RECONCILING SCIENCE AND RELIGION

SCIENCE AND ITS CONCEPTUAL FOUNDATIONS

A SERIES EDITED BY David L. Hull

RECONCILING SCIENCE AND RELIGION

THE DEBATE IN EARLY-TWENTIETH-CENTURY BRITAIN

PETER J. BOWLER

THE UNIVERSITY OF CHICAGO PRESS
CHICAGO AND LONDON

Peter J. Bowler is professor of the history of science at the Queen's
University of Belfast. He is the author of eleven books, including
*Life's Splendid Drama: Evolutionary Biology and the Reconstruction of Life's
Ancestry*, published in 1996 by the University of Chicago Press.

The University of Chicago Press, Chicago 60637
The University of Chicago Press, Ltd., London
© 2001 by The University of Chicago
All rights reserved. Published 2001
Printed in the United States of America

10 09 08 07 06 05 04 03 02 01   1 2 3 4 5

ISBN: 0-226-06858-7 (cloth)

Library of Congress Cataloging-in-Publication Data

Bowler, Peter J.
    Reconciling science and religion : the debate in early-twentieth-
century Britain / Peter J. Bowler.
        p. cm. — (Science and its conceptual foundations)
Includes bibliographical references and index.
    ISBN 0-226-06858-7 (alk. paper)
    1. Religion and science—Great Britain—History—20th cen-
tury. I. Title. II. Series.
    BL245 .B69 2001
    215'.0941'09041—dc21

                                              2001000719

♾The paper used in this publication meets the minimum require-
ments of the American National Standard for Information
Sciences—Permanence of Paper for Printed Library Materials,
ANSI Z39.48-1992.

CONTENTS

If the theme of this book seems rather widely removed from my usual research in the history of evolutionism, it is fairly easy to show how I was led into this new territory. I have long been interested in early-twentieth-century evolutionary theory, and had been vaguely aware that the controversies over the religious implications of the theory did not simply die away as the Victorian era ended. Work on E. Ray Lankester and Arthur Keith led me to the debates between H. G. Wells, Hilaire Belloc, and others on the validity of Darwinism. Having decided to explore these controversies in more detail, I found a wealth of literature largely unknown, at least to historians of science, and realized that a serious study was needed. Given that one area of scientific theory often articulates with another, I was led to equivalent debates over the theological significance of other areas of science, and decided that there was room for a general survey of the whole topic of science and religion in early-twentieth-century Britain. In the course of my research I became aware that one of the most prominent features of their interaction was an attempt to forge a reconciliation based on claims that science had now turned its back on Victorian materialism. The rise and fall of this attempted reconciliation form the central theme of this book.

In taking on so broad a topic, I was acutely aware that I was moving into unfamiliar territory. I am a historian of science by training and have only limited knowledge of religious history—although most historians of Darwinism have at least some understanding of the theological issues raised by that theory. Since I am not myself a religious person, I had no prior knowledge of the details of Christian theology from personal experience. I have done my best to understand the principles underlying the various theological and philosophical positions taken up by those whose commentaries I was reading. I am sure that I have to some extent oversimplified, but I have tried to approach these religious beliefs as sympathetically as possible. I hope that my inexperience in these areas has neither undermined the validity of my interpretations nor given offense to anyone whose beliefs do not coincide with my own. I am grateful to my referees, especially Professor John Brooke, for invaluable advice on this and other matters.

A few words of explanation may be needed concerning the format of this study. After much reflection, I decided to present it in three parts, dealing successively with the sciences, the churches, and the wider debate. This format has entailed a certain amount of repetition between the three parts, because the same debates were going on in all three areas. The implications of evolutionary theory, for instance, were discussed by scientists, by theologians, and by popular writers, so the same set of issues reemerges in each section. But the approach I have chosen has the advantage of introducing the scientific and the theological issues as coherent wholes, and I have included numerous cross-references to help the reader trace the issues through later chapters. A biographical appendix has been provided so that readers have a quick way of identifying personalities who crop up in all three sections.

Research for this study has been helped by leave of absence granted by The Queen's University of Belfast for the first semester of the academic year 1998–1999. The staff of the library at Queen's have been particularly helpful, especially with the large number of interlibrary loans required by so broad a project. As with most of my previous studies, I have focused mainly on published writings because it is the public debate, more than the private thoughts of the individuals involved, that interests me most. The research has, however, been supplemented by a certain amount of archival research at the libraries of the University of Birmingham (E. W. Barnes and Oliver Lodge papers), the University of Bristol (Lloyd Morgan papers), the University of Cambridge (Joseph Needham papers), and the National Library of Scotland (J. S. and J. B. S. Haldane papers, Patrick Geddes papers). I have also had access to the papers of E. Ray Lankester, in private hands, and to the Lankester–H. G. Wells correspondence (photocopies kindly supplied by Prof. Diane Paul, originals held at the University of Illinois at Urbana-Champaign).

Professors John Brooke and David Livingstone have provided advice at various stages in the project. Comments from David Hull and other referees, including Prof. Brooke, have helped me to improve both the structure and the content of the manuscript. Some of the material on Bishop E. W. Barnes has appeared in an article published in the *British Journal for the History of Science*, vol. 31 (1998), and a digest of the work on Charles Raven has appeared in the proceedings of a conference organized by the John Ray Trust.

A Legacy of Conflict?

Historians of science have devoted much attention to the Victorian era, and one of their most prominent themes has been the changing relationship between science and religion during the period when science first became a dominant force in Western society. The scientific naturalism promoted by figures such as Thomas Henry Huxley and John Tyndall represented a challenge to Christian faith and to the authority of the church. The Victorian debates over Darwinism and related topics represent a major focus of attention, equivalent in significance to earlier episodes such as the rise of Copernican cosmology in the seventeenth century. In contrast, the early twentieth century is seldom portrayed as a period in which the relationship between science and religion figured prominently, at least in Britain. It is as though the great debates of the Victorian era fizzled out after 1900, the underlying issues either having been resolved or no longer holding the interest of the majority of scientists or religious thinkers. James R. Moore's study of the Darwinian debates ends around 1900, as do most other studies of that topic, while John Brooke's survey of the changing relationship between science and religion devotes only part of a short epilogue to the early twentieth century.[1] The great exception, of course, is the rise of Fundamentalist opposition

---

1. Moore, *Post-Darwinian Controversies*; Brooke, *Science and Religion*.

to evolutionism in America in the 1920s, leading to the "monkey trial" of John Thomas Scopes in 1925, which has received a great deal of attention.[2] But there is a general impression that nothing of equivalent significance happened in Britain at the same time. In his history of the Victorian church, Owen Chadwick notes that "in 1900 men talked as though the conflict [with science] was over," leaving the reader to assume that the subsequent events are of little interest.[3] The British were steadily losing interest in religion, whatever happened on the other side of the Atlantic, and so the relationship between science and religion ceased to be a focus of attention.

This book is an attempt to show that the apparent lack of interaction between science and religion in the early twentieth century is an artifact of historians' neglect. Without claiming that events in early-twentieth-century Britain match those of the nineteenth century in significance, I argue that the issues were not as dead as the lack of historical emphasis might imply. Old topics such as the implications of evolutionism still attracted the attention of scientists and religious leaders and could still generate headlines in the popular press. New areas of interest emerged, especially as religious thinkers tried to incorporate the wider philosophical transformations stimulated by the revolutions in twentieth-century physics. When the BBC liberalized its policy on coverage of religious questions in 1930, the first topic to become the subject for a major series was science and religion. Popular writers such as H. G. Wells, Hilaire Belloc, and G. K. Chesterton debated the issues, Wells pushing a softened version of naturalism while Belloc and Chesterton resisted the rising materialism represented by science. George Bernard Shaw defended Lamarckism as part of his new religion based on "creative evolution." The novels of Aldous Huxley frequently reflected on the moral and religious implications of science, while his brother, the biologist Julian Huxley, wrote extensively on religion. Connie and Sir Clifford argue about Alfred North Whitehead's views on science and religion in D. H. Lawrence's *Lady Chatterley's Lover*, while Lord Peter Wimsey encounters a debate on the moral significance of biology between a slum vicar and a materialist scientist in one of Dorothy L. Sayers's popular detective stories.[4]

These examples show that we cannot so easily assume that the debates

2. The extensive modern literature has led to a substantial reassessment of this episode; see, for instance, Larson, *Summer for the Gods*, and Numbers, *Darwinism Comes to America*.

3. Chadwick, *Victorian Church*, part 2, 35.

4. Lawrence, *Lady Chatterley's Lover* (first published in 1928), 243–45; Sayers, *Unpleasantness at the Bellona Club* (also first published in 1928), 143. Wells, Huxley, and the other popular writers mentioned are discussed individually below.

between science and religion were over, or that they did not reach the now massively expanded middle-class audience. My own introduction to these issues came from my interest in evolutionary theory, but in extending my research to the more general interaction between science and religion, I uncovered an enormous volume of literature by scientists, theologians, and popular writers. If sheer volume is anything to go by, the debate over science and religion was anything but dead in early-twentieth-century Britain.

What emerged from a survey of this literature was a realization that this period saw a concerted effort to bring about a reconciliation between science and religion after the alienation of the Victorian era. A body of intellectually conservative scientists, liberal religious thinkers, and popular writers sought to convince the reading public that science had turned its back on materialism while religion had become more open to the kinds of changes that were consistent with the new understanding of nature. This attempted reconciliation was promoted most actively in the 1920s, but it fell apart in the course of the 1930s. By the time of World War II, its protagonists were marginalized, at least in the churches. The title of this book is derived from this model of what happened: a reconciliation was offered and widely debated, although interest gradually declined in the 1930s.

Not everyone accepted the proposed synthesis of science and religion, of course. Secularists were still active and wanted to use science as a weapon against all forms of what they regarded as superstition. Many conservative Christians, both Catholic and evangelical, reacted with suspicion to the claim that their faith could be adapted to the idea that human beings were the product of a natural process, even when that process was portrayed as the unfolding of a divine plan. It was the resurgence of this more conservative attitude that did most to undermine the reconciliation in the late 1930s. But enough scientists and religious thinkers had endorsed the proposed synthesis in the earlier decades of the century to create a short-lived expectation that it would become the permanent basis for a new kind of relationship between the two areas.

This pattern of events may seem surprising to readers used to thinking of the 1920s as the period in which the confrontation between science and Fundamentalist religion was at its height. Americans may also be shocked to realize that the whole sequence of events in Britain took place against a growing indifference to institutionalized religion—a trend that has affected British life through the whole twentieth century. The collapse of the proposed reconciliation coincided with the only period in the century when that trend was temporarily reversed. The fact that two English-speaking countries could experience so different a chain of events during the same period raises the prospect that this study may offer insights of wider significance to those currently interested in the relationship between science and religion. In effect, we have here a case study that may help us to understand what drives that relationship. We can ask which areas of science were most amenable to becoming the basis for a new, nonmaterialistic worldview, and which were the most difficult to incorporate into the synthesis. On the opposite side, we can ask which theological positions gave the greatest leeway to those seeking a reconciliation.

Here, at least, there are clues available in advance: the proposed synthesis required the modification, if not the actual suppression, of many aspects of traditional Christian belief, reducing religion to a more generalized theism. The synthesis thus depended on a degree of theological liberalism that many orthodox Christians regarded as a complete betrayal, and their fears were highlighted by the fact that it was endorsed by some openly non-Christian writers such as Shaw. This study will throw light on the intellectual links between the various scientific theories and theological opinions. But it will also reflect the ways in which the changing balance between the various positions was affected by events in the social and cultural arena. In the course of the 1930s, the synthesis was rejected because a growing body of opinion inside and outside the churches became suspicious of the ideology of progress with which it was associated. It was this cultural transformation, not the resurgence of materialistic theories such as Darwinism within science itself, that did the most to undermine the credibility of the proposed synthesis.

Do these long-dead debates have anything to tell us about the current state of the relationship between science and religion? History has a way of repeating itself, and it is not improbable that lessons could be learned from earlier episodes in which issues similar to those now stirring controversy were discussed. There have been three major episodes in the twentieth century when interest in the possibility of constructing a reconciliation between science and religion has flared. The first occurred in the early decades of the century and forms the theme of this book. The second wave (discussed briefly in the epilogue) began in the aftermath of World War II and lasted into the 1960s. The third seems to have arisen quite recently, at least according to the American humanist journal *Free Inquiry*, whose Spring 1998 issue proclaimed the need to resist a new effort to depict science as the basis for a religious view of the human situation.[5] The tensions of the Victorian era

5. See Cherry's introduction, "Science vs. Religion: War or Peace?" Curiously, Cherry lists no examples of the new spirit of reconciliation, and the articles that follow focus on traditional humanist concerns, such as the continued strength of creationism. Examples of recent efforts at reconciliation from the British side of the Atlantic include Peacocke, *Theology for a Scientific Age*; C. A. Russell, *Earth, Humanity, and God*; and K. Ward, *God, Chance, and Necessity*.

have thus been sustained throughout the twentieth century, each episode of challenge being followed by one of attempted reconciliation. These episodes seem to reflect the fluctuating balance of power between secularizing and traditional forces within our society, and if this is so, we can surely learn something of value from the debates of earlier decades—if only the futility of expecting the underlying issues ever to be resolved.

Scholarly interpretations of the relationship between science and religion have been transformed over the past few decades. The old model of inevitable conflict (still visible in the writings of extremists on either side) has been heavily qualified, if not abandoned. The diversity of theological positions offered by religious thinkers makes it clear that we cannot think of "religion" as a single category, even in a relatively limited time span such as that covered by this book. "Science," too, is a term that has been understood in different ways, covering rival visions of the nature and scope of the scientific method. One scholar has recently argued that this diversity of meaning is so great that we should avoid talking about the relationship between science and religion as though it were a unitary thing.[6] There is a serious point in this suggestion, however impractical it would be for those of us who want to paint with a fairly broad brush. In what follows, I take the term *religion* to include all the vast diversity of Christian belief articulated in Britain during the period under study, as well as a variety of implicitly and sometimes explicitly non-Christian forms of spirituality. It is precisely because there is such a wide spectrum of religious beliefs that we need to look for a complex rather than a simple model of how science and religion interact.

The scientists' own involvement in religious debates also needs to be monitored. In the early twentieth century many younger scientists became indifferent to religion—but not all shared this attitude, and an older generation with more conservative views was still active at the level of popularization. The churches themselves offer an obvious vector for analysis, with wide differences among Free Church, Anglican, and Roman Catholic writers—and wide debates within the Anglican Church, still an important focus of national religious consciousness despite declining attendance. The debates were also articulated within popular culture, including nonspecialist books, magazines, the increasingly ubiquitous newspapers, and the BBC's efforts to bring culture to the middle classes. The new century witnessed an explosion of interest in nonmaterialistic ways of thinking, many of which impinged on what can legitimately be called religious ideas. There were rival forces battling for

6. D. B. Wilson, "On the Importance of Eliminating *Science* and *Religion* from the History of Science and Religion."

the attention of the literate masses: various shades of opinion opposed to materialism, some of them quite unorthodox, confronted the rationalists, socialists, and exponents of scientific management. The changes taking place within early-twentieth-century British society and culture affected the way people thought about the same deep issues we are confronting today. At the same time, we must be aware of the sensitivities that cultural historians are now urging in the interpretation of these changes: all too easily we construct models of the past based on our own ideas about the outcome of the events we describe.

CONFRONTATION, COOPERATION, OR COEXISTENCE?

As John Brooke remarks, Alfred North Whitehead believed that the future course of history would be shaped by his generation's decisions on the relationship between science and religion.[7] Whitehead was not the only scholar of his period to look to history for an understanding of how science had developed the atmosphere of materialism that seemed to surround it, nor was he the only one to suggest that now was the last chance to dispel it. Lecturing in the darkest days of World War II, when European civilization itself seemed threatened, Charles Raven linked the failure of culture as a whole to the rise of a materialistic interpretation of science.[8] Raven's work as a historian of science reminds us that history has always been a resource for those wishing to comment on the relationship between science and religion in their own time, and that by the same token, few have written on the history of the topic without having a position of their own to defend. Modern scholarship, however, tends to be suspicious of any one-dimensional account of the past. Those who have axes to grind will reinterpret past events to ensure that they fit a preconceived scheme. All too often, we are left with the image of a single main focus of interaction—conflict and cooperation are the two most common—which, it is alleged, underlies how science (as a relatively new force in our culture) has dealt with the religious traditions it has encountered. It now seems more reasonable to suppose that there are several different ways in which science and religion can be related to each other. They may interact in various ways, including conflict and cooperation, or they may come to an amicable agreement to stick each to its own territory. But there has never been a

---

7. Brooke, *Science and Religion*, 1. On Whitehead, see chap. 11 below.

8. Raven, *Science, Religion, and the Future.* The reaction of historians of science to the crisis of the early 1940s was stressed by James Moore in a talk at a meeting held to discuss Brooke's *Science and Religion* in May 1992.

single attitude in play at a single time, and the balance between the various possibilities is always shifting.

Such an approach falls very much in line with contemporary studies in the sociology of scientific knowledge. One-dimensional accounts are a product of the decision to privilege religion by giving it the status of the key cultural force encountered by science. According to these accounts, science either arose from an essentially religious quest for knowledge or set itself up as the rival source of authority to religion, depending on the author's point of view. (I speak here of the history of science—a similarly one-dimensional account could be written by a historian of religion who saw science as either the greatest benefit or the greatest threat that religion has encountered.) The position of Whitehead referred to above arose from the belief that science ought to seek a vision of nature compatible with the universe's divine origin, a quest from which it had been deflected but which it might now hope to recover. The rival view, in which science and religion come necessarily into conflict, was based on the assumption that knowledge of the real world could supplant the invention of the supernatural as a guide to human behavior. The sociologist of science—who, of course, does not privilege scientific knowledge—sees science as the generation of knowledge-claims that are invariably influenced by the values of those who do the generating. Religion must then become but one facet of the system that provides those values, itself shaped and influenced by all the other social and cultural activities of the time. Under such a model it would not be at all surprising to find rival positions being staked out at any one time on the question of how science should relate to religion, nor to see the influence of those positions wax and wane with changing social conditions.

I do not intend to defend this more sociological approach to the study of science (or, for that matter, of religion) here.[9] I do believe that the production of scientific knowledge is influenced, though not determined, by its cultural environment, and to me, as an essentially nonreligious person, religion is just another facet of that environment. I am interested in how science is related to religion for exactly the same reason I am interested in how it is related to ideological values. I have studied the involvement of scientists and popular writers on science in the eugenics movement (the original political expression of the ideology of genetic determinism), and I now propose to study their involvement in debates over religious questions. The fact that the same persons are often involved, some of the popular writers on eugenics having been churchmen, makes the transition seem all the more natural. I am

9. For a detailed modern defense, see Barnes, Bloor, and Henry, *Scientific Knowledge.*

equally interested in the question of how a theologian or a scientist could come to support the claim that the breeding of "unfit" persons should be limited. If there were some who thought that science and religion must necessarily come into conflict, and some who hoped for a reconciliation, their attitudes are all grist to my mill. I do not intend to treat the one any more than the other as an expression of the "true" relationship that ought to obtain. My aim is to understand why people thought as they did, and to relate their views to the complex and ever-changing social and cultural life of the period under study.

Having laid these cards on the table, let me now progress to evaluate the three chief modes of relationship that have been suggested: conflict, cooperation, and coexistence. My intention is not to show that one is more natural than the others, but to uncover the various strategies that have been used by those seeking to create a discourse between science and religion. All of these strategies involve history, in the sense that they all seek to create an image of past interactions between science and religion that is compatible with how their supporters think things ought to develop in the present. In many cases, the interpretations articulated in the early twentieth century anticipate those still in use today, although modern historians are less likely to be trapped into the assumption that the past can be fitted into any single model. In the early twentieth century, models based on conflict, cooperation, and coexistence were all in use. Some of their supporters may have hoped that a single model might yet prevail, but from a perspective of hindsight, it looks much more plausible to see each interpretation as the strategy of a particular interest group in a world where old and new value systems were jockeying for influence.

Before exploring these rival models, it is necessary to comment on the possibility that the whole idea of treating science and religion as independent entities needs to be questioned. Led by Robert Young, historians focusing on the nineteenth century have protested that religion and science are better seen as complementary facets of a common context within which all knowledge was articulated.[10] Whatever one's reaction to Young's assertion that this context was essentially ideological, there is much value in the claim that in the age of natural theology, science was viewed as the exploration of a divinely created universe, and hence as another aspect of religion. In early-nineteenth-century Britain, many scientists were actually ordained clergymen, and most

---

10.   See the essays in Young's *Darwin's Metaphor*; see also Livingstone, "Science and Religion: Towards a New Cartography."

had strong religious views. Even in the later part of the century, natural theology was reshaped to include evolutionary ideas, while scientific naturalism was presented by Huxley and his followers as a substitute for religion, fulfilling essentially the same human needs (a function paralleled in the twentieth century by Marxism). The claim that science and religion interacted harmoniously at certain times can thus be translated into the more revisionist view that they are simply aspects of the same cultural or ideological forces. Why, then, preserve the old-fashioned image of two distinct entities, if the revisionist perspective has given us a deeper understanding of the relationship? The answer to this question has to be that the twentieth century is not the nineteenth: if it is inappropriate to impose a template based on modern ideas about science and religion onto the discourse of the nineteenth century, it may not be inappropriate to impose it onto the debates of the early twentieth.

The difference lies in the professionalization of both the scientists and the clergy and in the emergence of a new profession devoted to journalism and popular writing. It is possible to assign most of the people writing on the issues that concern us in the early twentieth century to one of these three categories unambiguously (the exceptions would be academics in other fields, such as philosophy). Very few scientists were now in holy orders, and very few clergymen were research scientists. Members of both groups wrote for the popular press, but are fairly easily distinguished from those who wrote on these and other topics for a living. Traces of the common context remain, especially among those advocating a new natural theology, but one can almost always say whether the advocacy is coming from someone whose primary identification is that of a scientist or of a clergyman. Under these circumstances, we are justified in speaking of science and religion as two separate discourses, which had to be related to one another by those practitioners who expressed an interest in so doing. Clergymen and popular writers got their information about science from those scientists who wrote for a wider audience, but they did not "do" science in the sense of contributing to scientific research. Some scientists wrote on religious matters, but many of those who were religious did not, and those who did often represented no organized religious group. The Social Relations of Science movement, which urged greater political and social involvement on the part of scientists, emerged in the 1930s, but concern for the ethical implications of science was expressed by some scientists who had strong religious views, by others who became Marxists, and by still others who had no commitment to any deeper worldview. Popular writers would have been very conscious of getting their information from two different sources—scientists and clergymen—however hard they (or the sources) might have been trying to reconcile the ideas thus derived.

How might these two increasingly separate areas of discourse be related? The claim that the advance of science necessarily bring it into conflict with established religious beliefs was advanced most energetically in the late nineteenth century by those who believed that science was the vehicle by which a new, secular view of the human situation would be established. T. H. Huxley was one advocate of this philosophy of scientific naturalism, but the metaphor of a "war" between the two areas was projected most explicitly by J. W. Draper's *History of the Conflict between Religion and Science* (1874) and A. D. White's *History of the Warfare of Science with Theology* (1896). The exponents of scientific naturalism believed that conflict was inevitable because religion was wedded to traditional dogma while science offered a new route to the truth that inevitably exposed the inadequacies of past ideas. This was a war that science was bound to win because it was the only reliable source of information.

Even at the time, there were critics of the assumptions built into this interpretation. They insisted that episodes of conflict occurred only because religious thinkers in the prescientific age had shortsightedly taken Holy Writ as a guide on matters of scientific fact, or had too hastily invoked the Deity to explain what was currently inexplicable by the sciences. They believed that once these items of outdated cosmology were purged, the source of conflict would be removed. But as Sigmund Freud argued in 1917, there was more to it than that. Christianity had begun with a particular view of humankind's significance in the world, and three great revolutions in thought had challenged its fundamental assumptions. The Copernican revolution had shown that humankind was physically insignificant in the universe, Darwin had shown that we were only highly developed animals, and now he, Freud, had undermined our claim to be rational beings aware of our own motivations.[11] The episodes of conflict were inevitable because science was establishing a new view of human nature in which traditional religious values had little place.

Freud's psychology was indeed seen as a new challenge to religion. But most of those who rose to the challenge believed that in this case, science had gone too far and had overexposed itself, while the earlier revolutions could, in fact, be absorbed even by sincere Christians. From their perspective, con-

---

11.   Freud, *Introductory Lectures on Psychoanalysis*, vol. 16 of *Complete Psychological Works*, 285.

flict arose only when science stepped outside its natural boundaries and tried to establish a completely materialistic metaphysics. Such a position was possible because the Copernican and Darwinian theories could be interpreted in ways that did not challenge traditional views of human nature. No one believed any longer that humankind's physical insignificance meant that we were necessarily of no spiritual interest to the Creator. And Darwin's theory had been absorbed via the assumption that evolution was an essentially purposeful process aimed at the production of morally aware beings. It is in the area of the Darwinian revolution that modern historians have launched the most sustained attack (to preserve the military metaphor) on the "warfare" model. James Moore's analysis has shown how that model was created as part of the naturalistic ideology of Draper and his supporters. Frank Turner has shown that many scientists resisted the rise of scientific naturalism as a philosophy, while my own studies of late-nineteenth-century evolutionism have revealed how the materialistic implications of Darwinian selectionism were evaded.[12] Robert Young has argued that the Victorians negotiated a new synthesis of science and religion, while John Durant has written of the "new natural theology of providential progress" that emerged at the end of the nineteenth century.[13] As we shall see, this new natural theology still played a major role in the early twentieth century.

There *was* tension in the Victorian era, of course. Religious thinkers at first found the new scientific theories difficult to assimilate, and some scientists—of whom Huxley is the best example—went out of their way to present science as a source of knowledge that would supplant religious superstition.[14] But the latter policy was at least as much a tactic employed by professional scientists seeking recognition as it was the product of real antipathy, and many commented on the religious tone of Huxley's own pronouncements about the moral character of agnosticism. There were many scientists who remained deeply religious and still hoped that the study of nature would endorse their beliefs. Religious thinkers renegotiated their image of God's role in the world, accepting that the creation of new species might take place through His laws rather than by miracle. When Huxley died in

12. Moore, *Post-Darwinian Controversies,* chap. 1; Turner, *Between Science and Religion;* Bowler, *The Non-Darwinian Revolution.* See also Helmstadter and Lightman, eds., *Victorian Faith in Crisis.*

13. Young, "The Impact of Darwin on Conventional Thought," reprinted in his *Darwin's Metaphor,* 1–22; Durant, "Darwinism and Divinity," 20. On "providential evolution," see also Elder, *Chronic Vigour.*

14. Adrian Desmond's two-volume biography *Huxley* stresses the confrontational side of its subject's nature.

1895, he was in the middle of a confrontation with the philosopher-politician Arthur Balfour that made it plain that scientific naturalism was on the wane.[15] By the early twentieth century it was a commonplace observation that scientific naturalism (which its enemies equated with materialism) was an illegitimate extension of science into the territory of metaphysics. Writing in 1903, the Anglican priest Philip Waggett argued that there was now a truce between scientific naturalism and religion, both sides having agreed to ignore each other.[16] But Waggett also noted the many biologists who had challenged Darwinian materialism, and he saw in this a hope that science was turning its back on the naturalistic philosophy. The stage was set for a revival of the belief that science and religion could work hand in hand.

To many of those favoring such a revival, this move represented no more than a return to the normal mode by which science interacted with religion. Some made direct use of history to emphasize this point, including Whitehead, Raven, James Y. Simpson, and in a slightly different vein, Joseph Needham. According to this model, science had arisen as a project to investigate God's handiwork in nature. Its most basic philosophical tools, including the use of mathematics to describe nature and the concept of natural law, had been inspired by this preconception. Some of the most eminent Victorian physicists had retained this faith, most notably Lord Kelvin. Naturalists had always taken it for granted that living things displayed the designing hand of their Creator. This tradition of natural theology stretched from John Ray to William Paley and was now manifested in the work of those biologists who showed that evolution is an essentially purposeful process. As Whitehead argued, the materialistic trend that began in the eighteenth century and came to a head with Victorian scientific naturalism was an aberration, a philosophical blind alley into which some scientists had been tempted by the lure of mastering physical nature. Science could and should turn aside from this blind alley into the more fruitful main line of development in which nature was seen as a system with built-in values.

Many in the churches were happy to accept this offer of reconciliation from the prodigal son. In the words of the encyclical letter of the bishops of the Anglican Church to the Lambeth Conference of 1930:

> There is much in the scientific and philosophical thinking of our time
> which provides a climate more favourable to faith in God than has

15. See Lightman, "Fighting Even with Death," and more generally, the same author's *Origins of Agnosticism.*

16. Waggett, *Scientific Temper in Religion,* 10–11.

existed for generations. New interpretations of the cosmic process are now before us which are congruous with Christian Theism. The great scientific movement of the nineteenth century had the appearance, at least, of hostility to religion. But now, from within that movement and under its impulse, views of the universal process are being formed which point to a spiritual interpretation. We are now able, by the help of the various departmental sciences, to trace in outline a continuous process of creative development in which at every stage we can find the Divine presence and power. Thus scientific thinking and discovery seem to be giving us back the sense of reverence and awe before the sublimity of a Creator Who is not only the cause and ground of the Universe, but always and everywhere active within it.[17]

With hindsight, of course, these expectations seem somewhat overoptimistic. The anti-Darwinian and sometimes openly vitalistic biology on which they were based had been popular at the turn of the century but was now being abandoned by many younger researchers. Yet the willingness of older biologists such as J. Arthur Thomson and psychologists such as C. Lloyd Morgan to endorse and promote these older views in popular writings blinded religious thinkers to what was happening at the cutting edge of research. Somewhat more positive encouragement for the new synthesis came from the latest developments in the physical sciences, which had clearly overthrown the simple materialism of the previous century. With scientists such as Arthur Stanley Eddington and James Jeans openly proclaiming a new idealism that made a Creator seem more plausible, the churches can be forgiven for hoping that the old rift had been healed.

If supporters of the old rationalism and naturalism could be dismissed as out of date, they were still nevertheless active. In general, however, they were now less openly dismissive of religion. H. G. Wells, a prominent and popular exponent of antireligious values, began to treat the religious sense as an unconscious recognition of the higher values that humanity had formulated in the course of its development. Julian Huxley (T. H.'s grandson) turned his back on pure rationalism and began to argue that the religious sense was a valuable human asset, even if the objectification of its values in a God was illusory. Religious feelings, especially those of a mystical character, were increasingly taken as valid guides to reality, guides that coexisted with the rational and empirical faculties used by science. Here was a more subtle model of the relationship between science and religion: neither cooperation

17. Encyclical letter, in *Lambeth Conference, 1930*, 17–34, 19.

nor conflict, but coexistence based on a mutual recognition that each had its own legitimate questions to ask and territory to explore. Conflict occurred only when one side overstepped the mark and encroached on the other's territory.

The possibility that this attitude of mutual tolerance might be more natural than either conflict or cooperation had already been expressed some years before the 1930 Lambeth Conference by an Anglican intellectual, Canon B. H. Streeter. Streeter led a discussion group that generated a 1927 book stressing that both science and faith were "adventures" into new conceptual territory. As Streeter wrote in his introduction to the book:

> ... to us it seems that ... the period when the 'reconciliation' of Science and Religion was something which men could regard, according to their temperaments, as a matter either for hope or despair, is passing away— to be successed by a period in which they will be regarded as two diverse, but intrinsically connected *adventures* of the spirit of man.[18]

In fact, however, things were not quite so simple. In the same year, Streeter's own book *Reality* proclaimed in its subtitle *A New Correlation between Science and Religion.* These two intellectual adventures could exist in a state of mutual tolerance only if science did not promote a worldview that made it difficult to support the belief that the universe is a divine construct. Streeter himself wrote books on the problems posed by natural selection and the new psychology. For all his professed lack of belief in a personal God, Huxley continued to suppose that evolution was ultimately a purposeful process. And, unlike some more materialistic scientists, he accepted that religious experiences had a genuine value for human life. Without this kind of agreement, friendly coexistence was impossible.

VICTORIAN BACKGROUND

In the Victorian era, the metaphor of conflict had been promoted as part of the rhetoric by which the new generation of professional scientists sought to establish themselves as a source of authority in an industrialized society. There is no doubt that for T. H. Huxley, the opposition of religious thinkers to Darwin's theory reflected a worldview in which science was subservient to theology. He was determined to throw off the shackles, and in so doing he

18.   Streeter, ed., *Adventure: The Faith of Science and the Science of Faith,* ix.

may have said things that encouraged spectators to view the relationship as one of inevitable conflict. Rationalists and freethinkers certainly seized upon science as a stick with which to beat the churches, going further in this respect than Huxley himself. This rationalist or materialist trend survived into the twentieth century, where it continued to promote antireligious sentiment, sometimes in the less strident guise adopted by the younger Huxley. Yet many Victorian scientists had been deeply religious, and there were members of the younger generation who continued that tradition. Popular writing on science and religion in the early decades of the twentieth century was to some extent dominated by the generation trained in the 1880s and 1890s, which included many who still thought it important to preserve the link between the two areas.

It was in his review of Darwin's *Origin of Species* that Huxley wrote: "Extinguished theologians lie about the cradle of every science as the strangled snakes beside that of Hercules."[19] To Huxley, the promotion of Darwinism was indeed a struggle, representing another step in science's advance toward the truth, another nail in the coffin of religious dogma. In the course of the battle, he would take on opponents as diverse as the Roman Catholic Church and W. E. Gladstone's defense of biblical literalism. John Tyndall's "Belfast Address" of 1874 made similar demands for the authority of science and also provoked much critical response from church leaders. These demands constituted the program of scientific naturalism; they were based on the claims that the natural world revealed by our senses is the only true reality and that science (the rationally organized use of the senses) is the only way of gaining reliable knowledge of that reality. By its critics, scientific naturalism was often identified with materialism—indeed, with a simplified materialism that supposed the universe to be composed solely of hard, indestructible atoms, something like a vast collection of billiard balls. But true materialism holds that there is no mental level of existence at all—all references to mental states are merely shorthand ways of describing material events in the brain. Huxley certainly did not subscribe to such a position, and was quite willing to accept that mental events were real. But he did believe that those events exert no control over the natural world, so that the mind is a mere passive spectator of events it cannot control (this reduces the mind to what is called an epiphenomenon). Scientific naturalism thus implied determinism—the rigid determination of all events, including mental events, by natural laws. And while it did not subscribe to the billiard-ball model of nature,

19.   T. H. Huxley, "The Origin of Species," in his *Darwiniana,* 52.

it did hold that the only way of understanding the world is to treat it as a mechanism. In biology, this mechanist program was a particularly effective, and ultimately successful, product of the naturalistic philosophy. By definition, mechanism excluded any claim that life was something more than the physical activity of a complex material structure.

Scientific naturalism was certainly materialistic in the sense that it denied the existence of a spiritual world apart from the universe revealed by the senses. There was no soul that could survive the death of the physical body. Huxley was scathing about the popular craze for spiritualism, which was based on the assumption that the soul does continue to exist on a spiritual plane after bodily death. But spiritualism was dangerous to the exponents of scientific naturalism precisely because it blurred the distinction between the material and the spiritual worlds—the medium claimed to be able to contact the spirits of the dead and hence provide material evidence of their continued existence. Some mediums also claimed to be able to manipulate an ethereal kind of matter known as ectoplasm, which could be observed in certain circumstances by normal observational means such as photography. Here again scientific naturalism adopted a materialistic stance, denying the validity of the evidence for these effects. The scientific naturalists also denied any evidence for ghosts or spirits, holding that these ancient superstitions had been created to bolster the claim that the material world was not the whole realm of existence. Much of the evidence for the claims of orthodox religion was dismissed in the same way.

On the question of the existence of God, the scientific naturalists were less dogmatic. They certainly did not believe that observing the material world gave clear evidence that it had been designed by a wise and benevolent Creator. This was why they welcomed evolutionism as a weapon to use against the old tradition of natural theology. But to make a positive claim that nothing existed beyond the world of the senses went too far; all that could be said was that we have no direct evidence for such a Creator. Searching for a name that would provide an adequate label for his beliefs, Huxley coined the term "agnosticism" in 1869. Agnosticism symbolized the scientist's refusal to dogmatize on questions that could not be answered by rational means and the willingness to search open-endedly for the truth in those areas in which investigation was possible. In a very real sense, Huxley saw this search for the truth as a moral crusade, giving science a key role in the value system that would replace theological dogma.

It was this element of moral earnestness that allowed agnosticism and the philosophy of scientific naturalism that accompanied it to take on the atmosphere of a religious movement. Huxley gave "Lay Sermons" in which he preached his philosophy to a wide audience.[20] Herbert Spencer's philosophy taught that behind the material universe lay the "Unknowable," which for many Victorians represented a veiled Deity. Far from being linked to atheism, Huxley's agnosticism was widely perceived as being compatible with a theistic position.[21] It was by no means incongruous for scientists such as Francis Galton to speak of a "scientific priesthood," which, through its dedication to the truth and to the moral betterment of the race, could legitimately take the place of the church.[22] So strong was Huxley's commitment to this moral crusade that he resented efforts by militant secularists to take over the use of the label "agnostic." Now that he himself had become part of the political establishment, he wanted no truck with those who still used science as a weapon in a materialistic assault on the ideological foundations of a stratified society.[23]

Yet this more radical rationalist program continued, making extensive use of the new science. Through a number of socially militant scientists, it played an important role in setting the scene for the debates of the early twentieth century. Under Charles Albert Watts, the Rationalist Press Association was created in the 1880s to publicize what was, in effect, scientific naturalism as the key to social reform. This was not rationalism in the sense adopted by those seventeenth-century philosophers who thought that all knowledge could be constituted on an a priori basis. The RPA was firmly behind the campaign to base knowledge on empirical science—but it stressed the need for a rational approach to the study of nature as a means of undermining belief in the supernatural. By popularizing the philosophy of scientific naturalism, they would convince ordinary people that the churches should be swept away, and with them the old social hierarchy. Writers such as Samuel Laing produced readable summaries of sciences such as evolution to bolster the claim that social reform would follow the emergence of a more rational view of human nature.[24] The RPA continued into the early twentieth century and issued vast numbers of cheap reprints of books by Laing, Spencer, and other heroes of what was perceived as an essentially antireligious philos-

---

20.   See Desmond, *Huxley: The Devil's Disciple*, 344–45, and Moore, *Post-Darwinian Controversies*, 58–68.

21.   Lightman, *Origins of Agnosticism*.

22.   Brooke, *Science and Religion*, 305–6.

23.   Lightman, "Ideology, Evolution, and Late-Victorian Agnostic Popularizers."

24.   Laing, *Human Origins*. For an outline of the same author's agnosticism, see his *Modern Science and Modern Thought* and "Religion of the Future," essay 8 in his *Problems of the Future*.

ophy.[25] As we shall see in chapters 2 and 10 below, important scientists and philosophers remained associated with the RPA through the early twentieth century, including E. Ray Lankester, Arthur Keith, and Bertrand Russell.

The link between scientists and the RPA shows that the metaphor of conflict can be appropriate in certain circumstances, but these scientists represent the far end of a spectrum of attitudes. Perhaps feeling more secure by the end of his career, even T. H. Huxley tried to distance himself from this kind of outright hostility to religion. He remained on amicable terms with the religious thinkers who frequented the Metaphysical Society's after-dinner discussions.[26] An even clearer indication of the restricted applicability of the conflict metaphor is provided by the vast number of scientists who retained their religious beliefs and struggled to form a workable compromise. Francis Galton's 1874 survey showed that seventy percent of English scientists still regarded themselves as Anglicans, and several later surveys still confirm a surprisingly high level of religious belief.[27] It was these scientists, and like-minded religious thinkers, who created the new natural theology that would survive into the early twentieth century. The most significant change in the balance between the opponents and supporters of religion is that the latter expressed a growing self-confidence in the early decades of the new century. Indeed, they created their own mythology of the Victorian era, presenting it as a period of unbridled materialism that had now been swept away. I shall argue below that it was this myth of Victorian materialism that created the artificial image of an "eclipse of Darwinism" in the decades around 1900. In fact, Darwinism (in anything like the modern sense of the term) had never been popular among Victorian biologists. Those associated with the myth, especially Huxley and Spencer, came to be regarded as outdated and scarcely worthy of mention.

By the end of his life, Huxley himself had come into conflict with this more self-confident opposition to materialism in the form of Arthur Balfour's *Foundations of Belief.*[28] Balfour argued that the scientific method was not enough—that we do have intuitive knowledge of a world beyond the senses. The academic philosophy of the late nineteenth century was increasingly idealist: it held that it was the mental world, not the physical, that was the true reality. After all, it argued, our knowledge of the natural world can only be

25. Whyte, *Story of the RPA;* see also Budd, *Varieties of Unbelief;* Berman, *History of Atheism in Britain,* 221–33; Royle, *Radicals, Secularists, and Republicans.*

26. Moore, *Post-Darwinian Controversies,* 97–99.

27. Galton, *English Men of Science,* 126–27; later surveys, discussed below, include Tabrum, *Religious Beliefs of Scientists,* and Drawbridge, *Religion of Scientists.*

28. Lightman, "Fighting Even with Death."

built on sense impressions, which are by definition mental. But by acknowl-edging the primary reality of the mental, idealism opened the doors to a wider supernaturalism in which spiritual forces once again became acceptable. By 1918, a pessimistic Lankester could write to Wells that Huxley was now al-most forgotten.[29] Anti-Darwinian evolutionary theories postulated "cre-ative" forces in nature driving it toward morally significant goals. Vitalistic physiology flourished again, reviving the old idea that life was a force exist-ing apart from the mechanical functioning of the body. These theories in-voked what the rationalists had dismissed as the supernatural, and they were regarded as the key to a new philosophy in which nature was directed by the nonphysical agents of life and mind. In physics, the theory of the ether was regarded as the basis for a nonatomistic worldview that would sweep away the foundations of materialism by showing that the whole universe was a har-monious cosmos. Such ideas were promoted by a conservative segment of the academic community that regarded scientific naturalism as a product of vul-gar utilitarianism. This was the era of the "revolution of the dons," in which idealism became the symbol of the universities' determination to resist the move to regard applied science as the basis for education. It was also an era that saw a renewed interest in spiritualism and psychic research.[30] Both Bal-four and the physicist Oliver Lodge supported the latter movement. Mean-while superstition, astrology, and occult religions such as Theosophy gained a new lease of life in popular culture.

SCIENCE AND RELIGION IN THE NEW CENTURY

The last two sections have hinted at a number of ways in which the Vic-torian era laid the foundations of the debates that continued into the twen-tieth century. But circumstances were changing within both the sciences and the churches, while major social and cultural transformations were creating a new arena in which their interactions had to be played out. Details of these transformations are provided at the beginning of each of the three main parts into which this book is divided. This last section of the introduction will provide a brief overview of the new situation, leaving the details and the rel-evant documentation for later.

The growing professionalization of early-twentieth-century science in-evitably affected the way it interacted with other areas of society and culture. To some extent, Huxley's opposition to organized religion had been based on

29.   Lankester to Wells, 23 September 1918, Wells papers, University of Illinois.
30.   Rothblatt, *Revolution of the Dons*; on spiritualism, see Oppenheim, *The Other World*.

the demand that science replace the churches as a source of knowledge and expertise. Almost everyone agreed that science was becoming more powerful through its provision of practically useful information, and some scientists sought to capitalize on this power to boost the status of their profession. In fact, their advance toward greater prestige and influence was slower than they had hoped, and there was a temptation for more radical scientists to become involved in calls for a planned society that would apply the benefits of science in a rational way. Those who continued on the path established by the RPA saw this program as part of a moral crusade to reform society and presented the churches as obstacles to its progress. Their heirs, in this respect at least, were the Marxists of the 1930s.

Many scientists, however, retreated into a kind of intellectual isolation, depending on their technical specializations to insulate them from all moral or theological involvements. Others avoided comment on wider issues in order to preserve a sense of objectivity, even when they felt quite deeply one way or another. Any effort to present an overview of scientists' thinking has to confront the difficulty presented by this attitude of indifference or aloofness. Some scientists who retained deeply religious feelings expressed them only in private, refusing to participate in the public debates. Several surveys conducted to determine the religious beliefs of scientists reported surprisingly high proportions who claimed some degree of faith, although this was increasingly less likely to be an orthodox form of Christianity (see chapter 1). In addition, there was a significant group that campaigned actively to overthrow the image of science as the handmaiden of materialism and rationalism. Although many of these scientists were by now senior figures and no longer active in research, they were able to use their influence to gain access to editors and publishers in order to mount a public campaign of considerable effectiveness. It was in part their ability to convince church leaders that they represented the opinion of the scientific community as a whole that created the hope of a more general reconciliation. Younger scientists were on the whole less likely to be supportive of this move, and to this extent, the proposed rapprochement was based on a misapprehension. Even so, there were some eminent members of the active research community who still held conventional religious views—although they were aware that their position did not reflect that of their generation as a whole.

If science did not gain in prestige as rapidly as it had hoped, the churches were now losing influence at a rate they could hardly have expected (see chapter 6). Historians have charted a steady decline in church attendance for all denominations except the Roman Catholic, beginning in the 1890s for the Anglican Church and a decade or so later for the Free Churches. Much of the rationale for the proposed reconciliation with science was based on the hope that this decline could be halted or reversed if religion could be made more amenable to modern values.

Despite its ebbing support, the Church of England was still the established church of the nation, and it represented a powerful symbol, any open challenge to which produced an outcry even among those who remained Christian in little more than name only. Yet this symbol meant different things to different people. Much of the church's difficulty lay in the fact that it attempted to cover so broad a spectrum of theological opinion that the resulting tensions threatened to pull it apart. The evangelical wing jealously guarded the church's Protestant heritage, focusing on the Bible and the need for personal salvation. Few evangelicals engaged in detailed evaluation of new scientific ideas, although there was a small antievolution movement. At the opposite extreme were the Anglo-Catholics, determined to retain the Anglican Communion's historic claim to be part of a spiritual heritage derived from the very origins of Christianity. Their preference for a form of service using ritual similar to that used by the Church of Rome was anathema to the evangelicals and led to a major public debate over revision of the Prayer Book. Some Anglo-Catholics adopted a liberal position with respect to new ideas such as the theory of evolution, but they remained suspicious of any scientific attempt to impose a materialistic view of life or mind. Most sympathetic to the prospect of a reconciliation with nonmaterialistic science was the Modernist movement, whose supporters openly doubted the veracity of the miracles reported in the Bible and proclaimed the hope that following Christ's moral teaching would produce salvation in this world as well as the next. By focusing on the idea of progress, they opened the way to a general acceptance of evolutionism, but were forced to repudiate the traditional interpretation of original sin and the need for redemption. Their position appealed to many scientists, but was regarded as a betrayal of Christianity's basic teachings by both evangelicals and Anglo-Catholics.

The Free Churches presented a spectrum of beliefs allied to the evangelical wing of the Church of England. The Congregationalists experienced a brief surge of Modernism in the form of the so-called "New Theology" of R. J. Campbell. The Roman Catholic Church retained a small but stable membership and remained largely hostile to liberalizing trends. In the early decades of the century, when much of the nation's intellectual elite seemed to have abandoned religion altogether, the Catholic Church remained a vocal source of support for traditional beliefs and enjoyed a degree of influence out of all proportion to its numbers, thanks to popular writers such as Hilaire Belloc. In the pessimistic atmosphere of the 1930s, it began to attract some support from intellectuals precisely because it had held out against the materialism that was now seen as a threat to civilization itself.

The declining membership of the churches seems to imply a wholesale secularization of British culture, but some historians have warned that this may be an illusion. While institutional religion declined, there is evidence of a continuing demand for some kind of spiritual engagement, which manifested itself in a variety of ways within popular culture. The RPA and the Marxists were pressure groups rather than harbingers of a mass conversion to materialism. Some people turned to spiritualism, especially in the traumatic years of the Great War and after, while astrology, Theosophy, and various other nonmaterialistic ways of thought flourished. These movements sometimes intersected with the debate over science and religion, as with Lodge's support for spiritualism and Shaw's "creative evolution." Nor were the leading intellectuals the only ones attempting to spread such a message: there was a network of popular writers catering to the public's vague feeling that there must be a spiritual element to life. Some of the resulting ideas were presented as closely allied to liberal Christianity (although more traditional believers repudiated them along with Modernism), while others were offered as alternative forms of spirituality.

The last paragraph has taken us from organized religion into the wider realm of British culture, at both the high and the popular levels (see chapter 10). The two levels interpenetrated to some extent, since the elites in the sciences, the churches, and the academy all included some figures who were both able and willing to write for a wider audience. This was an age in which better education for the lower middle and upper working classes had produced a substantial market for "middlebrow" literature, which was satisfied by a wide range of book publishers and magazines. Books and periodicals were cheaper now, and best-sellers—even on the subject of science—could sell by the tens of thousands of copies. This was the audience to which much of the popular literature studied below was targeted. Writers such as Wells and Huxley were explicitly trying to educate the general reading public in scientific matters. They also hoped to convey a rationalist message, and were resisted by equally popular literature written by Belloc, Chesterton, and others. But there were also new avenues by which the public could be addressed. The mass-market newspaper industry was firmly in place, and while ostensibly scorned by the elite, it did occasionally tackle serious issues, provided they could be given a "human interest" element. The radio was an entirely new medium, tightly controlled by the director of the BBC, Lord Reith, who insisted that serious matters such as religion be approached with respect. Even so, the 1930s saw a relaxation of this attitude that permitted some critical discussion, although outright atheism was still banned from the airwaves.

The fact that some members of the intellectual elite attempted to disseminate their views more widely should not, however, incline us to believe that there was a unified British culture. Even the elite was divided, as Noel Annan reminds us, with the Bloomsbury group at loggerheads with the Oxford wits.[31] The literary avant-garde despised those such as Wells who wrote for a wider audience. This division also highlights an important message for any study of British life as a whole, because the modernism of the avant-garde was very different from the Modernism that was being promoted by liberal religious thinkers. The atheism that the Victorians had seen as a threat on the horizon had now become the dominant attitude of the intellectual elite. But while the high intellectuals welcomed the collapse of nineteenth-century values and beliefs, symbolized by the new literature and the music of Stravinsky's *Rite of Spring*, the middlebrow debate still took those values for granted. Despite the horrors of the Great War, faith in the possibility of progress remained intact for all but a few intellectuals, as Bertrand Russell acknowledged.[32] The argument developed in this book depends crucially on this point: the reconciliation proposed between nonmaterialistic science and liberal Christianity was based on a continued belief in progress and in the purposefulness of the material universe. It was taken seriously only because a large proportion of the educated public—to say nothing of the scientists and the Modernist clergy—still hoped for progress. Curiously, the literary elite paralleled more traditional Christian thinkers, both evangelical and Catholic, in rejecting this faith, although for very different reasons.

Modern images of the past are all too open to distortion through a concentration on literary or artistic innovation, which focuses the historian's attention on an avant-garde that is by no means representative of its generation. This is certainly true for the 1920s and the 1930s: we have constructed our models of what those decades stood for out of what we take to be their most important legacies for the present.[33] A study that hopes to gain insight into the wider attitudes of the time must take account of the literary second-rate as well as the elite. Geographic differences are also important; this book focuses mostly on England, and even here the elite was mostly concentrated in

31. Annan, *Our Age: Portrait of a Generation.*

32. B. Russell, "Eastern and Western Ideals of Happiness," in his *Sceptical Essays,* 99–108, see 100; reprinted in Russell, *Basic Writings,* 555–61, see 555.

33. See, for instance, Baxendale and Pawling, *Narrating the Thirties.*

the "golden triangle" of London, Oxford, and Cambridge. But attitudes toward religion were very different in Scotland, Wales, and Ireland, and writers from those regions participated in the national debates.[34] We must also be alert for changes in the popular consciousness, and here there is another important point crucial for the argument below: If the middlebrow culture of the 1920s retained the hope of progress, that hope was dashed by the events of the following decade. As the economic depression took hold and Fascism reared its head in Europe (and to a limited extent in Britain itself), it became increasingly difficult to retain the belief that the human race could rise to new moral heights.

Developments in science may have threatened the proposed reconciliation with religion—certainly in biology there was a sustained movement away from the nonmaterialistic and anti-Darwinian theories favored by liberal religious thinkers. But even without these developments, the reconciliation was doomed by a more general loss of faith in the idea of progress, which had been a key component of the claim that science revealed evidence of a divine plan. In this more pessimistic climate of opinion, only the Marxists still hoped for a perfect society, and then only after a great struggle with the forces of reaction. For many religious thinkers, Modernism was discredited, and the more traditional interpretations once again seemed plausible: humanity was indeed deeply flawed and needed a divine source of salvation, just as Christians had always maintained. The existential theology of Karl Barth began to penetrate the English-speaking world, bringing with it an indifference to science based on the assumption that a natural theology was impossible. Religious thinkers simply lost interest in maintaining any relationship with science, believing that the study of nature was a distraction from the contemplation of humanity's alienation from God. Science was not so much an enemy as an irrelevance, except to the extent that it was still used by those calling for a centrally planned society. In this harsher climate, the liberal theology that sustained the hope of reconciliation wilted even before the majority of its supporters had recognized that the latest developments in biology were no longer supporting the claim that "Darwinism is dead." Science and religion might continue to coexist, but with a spirit of mutual distrust more akin to a cold war. Only when the real Cold War began in the 1950s did some British scientists and religious thinkers begin to reexamine the possibility of tolerant coexistence or even active cooperation.

34. Historians are increasingly conscious of the significance of local cultural climates; see, for instance, Livingstone, "Science and Religion: Towards a New Cartography."

PART ONE

The Sciences and Religion

CHAPTER 1

The Religion of Scientists

S tudies of the Victorian period have encouraged us to
think in terms of a newly professionalized scientific
community seeking to wrest authority from the churches.
But there was a major reaction in the 1890s against the scientific naturalism promoted by Huxley and Tyndall, and
by the turn of the century there were many scientists who
openly resisted the attempt to portray the search for natural knowledge as a rival to religious belief. The more articulate among them engaged in a major effort to convince the
public that science was not the enemy of traditional religion. On the contrary, they believed that if only religion
could discard those aspects of its origins that did not fit
the new worldview, its underlying foundations would be
strengthened by the realization that we live in a universe
that cannot have been the product of chance. In this way,
an evolutionary natural theology set itself up in opposition
to naturalism and rationalism.

Surveys of scientists' opinions (outlined below) were
published with the obvious intention of bolstering the
claim that the majority did not support an antireligious
stance. These surveys confirmed that many scientists who
did not express their opinions in public were nevertheless
sincerely religious. The British were notoriously reluctant
to bare their souls, and there were eminent figures, such as
the physicist Joseph John Thomson, who retained a deep

faith but kept largely quiet about it. Clues to the views of those scientists who did not write openly on religion can be obtained from their reactions on related issues. The reactions of biologists to antimechanistic philosophies such as that proposed by Henri Bergson tell us something about their deeper feelings, while the involvement of several physicists, including Thomson, in the study of the paranormal is also significant. These clues must be interpreted with caution, however, because some scientists opposed materialism more on moral than on religious grounds. Nor can we equate an interest in spiritualism with formal religious belief, although it is a good indication of a concern that the material world may not be the only focus for human life. Of those who did express such a concern, some were religious in the conventional way (Thomson was an Anglican), while others adopted a vaguer and less explicitly Christian view of spiritual matters.

Better known to the general public was a cohort of scientists who had gained a degree of eminence and now used their positions to argue actively for the new natural theology. Their names were routinely cited by religious thinkers anxious to stress that science was becoming more amenable to a reconciliation with religion. Their names will appear over and over again throughout this study: Robert Broom, J. S. Haldane, Oliver Lodge, Conwy Lloyd Morgan, E. W. MacBride, William McDougall, and J. Arthur Thomson. Around 1930, two new names associated with the implications of the latest developments in physics leapt to prominence: A. S. Eddington and James Jeans. We can only guess at the balance of opinion among the lesser figures whose views would seldom have been canvassed or recorded, but the activity of these popularizers was enough to convince many nonspecialists that the bulk of the scientific community was now on their side.

To the rationalists who still opposed the link between science and religion, there was a major weakness in the claim that the majority of scientists were now more sympathetic to religion. In 1933 Joseph McCabe published a new edition of his *Existence of God*, revised to expose the shallowness of the supposed reconciliation. He insisted that few of the scientists mentioned in the surveys were really distinguished, and that the well-known names were mostly veterans more representative of the late nineteenth century than of current views. According to McCabe, the biologists mentioned were "a lingering group of elderly men—the late Sir A. Thomson, Dr Haldane, etc.—whose watches stopped forty years ago."[1] Morgan's theory of emergent evolution was supported only by "the small group of scientists who admit some mysticism." Oliver Lodge's views were suspect because of the gullibility re-

---

1. McCabe, *Existence of God*, 77; see also 142–43. On Arthur H. Tabrum's survey, see 6.

vealed by his acceptance of spiritualism.[2] McCabe was particularly scathing about the "Jeans-Eddington outbreak," which he portrayed as untypical of modern physics.[3]

J. B. S. Haldane—although sympathetic to idealism himself in the 1920s—agreed with this assessment. He explicitly challenged Bertrand Russell's acceptance of the claim that "the bulk of eminent physicists and a number of eminent biologists have made pronouncements stating that recent advances in science have disproved the older materialism, and have tended to re-establish the truths of religion." Haldane insisted that Russell was wrong to portray Jeans and Eddington as representative of the scientific community as a whole.[4] Other anecdotal evidence suggests that the younger generation of scientists were less likely to be religious—Joseph Needham, for instance, had difficulty in getting his Cambridge colleagues to take his religious views seriously. Those scientists who, like Needham, began to take a stronger interest in social affairs during the 1930s complained that many of their fellows were reluctant to discuss any wider issues (see chapter 2).

Biographical details suggest that McCabe was right in his assessment of the biologists who wrote openly in support of religion. Almost all of the activists were born before 1875 and now occupied senior professorships and administrative positions. The same is true of J. J. Thomson and other senior physicists known to be deeply religious. Eddington and Jeans were slightly younger, but were still very senior by the time they made their mark on the public debate.[5] The majority of the younger generation tended to express more skepticism, if they wrote at a nonspecialist level, or in a few cases remained silent (R. A. Fisher was an Anglican, but wrote nothing on his religious views until the 1950s). Needham is the great exception here, although his views were a good deal more subtle than those of the elder statesmen criticized by McCabe. The older generation thus continued to have influence on public affairs long after it had been replaced by the new at the level of original research. The scientists who gained their reputations in the last decades of the nineteenth century were still highly influential and articulate in the early decades of the twentieth, but they no longer reflected the ideas that were transforming science at the time. By the 1930s their publications were drying

2. Ibid., 112–13.

3. Ibid., 144–45.

4. J. B. S. Haldane, *Inequality of Man*, 158–59; see B. Russell, *Scientific Outlook*, 105.

5. The relevant dates of birth are as follows: Bragg, 1862; Broom, 1866; Haldane, 1860; Lodge, 1851; MacBride, 1866; McDougall, 1871; Morgan, 1852; Rayleigh, 1848; Sherrington, 1857; J. A. Thomson, 1861; J. J. Thomson, 1856. Eddington was born in 1882 and Jeans in 1877. For details of careers, see the biographical appendix.

up as they became less active, and there were few members of the younger generation willing to speak out in favor of the link with religion.

CHANGING PATTERNS OF BELIEF

The aggressive scientific naturalism of Huxley and Tyndall had never reflected the beliefs of the whole scientific community, and the last years of the Victorian era witnessed a resurgence of support for religious values. The new century thus began on a note of optimism for those who wished to reconcile science and religion. As science became more professionalized, it did not necessarily maintain its opposition to traditional culture, in part because the scientists themselves now constituted part of the elite. Even the tragedy of the Great War did not altogether dim the hopes of restructuring Western civilization in a way that would transform the old in the light of the new. Only in the 1930s, as economic depression and the rise of Fascism in Europe began to undermine confidence in human progress, did scientists—like many other thinkers—begin to explore a more critical approach to the relationship between science and society. Historians' attention to this period has naturally focused on the surge of support for Marxism (itself an emotional substitute for religion), but a few scientists jumped in the opposite direction, seeing a return to the Christian faith as a way of acknowledging the failure of the idea of progress.

As Frank Turner has shown, there was always a substantial level of resistance to scientific naturalism from those who wanted to defend a modernized version of traditional religious values.[6] The preference of many biologists for less materialistic versions of evolutionism than Darwinian natural selection makes the same point.[7] Some saw life as an active force in nature and believed that it drew its power from a Creator who intended the universe to develop in a meaningful way. Most Victorian physicists did not share the materialistic opinions articulated in Tyndall's notorious Belfast address of 1874; figures as eminent as Lord Kelvin, James Clark Maxwell, and Lord Rayleigh retained strong religious beliefs.[8] As early as 1864–1865, a group of minor scientists had circulated a declaration that science must be conducted in a way that supports religious belief. This view had been resisted even by some religious scientists on the grounds that it seemed an assault on freedom of thought, but one consequence was the founding of the Victoria Institute in 1865 as a fo-

6.   Turner, *Between Science and Religion.*
7.   Bowler, *Eclipse of Darwinism;* J. R. Moore, *Post-Darwinian Controversies.*
8.   On Kelvin's beliefs, for instance, see Smith and Wise, *Energy and Empire.*

rum within which the intersection of science and Scripture could be discussed in a sympathetic way.[9] In 1875 the physicists Balfour Stewart and P. G.
Tait published an anonymous book, *The Unseen Universe*, arguing openly for a
parallel world in which unknown forces expressed a higher purpose in ways
that could somehow impinge on the world studied by science.[10] Belief in miracles and survival of bodily death was justified on what were claimed to be
scientific principles: the principles of continuity and conservation were valid
only if the nonphysical universe was taken into account. Such ideas were
made more concrete in the late-nineteenth-century articulations of the theory of the ether made by physicists such as Lodge. The ether was a tenuous
material spread through the whole of space, capable of linking and energiz-
ing material structures. Lodge also followed William Crookes in open support of spiritualism, firmly linking his views into a philosophy designed to
challenge scientific naturalism and reinstate a role for religious belief.

The new century thus began on an optimistic note for those who wished
to portray science as a force that could be used to modernize religion rather
than replace it. One illustration of this optimism is a survey conducted for
the North London Christian Evidence League by Arthur H. Tabrum, published in 1910. The survey was begun in the hope of refuting the Rationalist
Press Association's frequently expressed view that the majority of scientists
opposed religion.[11] It reprinted letters from scientists, some gathered four-
teen years earlier and others received in response to a currently circulated re-
quest, to show that a large proportion of the scientific community had strong
religious beliefs. Some of the names cited were members of the previous generation collected in the earlier trawl, including Kelvin, but most were still active either in science or the dissemination of scientific ideas. Several were eminent figures who had not expressed their ideas on this topic in print but were
prepared to commit themselves when challenged. They included the astronomer Robert Stawell Ball, who was an Anglican. The survey notes that
in 1905, Ball joined with Kelvin and a number of other scientists in an appeal
to the Local Education Authorities of England for moral training on Christian principles in the schools.[12] Frank Cavers, professor of biology at a
Southampton college, complained about the influence of agnostic lecturers

9. See Brock and MacLeod, "The Scientists' Declaration."

10. [Stewart and Tait], *The Unseen Universe: or Physical Speculations on a Future State*; see Heimann,
"The Unseen Universe."

11. Tabrum, *Religious Beliefs of Scientists*; see the preface and the Rev. C. L. Drawbridge's introduction, xi–xiii.

12. Ibid., 66–67. Ball is an interesting case because his popular works, such as *The Story of the
Heavens*, made little effort to present cosmology as a vehicle for the argument from design.

(a significant concession to the RPA's claims) and stressed the importance of countering that movement in education.[13] Gerald Leighton, professor of pathology and bacteriology at the Royal Veterinary College, Edinburgh, thought that most scientists were religious, although "few, if any, could be called precisely orthodox."[14] Other biologists who responded positively included George Carpenter, W. Kitchen Parker, and Sidney F. Harmer. The geologist William J. Sollas argued that there was no conflict between science and religion because most men of science saw the mystery behind it all and regretted the conflicts of the Victorian era.[15] When Cambridge dons presented a memorial to the archbishops of the Church of England four years earlier on a doctrinal question, twenty of the names had scientific connections.[16] Overall, Tabrum's survey makes no attempt to quantify the proportion of scientists who were still religious in some sense, making its impression simply on the basis of the substantial number of individuals cited. Most of those contacted were senior figures, but the editor clearly felt that he had enough evidence to challenge the assertion that the majority of scientists were irreligious.

Another survey was published in 1932 by the clergyman C. L. Drawbridge, who had written the introduction to Tabrum's book. This survey was explicitly aimed at the elite, since it solicited the response of two hundred Fellows of the Royal Society, but it made some effort to present a balanced view by listing both positive and negative responses to questions about the relationship between science and religion. The results still came out very much in favor of scientists' continued support for religion, at least in the informal sense. It was probably too soon for the effects of the social and cultural changes of the 1930s to have manifested themselves, especially on senior figures, but the clergy who still hoped to find science on their side would not have been disappointed. Drawbridge did note some resistance to his project even from those sympathetic to religion. One (unnamed) FRS, an orthodox Anglican, thought the survey was a waste of time on the grounds that many would not reply. Others thought the questions simplistic or too personal, and Drawbridge was honest enough to quote the letters he had received expressing such misgivings.[17] In fact, a very large proportion of his subjects did

13.   Ibid., 75–78.
14.   Ibid., 80–81.
15.   Ibid., 140–42.
16.   Ibid., 146–49.
17.   Drawbridge, *Religion of Scientists*, 18–19. It is worth noting that similar surveys are still being carried out, but with conflicting results; see Larson and Whitman, "Scientists Are Still Keeping the Faith," and "Leading Scientists Still Reject God."

reply in a form that allowed their responses to be tabulated for some of the questions asked.

The first question, "Do you credit the existence of a spiritual domain?" was introduced in the book with a quotation from A. S. Eddington, who had now emerged as a leading exponent of the view that the new physics made a religious view of the word possible. Some of those contacted were worried that a positive response implied belief in spiritualism, and 66 refused to answer the question. But of those who did respond, 13 answered in the negative and 121 in the positive.[18] Those who accepted a spiritual dimension included Bishop E. W. Barnes (originally a mathematician), the paleontologists Robert Broom and W. D. Lang, the physiologist J. S. Haldane, the biologists E. W. MacBride and A. C. Seward, Sollas the geologist, and of course, Lodge.

The second question was less directly related to religion: "Do you consider that man is in some measure responsible for his acts of choice?" There were 7 negative responses and 173 affirmative. Next came the once thorny question of evolution: "Is it your opinion that belief in evolution is compatible with belief in a Creator?" There were only 5 negatives and 142 positive responses. Then, "Do you think that science negatives the idea of a personal God as taught by Jesus Christ?" The number of positive (i.e., antireligious) responses now increased to 26, while 103 felt that science and the idea of a personal God were compatible. Of the 71 who gave no clear answer, most indicated that they thought science was irrelevant to the question, not that they were atheists.[19] A question about the survival of the personality after the death of the physical body indicated 47 who believed in its survival and 41 who did not—the closest to parity in any of the questions asked. The final question was, "Do you believe that the recent remarkable developments in scientific thought are favourable to religious belief?"—the point urged directly by Eddington. There were 27 negative responses, 77 affirmative, and 99 who gave no clear answer. In summing up his efforts, Drawbridge suggested that many scientists retained some form of religious belief even though they were suspicious of dogmatic theology. He believed that his survey offered remarkable evidence of a trend away from the materialism and determinism of the previous generation.[20]

For all its limitations, Drawbridge's book does provide evidence that among senior figures in the scientific community, there was as yet no major trend away from religion. On almost every question there was a significantly

18. Drawbridge, *Religion of Scientists*, 28–32.
19. Ibid., 84–95.
20. Ibid., 133.

higher number of positive responses, indicating that relatively few scientists were militantly opposed to a generalized form of religious belief. Whatever the general cultural trends in British society, scientists were aware that the credibility of the old mechanistic worldview had been undermined, and some, at least, shared Eddington's opinion that it was now possible to believe in a spiritual dimension again. The thirties would bring a new wave of problems, of course, and we have no surveys equivalent to those of Tabrum and Drawbridge to provide any overview of how the opinions of the scientific community changed at that time. Anecdotal evidence suggests that the new generation was less concerned with religion than the senior figures surveyed by Drawbridge. The rise of Marxism must have had some effect in turning younger scientists away from religion (although Joseph Needham managed to combine the two). A very few scientists joined those intellectuals who reacted to the crisis of civilization by returning to orthodox religion.

SCIENTISTS AND CHRISTIANITY

Tabrum and Drawbridge both concluded that scientists were more likely to have a generalized sympathy for religion than to accept the position of a particular church or faith. Some scientists did have a Christian faith, however, and although they did not always express that faith in public, their contemporaries may well have been aware of their feelings. Their faith was often of the more liberal kind known as Modernism within the Anglican Communion, which minimized the role of miracles and placed less emphasis on original sin. Nevertheless, these were Christians who acknowledged Christ as a transforming spirit in their lives, and their membership in the Anglican and other churches marked them off from the vaguer kind of theism that the surveys suggested was more popular among the scientific community. A few scientists held to a more conservative faith, as in the case of those who still found it difficult to accept evolutionary theory.

Several eminent scientists remained members of the Anglican church throughout their careers. This is especially obvious in the physical sciences, reflecting the role played by physics in the reaction against the naturalistic philosophy of Victorian times. Even in the 1890s, the theory of the ether had been developed in a way that challenged the mechanistic viewpoint, and the physicists who built this worldview were still influential figures in the early decades of the new century.[21] William Barrett, professor of physics at the

21. D. B. Wilson, "The Thought of Late Victorian Physicists"; Wynne, "Physics and Psychics."

Royal College of Science in Dublin until 1910, shared Lodge's commitment to spiritualism and, like Lodge, wrote extensively to argue that the ethereal world provided knowledge that would put religion on a new foundation. But while Lodge's interpretation of Christ's teachings did not fit the orthodox Christian position, Barrett remained a communicant at St. Martin-in-the-Fields in London after his retirement.[22]

Both Lord Rayleigh and J. J. Thomson were Anglicans, and although neither wrote on science and religion, they were both openly interested in spiritualism. John William Strutt, third baron Rayleigh, was a close friend of the politician and philosopher Arthur Balfour, marrying his sister Evelyn in 1871. He became widely known for his isolation of argon in 1894 and was still active and influential after 1900, being awarded the Nobel Prize for physics in 1904 and serving as president of the Royal Society in 1905. He believed strongly in the value of prayer and was a regular churchgoer, although he disliked Anglo-Catholic ritualism. Rayleigh clearly believed in the "unseen universe" and saw Christ as a great moral teacher:

> I have never thought the materialist view possible, and I look to a power
> beyond what we see, and to a life in which we may at least hope to take
> part. What is more, I think that Christ and indeed other spiritually
> gifted men see further and truer than I do, and I wish to follow them as
> far as I can.[23]

The latter sentiment echoes the undermining of the Savior's unique status that many traditionalists saw as the most disturbing outcome of Modernism.

Joseph John Thomson, whose discovery of the electron in 1897 marked an important step toward the new physics (although Thomson himself remained committed to the ether theory), was also a devout Anglican and an opponent of ritualism.[24] Like Rayleigh, he was reticent on such matters in public, his only published indication of his faith being the concluding sentence of his 1909 presidential address to the British Association for the Advancement of Science: "Great are the Works of the Lord."[25] The crystallog-

22.   See R. J. Campbell's preface to Barrett's *Personality Survives Death*, vii. (This is the R. J. Campbell who, as a Congregationalist minister, stirred up the "New Theology" controversy in 1907; see chap. 7 below.)

23.   From a letter on a family bereavement, quoted in R. J. Strutt, *John William Strutt*, 361.

24.   R. J. Strutt, *The Life of Sir J. J. Thomson*, 283–85. Thomson was consulted by Lord Saunderson, who was drawing up an account of relativity for the archbishop of Canterbury, Randall Davidson; see 203–4.

25.   Thomson, "Inaugural Address," 257.

rapher Sir William Henry Bragg was similarly a devoted Anglican who made few references to religion in his publications, although he did include a paragraph on the topic at the end of his popular *The World of Sound* in 1920.[26] He began to speak more openly about the link between his science and his faith in the last years of his life, giving the Riddell Memorial Lectures on this theme in 1941.[27] Bragg's argument was that science itself did not offer certainty, so the religious search for faith was not as far removed from the scientific quest as most people supposed. For him, Christianity had worked as a means of transforming his personal life, and it had thus passed the test of experience.

Rayleigh and Thomson shared Lodge's commitment to the theory of the ether and his interest in the paranormal (although both were more skeptical about the evidence). For most early-twentieth-century thinkers, it was the new physics of relativity and the quantum theory that symbolized science's destruction of the old mechanical universe and opened the gates to a synthesis with the search for spiritual values. Arthur Stanley Eddington most clearly articulated this position from the scientific side, proclaiming in his *The Nature of the Physical World* that "religion first became possible for a reasonable scientific man" in 1927.[28] Whatever the significance of that year as a turning point in the destruction of the old mechanistic philosophy, Eddington himself was a lifelong member of the Society of Friends (Quakers) and had thus no need of science to create a foundation for his faith. His faith was certainly apparent to his readers—his *Science and the Unseen World* was based on his Swarthmore Lecture of 1929 for the Society of Friends—although he did not go out of his way to promote a specifically Christian view of religion in his popular writings. Yet his whole approach threw the emphasis back onto religious faith. He claimed only that science had made religion *possible* once again; it did not prove that there was a God, or reveal any insights into the nature of God. It was up to the individual to search his or her heart for those insights—a position that eminently fitted the quiet mysticism of the Quaker faith.

A few eminent engineers made reference to Christian principles in the course of the early-twentieth-century debates. In 1932 Sir Alfred Ewing gave his presidential address to the British Association, lamenting the harmful effects of modern science and technology being highlighted by the Great De-

26.   Bragg, *World of Sound*, 195–96. See Caroe, *William Henry Bragg*, chap. 11, and Hiebert, "Modern Physics and Christian Faith."

27.   Bragg, *Science and Faith*.

28.   Eddington, *Nature of the Physical World*, 350. On Eddington's religious beliefs, see Douglas, *Life of Arthur Stanley Eddington*, 31 and 128.

pression. In a committee set up to consider the problem, Ewing expressed the view that "the only way we should find our salvation was that industry should be carried on in the spirit of the teaching of Christ."[29] The electrical engineer Sir J. Ambrose Fleming, inventor of the thermionic valve, was well known for his evangelical Anglican faith, which led him to undertake a critique of the theory of evolution. Beginning in 1928 and continuing through the 1930s, Fleming published a series of attacks on the theory, drawing a response from no less a figure than Sir Arthur Keith. He was a leading figure in the Victoria Institute, contributing to the sustained attack on evolution mounted by that institution.[30]

According to William Osborne Greenwood, whose *Biology and Christian Belief* was published by the Student Christian Movement in 1938, the physicists had done much to turn back the tide of materialism promoted by biologists.[31] Greenwood admitted, however, that the perception of biology as a leading source of materialism was generated by the propaganda of a few activists. There were certainly some biologists and physicians who were Christians or who brought a Christian message into their writings. Perhaps the best known is the Scottish-born paleontologist Robert Broom, who did important work on the origin of mammals from his base in South Africa, but visited Britain regularly. Broom remained a Presbyterian throughout his life and corresponded with Bishop E. W. Barnes, a Modernist Anglican whose antiritualist views appealed to many Nonconformists. Despite his important contributions to phylogenetic research, Broom was an old-fashioned thinker who still believed that evolution unfolded in accordance with a divine plan.[32] A new slant on this theme was developed in a book entitled *Evolution and Redemption* by the medically trained H. P. Newsholme. Here progress was interpreted as the struggle of spirit against matter—but something had gone wrong at the beginning and thrown progress off course, requiring Christ to step in as the "rescuer." Newsholme was sympathetic to Lodge's views on the ether as the medium through which spiritual activity took place.[33] The entomologist Charles J. Grist, a friend of the Darwinian evolutionist Edward B.

29.   Quoted in the *Manchester Guardian*, 5 September 1932, 12; see McGucken, *Scientists, Society, and State*, 32, where this remark is dismissed as illustrating the inadequacy of Ewing's response.

30.   For details of Fleming's arguments and the responses, see chap. 4 below.

31.   Greenwood, *Biology and Christian Belief*; see, e.g., 11–12. See also Greenwood's *Christianity and the Mechanists*, 9, which more explicitly condemns the biologists.

32.   See Findlay, *Dr. Robert Broom*. Broom's correspondence with Barnes is preserved in the E. W. Barnes papers, Birmingham University Library; see especially the letter of 16 January 1929, EWB 12/5/104. Broom's views on evolution were published in his *Coming of Man: Was It Accident or Design?*

33.   Newsholme, *Evolution and Redemption*, 90–94.

Poulton, wrote *Science and the Bible* in 1941, noting that he had returned to a belief in the Bible's fundamental truths after a long episode of unbelief.[34] He
argued that the texts should be treated as allegory and metaphor, and developed the theme that evolution was a spiritual process. In 1945 the eminent
physician Lord Horder contributed to a brief booklet, *The Philosophy of Jesus.*[35]

A major current of opposition to Darwinism centered on the Lamarckian theory of the inheritance of acquired characters, long associated with the
claim that evolution was a purposeful process directed by the mental powers
of animals. In the late nineteenth century, the botanist George Henslow, an
Anglican clergyman, had written widely in favor of Lamarckism. In the early
twentieth century, Henslow joined Lodge in the chorus of support for spiritualism, arguing that it was vital for the future of the church to overthrow
materialism by demonstrating the survival of the personality after death.[36]
Another Lamarckian who expressed a Christian faith openly was the pathologist J. George Adami, who ended his career as vice-chancellor of Liverpool
University. Adami was an Anglican who preached the occasional sermon, including one at the British Association meeting in 1923. He played a major role
in a church congress organized by Charles Raven in 1926 to foster the link
between science and religion, and his address there on "The Eternal Spirit of
Nature" was published by *Modern Churchman.* Adami also wrote a booklet on
the unity of faith and science for the Anglican Evangelical Group.[37]

Biologists also figured prominently in the socialist-inclined Social Relations of Science movement of the interwar years, which called for scientists
to show greater awareness of the practical and moral implications of what
they were doing. The more radical became Marxists, and some—most
prominently J. D. Bernal—outright opponents of religion. But the link between socialism and atheism can be overstated, and several prominent members of this group retained some form of Christian faith. Lancelot Hogben,
an outspoken materialist and opponent of Eddington-style idealism, described himself as an atheist, yet he had joined the Society of Friends as a
student and retained the connection through much of his career, resigning

34. Grist, *Science and the Bible,* preface.

35. Roberts and Horder, *Philosophy of Jesus.*

36. Henslow, *Proofs of the Truths of Spiritualism,* 3; see also *Religion of the Spirit World.* On Henslow's
Lamarckism, see Bowler, *Eclipse of Darwinism,* 85–86, and on his religious views, see Moore, *Post-
Darwinian Controversies,* 221 and 233.

37. See M. Adami, *J. George Adami,* 99–101. Adami's contribution to the church congress is
noted in the *Nature* report "Science and Religion." See J. G. Adami, "The Eternal Spirit of Nature"
and *The Unity of Faith and Science.* For his evolutionary ideas, see J. G. Adami, *Medical Contributions to the
Study of Evolution.*

only with reluctance as late as 1955.[38] Although he claimed that socialism filled the vacuum left by his rejection of theism, one of his characteristic sayings was "I'm an atheist, thank God." His connection with the Quakers reflected an ongoing commitment to their moral values, but Hogben's materialism seems to have left him enough flexibility (in his private mind, at least) to preserve some links with their undogmatic form of Christianity.

Joseph Needham's intellectual development also illustrates that a move to the left did not necessarily entail a rejection of Christianity. Needham is best known to historians of science for his work on the history of Chinese science, but he began his career in the 1920s as a chemical embryologist deeply concerned with the relationship between the physical and biological sciences. As a student at Cambridge he was a member of an Anglican order, the Oratory of the Good Shepherd, and he remained active in the church through the next couple of decades. He belonged to a High Church socialist movement, the Order of the Church Militant, and worked with Anglican intellectuals such as the Rev. J. H. Oldham and Canon B. H. Streeter, who were trying to make the faith more acceptable to the modern climate of opinion. Needham has often been described as an Anglo-Catholic, and he was certainly High Church (that is, he welcomed the traditional ceremony surrounding the sacraments), but in other respects he was a Modernist who wanted to bring the church's beliefs into line with the ideas of science. No Anglo-Catholic (in the doctrinal sense) could have written, as Needham did in 1934: "I am only a Christian because I happened to be born in the European West in 1900, and Anglican Christianity was the typical form religion took for my time and race."[39] He saw no conflict between his interpretation of Christianity and his growing commitment to Marxism after 1927, and was in close contact with Conrad Noel, the "Red Vicar" of Thaxted in Essex. Religion was important to him because it gave a sense of the holy to those aspects of human life that involved ethical and other values. He was prepared in the end to admit that Christianity might disappear, although its spiritual strengths would be preserved in some other form—he would not complain if the spirit were taken away and the letter left behind. But in the meantime, he spoke for those

---

38. For Hogben's attack on Eddington's philosophy, see his *Nature of Living Matter,* chap. 10. On his atheism, see Werskey, *Visible College,* 105, but on his membership in the Society of Friends, see G. P. Wells, "Lancelot Thomas Hogben," 188–90. The "unauthorised autobiography," *Lancelot Hogben: Scientific Humanist,* notes his early links with the Quakers, but then stresses the break with religion around 1918; see 40, 68–69.

39. From a statement of his beliefs written in 1934, Joseph Needham papers, Cambridge University Library, F 177. On Needham's involvement with Oldham's group, see sections L 2–8. On Needham's religious views, see Goldsmith, *Joseph Needham,* esp. 40–41 and 55–65.

who wanted to preserve the old forms of liturgy, putting new wine into old bottles.[40]

Needham published widely on the relationship between science and religion, editing the collected work *Science, Religion, and Reality* in 1925. Significantly, however, he confessed that he had tried in vain to interest his Cambridge colleagues in his religious beliefs, and conceded to Lloyd Morgan that they regarded him as a mystic.[41] A decade later he joined Charles Raven on the editorial board for a book designed to show how Christianity and socialism could be combined to resist the challenge of Fascism, and also contributed to the book himself.[42] Needham was committed to the idea of progress, despite all the obstacles in the way. It was important to show that *History Is on Our Side*—the title of his 1946 book of essays. His own scientific work was in biochemistry, in which he strove to apply an organicist approach that negated the philosophical dangers of mechanistic materialism without requiring an appeal to vital forces. But he was aware of contemporary developments in evolutionary theory, and one aspect of his appeal to history was the claim that evolution's development of increasing levels of coherence in organisms and groups favors the development of a collectivist (but not totalitarian) society in the future.[43] Needham believed that this Marxist faith should also be the faith of any Christian aware of the egalitarian social movements that have constituted the faith's true contribution to social policy from the early church through to seventeenth-century England and now into the present.

Another younger biologist who retained a Christian faith was R. A. Fisher, a statistician and one of the founders of the genetical theory of natural selection. Like Needham, he was an Anglican throughout his career, although their political views were very different—Fisher being an ardent supporter of eugenics (planned control of human breeding). Also unlike Needham, Fisher wrote very little on science and religion until the later years of his life, when he gave the Eddington Memorial Lecture in 1950 and wrote a short article in 1955. He also preached sermons in his college chapel at Cambridge.

---

40.   Needham, "Science, Religion and Socialism," in his *Time, the Refreshing River*, 42–74; see 59–60. This is a modified version of the article under the same title in Lewis, Polanyi, and Kitchin, eds., *Christianity and the Social Revolution*, 416–41.

41.   See Goldsmith, *Joseph Needham*, 41; Needham to Morgan, 18 November 1925, Morgan papers, Bristol University Library, file DM 612.

42.   Lewis, Polanyi, and Kitchin, eds., *Christianity and the Social Revolution*. On the Society for Freedom in Science, see Werskey, *Visible College*, 281–92.

43.   Needham's essay "History Is on Our Side" was originally a paper for the Modern Churchmen's Union in 1937; see Needham, *History Is on Our Side*, 22–34; on evolution, see 24–27. See also his *Time, the Refreshing River*.

Fisher was a Modernist who argued that science was compatible with belief in a personal God and "a historical person, Jesus of Nazareth, [who] exhibited and taught the perfect way of life, which God desires human beings to endeavour to follow in a spirit of gratitude and benevolence."[44] His work on the genetical theory of natural selection was inspired by a desire to show that the theory, far from deserving its old image as a key plank in the case for a mechanistic universe, undermined determinism and was thus a plausible means by which a Creator might seek to encourage the development of higher forms of life with a degree of freedom of choice. He saw a close parallelism between the logic of the Darwinian theory and the Christian view that the service of God requires not only faith but also good works—good intentions are not enough, and the successful organism is one that achieves something in the real world.[45] Fisher's assumption that some individuals were better able to make such contributions than others led him to support eugenics, and he was involved with groups trying to encourage the church to take an interest in eugenic policies during the 1930s. His belief that Christianity imposes a moral duty on us to improve the race (or at least to prevent its degeneration) was shared by many Modernist Anglicans.[46]

A small number of scientists went through a phase of atheism or materialism and then returned to a religious faith. An early example was the Darwinian biologist George John Romanes, whose *Thoughts on Religion* was published posthumously in 1895 after he had been received back into the Anglican Church.[47] We have already noted the case of Charles Grist, who also returned to the Anglican fold. Others turned to Roman Catholicism. The anatomist Sir Bertram Windle was raised as an Orangeman (an Irish Protestant) and became an atheist during his medical training at Trinity College, Dublin. He soon became a convert to Roman Catholicism, however, insisting that it was the severity of his strict Protestant upbringing that had been responsible for his temporary rejection of religion. Through a distinguished career as a medical teacher and university administrator at Birmingham, Cork, and Toronto, he wrote extensively to defend the church against the charge that it was opposed to science. Hilaire Belloc wrote an introduction to Windle's *The Catholic Church and Its Relations with Science*, published in 1927. Windle supported a vital-

44. Fisher, "Science and Christianity," 995.

45. Fisher, *Creative Aspects of Natural Law*, 20; see also Box, *R. A. Fisher*, esp. 8, 193, 293–95.

46. See Fisher's correspondence with Bishop E. W. Barnes, Barnes papers, Birmingham University Library; see also Bennett, ed., *Natural Selection, Heredity, and Eugenics*, 34–35 and 181–82.

47. Romanes, *Thoughts on Religion*; see the concluding summary, 184. On Romanes's spiritual journey, see Moore, *Post-Darwinian Controversies*, 107–9, and R. J. Richards, *Darwin and the Emergence of Evolutionary Theories of Mind and Behavior*, 370–75.

ist stance in biology and rejected the Darwinian view of evolution. Writing to a friend in 1914, he argued that it was difficult for the scientist to acknowledge religion because to accept something on faith went against the scientific method.[48] Once converted, he had no difficulty retaining his faith and was prepared to write in support of the church's position on miracles, suggesting that the evidence from Lourdes had transformed the situation on this score.[49]

Two scientists of note converted to Catholicism in the traumatic days of the 1930s and early 1940s. The astronomer and mathematician Sir Edmund Whittaker was received into the church in 1930 and gave the Donnellan Lectures at Trinity College, Dublin, in 1946, linking modern science to the philosophy of St. Thomas Aquinas.[50] The chemist and historian of science F. Sherwood Taylor was raised as a Christian and found himself living in a schizophrenic state as his work seemed to commit him to a materialistic worldview. As late as 1937 he was commissioned to write a book on Galileo by the RPA, but by 1941 he had converted to Catholicism, having decided first that the evidence for Christianity itself was sound, and then that having taken one step toward belief, he might as well go the whole way.[51] This kind of reaction was, perhaps, more typical of those nonscientific intellectuals who, faced with what was perceived as a choice between Marxism and formal religion, opted for the latter.

SCIENTISTS AND THEISM

The surveys by Tabrum and Drawbridge found large numbers of scientists who were loosely sympathetic to religion. Without accepting the traditional Christian message of redemption through Christ, these scientists were nevertheless prepared to believe that there was a God who had created the universe, and perhaps that He was able to interact with that universe, at least through human spiritual faculties. They were thus theists if not actually Christians, and many would have accepted the moral teachings of Christ as an important step in the growth of civilization. To some extent, the latter position would have brought them close to the beliefs of the extreme Modernists within the Anglican church.

48. See the letter quoted in M. Taylor, ed., *Sir Bertram Windle*, 233–34; see also Neeson, "The Educational Work of Sir Bertram Windle."

49. Windle, *On Miracles and Some Other Matters*, esp. 25–47.

50. Whittaker, *Space and Spirit*.

51. F. S. Taylor, *Man and Matter*, esp. 22–29; see also Hastings, *History of English Christianity*, 290.

Most theistic scientists campaigned against materialism, although opposition to old-style materialism cannot automatically be equated with religious belief. One could be an antimechanist for moral rather than religious reasons, and several important figures associated with the antimechanist movement must be viewed in this light. Perhaps the best example of this moral antipathy to materialism is provided by the psychologist William McDougall, who moved from Britain to America in 1920 but continued to publish widely for a British readership. McDougall wrote tirelessly to promote the view that mind was a force that transcended the material world, and his writings were frequently cited by supporters of the new natural theology. Yet McDougall's autobiography reveals him to be an agnostic with only residual sympathies for religion. He expressed his core motivation thus:

> To reconcile Science with morals seems to me a more urgent need than its reconciliation with Religion. I have never yet been able to convince myself that religious belief of any kind is an important human need. . . . On the other hand, belief in the efficacy of moral effort and the reality of moral choice does seem to me an imperative human need.[52]

All talk of morality in a materialistic world was nonsense—hence the need to oppose that philosophy.[53] McDougall professed some sympathy with the religious outlook, however, accepting that its continued existence probably depended on resisting the onslaught of materialism and that its collapse would be "calamitous for our civilization."[54] He supported research into the paranormal, but claimed that (unlike many others) he had little personal interest in providing evidence of survival after death. If anything, he became gradually more skeptical about psychic effects, with the sole exception of telepathy.

McDougall was aware that his position was that of a minority. Most opponents of materialism wanted to see a wider manifestation of purpose in the universe consistent with the belief that it was created by an Intelligence wishing to achieve a moral end. Theistic scientists often joined with Christians to promote the evolutionary natural theology in which science revealed a universe designed to advance toward higher levels of organization and aware-

---

52. McDougall, "William McDougall," in Murchison, ed., *A History of Psychology in Autobiography,* 218.

53. McDougall, *Modern Materialism and Emergent Evolution,* vii.

54. McDougall, *Body and Mind,* xviii.

ness. A regular vehicle for promoting this position was the Gifford Lectures on natural theology, given at the Scottish universities of Edinburgh, Glasgow, Aberdeen, and St. Andrews. Founded by the will of advocate and judge Lord Gifford, the first of these lectures was delivered in 1888. Over the succeeding decades they provided a platform for a number of scientists, philosophers, and theologians to promote their views on the positive implications of the study of nature for religious belief. The scientists who contributed included the German biologist Hans Driesch, who provided an English-language account of his vitalistic philosophy in the Giffords for 1907–1908, while J. S. Haldane explored the same theme in 1927–1928. One of J. Arthur Thomson's most extended defenses of his progressionist view of evolution was his Giffords for 1915–1916, while Lloyd Morgan's theory of emergent evolution was proclaimed in the Giffords of 1922.[55] Eddington's *Nature of the Physical World* and Whitehead's *Process and Reality* both began their careers as Gifford Lectures in 1927. The lectures and the subsequent publications offered a high-profile opportunity for scientists with theistic views to contribute to the public defense of their position.

Some eminent scientists worked hard to promote the theistic worldview to the general public. Popular books, magazine articles, collected volumes, and radio talks were all employed, and certain names crop up over and over again as authors or contributors. Some of these were well-established figures now long past their prime as scientific researchers. Lodge, Thomson, and Morgan clearly devoted much of their time in the later parts of their careers to promoting a worldview that had become established by the reaction against scientific naturalism at the end of the previous century. They provided the new terminology (such as Morgan's "emergence") that helped to give progressionist, teleological evolutionism an air of having moved beyond the Victorian era—although often without the conceptual clarity that would have helped real scientific research. Only in physics did the latest developments in science seem to aid the antimechanist position, and here the idealism of James Jeans reinforced the work of Eddington to provide the foundation for a popular form of religious belief.

Several volumes of collected essays (some of the later ones based on radio talks) give us a series of snapshots of the move to create the new natural theology. As early as 1904, Lodge, Thomson, and Patrick Geddes wrote essays for a volume titled *Ideas of Science and Faith.*[56] Joseph Needham's *Science, Re-*

---

55. Driesch, *Science and Philosophy of the Organism;* Haldane, *The Sciences and Philosophy;* Thomson, *System of Animate Nature;* Morgan, *Emergent Evolution.* On the Gifford Lectures, see the introduction to Spurway, ed., *Humanity, Environment, and God.*

56. Hand, ed., *Ideas of Science and Faith.*

ligion, and Reality of 1925 brought together scientists, philosophers, and religious thinkers, with Needham himself and Eddington providing the main scientific component. Two Anglo-American books edited by Frances Mason provided the most explicit collected articulation of the new natural theology. The British contributors to *Creation by Evolution* (1928) included Charles Scott Sherrington, Thomson, MacBride, Morgan, and paleontologists D. M. S. Watson, F. A. Bather, and Arthur Smith Woodward. The successor volume, *The Great Design* (1934), addressed "the greatest question in the world to-day: Is there a Living Intelligence behind Nature, or does the great Cosmos somehow run itself, driven by blind force?" and sought to show "that science is not undermining or superseding religion, but is discovering for religion a vaster and sublimer universe and thus supplying a surer foundation for Faith."[57] The contributors included Thomson, Morgan, MacBride, and Lodge. Supporters of the new natural theology were even able to gain access to collected volumes planned by those with more positivist views. A 1923 volume called *Science and Civilization,* originating from an Adult School Union conference, included essays by Thomson and Whitehead, along with more radical pieces by Julian Huxley, Charles Singer, and others.[58] J. G. Crowther, science correspondent of the *Manchester Guardian,* edited a volume called *Science Today,* originally planned by Thomson, which included essays by Morgan, Hogben, and John Joly.

When the BBC's originally very strict policy on religious broadcasting was relaxed, one of the first products was a symposium on science and religion broadcast from September to December 1930. The relaxation was by no means complete, of course, and no explicitly antireligious views would have been tolerated—as a member of the establishment, Julian Huxley was perhaps lucky to get his very controversial views included. Other scientists included were Thomson, J. S. Haldane, and Eddington.[59] A few years later, a more general series called *Science in the Changing World,* which debated a wide range of philosophical and social issues, included contributions by Huxley and Lodge on religious questions.[60]

Several of the names cited above will appear frequently below because their writings were sufficiently prolific or influential to catch the attention of both the clergy and the general public. All shared an antipathy to materialism in biology, an enthusiasm for progressive or teleological evolutionism, and an

57.   Mason, preface to *The Great Design,* 5–6.

58.   Marvin, ed., *Science and Civilization.*

59.   *Science and Religion: A Symposium* [no editor listed].

60.   Adams, ed., *Science in the Changing World.*

interest in nonmechanistic approaches to physics. As McCabe complained, however, they were all members of the generation that had gained its reputation by the turn of the century and was now more absorbed in administration or popular writing than in original research. Their views are typical of the extremely liberal approach to religious belief, reflecting a general theism that was either explicitly hostile to the Christian message or required that message to be modernized to an extent that most orthodox Christians would have seen as a betrayal of the faith.

The physiologist John Scott Haldane was widely hailed by religious writers as a scientist who wanted to reintroduce a role for nonphysical forces, although his attacks on mechanistic biology had a more complex foundation. He had abandoned formal religion at an early age and adopted a philosophy that was certainly theistic, but offered little comfort for Christians. He resigned his membership in the Free Church of Scotland in 1883, disgusted by the factionalism within the churches.[61] But at the same time, he was disturbed by the naturalistic philosophy of Huxley and Tyndall and by its links to a liberal social policy in which society was seen as no more than an aggregate of atomistic individuals. Deeply influenced by the idealist philosophy of T. H. Green and F. H. Bradley, he soon developed an idealist viewpoint of his own, which allowed him to treat life as an autonomous entity to be studied by the biologist.[62] His twentieth-century writings on science and religion argued that consciousness was the key element of reality and suggested that God was, in effect, the consciousness of the whole universe.[63] Not surprisingly, Haldane was accused of being a pantheist by orthodox Christians. He responded by arguing that he could not belong to a Christian church because they were all infected by materialism. For Haldane, materialism included the belief that minds can exist independently of the universe revealed by the senses, for which reason he rejected both immortality and miracles produced by supernatural beings.[64]

Far more positive endorsement for a theistic position came from evolutionary biologists who were prepared to argue that the progress of life toward

61.   For an account of Haldane's philosophical and ideological positions, see Sturdy, "Biology as Social Theory: John Scott Haldane and Physiological Regulation."

62.   Haldane's idealism is already apparent in a lecture given to the Royal Medical School, probably just after 1882; see the manuscript notebooks, National Library of Scotland, MS 20656, and R. B. Haldane and J. S. Haldane, "The Relation of Philosophy to Science."

63.   On Haldane's later idealism, see his *Philosophy of a Biologist*, preface; his Gifford Lectures were published as *The Sciences and Philosophy*. See also his "Natural Science and Religion," "Biology and Religion," and the essays collected in his *Materialism*.

64.   See, e.g., Haldane, *Materialism*, 139–41.

humankind was the unfolding of a divine plan or purpose. Even the veteran Darwinist Alfred Russel Wallace adopted such a position in his last major survey of evolution, *The World of Life: A Manifestation of Creative Power, Directive Mind, and Ultimate Purpose* (1910). At an earlier stage in his career, Wallace had startled Darwin by making an exception for the origin of humankind, arguing that here at least, some supernatural intervention was needed to explain the appearance of the soul. His defection from the mechanist camp was prompted by his conversion to spiritualism as well as by his socialist political views.[65] Now, in his final years, he went further to build teleology into the whole history of life, arguing that its preordained goal was the production of the human mind.

J. Arthur Thomson, professor of natural history at Aberdeen, also endorsed an organismic philosophy and the claim that evolution was a morally purposeful process. While studying under Haeckel in the 1880s, Thomson had still thought that he might abandon science to become a minister in the Free Church of Scotland. He was converted to a less specifically Christian viewpoint by the sociologist and town planner Patrick Geddes, who favored an almost mystical philosophy verging on pantheism.[66] Thomson retained much respect for the church and its teachings, however—he sent the Missionaries' Furlough Society the sum of thirty shillings so that they could acquire a copy of his *System of Animate Nature.*[67] He wrote a number of works arguing that while science did not prove the existence of a Creator, it certainly made the theistic position seem highly plausible.

Thomson's views paralleled those of evolutionary psychologist C. Lloyd Morgan, who became vice-chancellor of Bristol University. Morgan's early career had been strongly influenced by T. H. Huxley, but he had soon turned away from the naturalistic philosophy.[68] He remained a deeply religious man, but had little interest in formal Christian theology. He seems to have shared Lodge's view of Christianity as a religion that had much to offer if only it

65. See Kottler, "Alfred Russel Wallace, the Origin of Man, and Spiritualism"; Turner, *Between Science and Religion*, chap. 4; Oppenheim, *The Other World*, 296–325.

66. See the letters from Thomson to Geddes, Patrick Geddes papers, National Library of Scotland, MS 10555, especially those of 2 January, 24 March, and 4 December 1883 and 10 August 1886. The latter stresses Geddes' influence on the young Thomson, equating it with his earlier regeneration through the Christian faith. The later letters illustrate the extent of Thomson's involvement in popularization—he was an editor of the Home University Library series. On Geddes' views, see Kitchen, *A Most Unsettling Person.*

67. J. Arthur Thomson to the Missionaries' Furlough Society, 8 February 1921, letter tipped in to the author's copy of *System of Animate Nature.*

68. See "C. Lloyd Morgan" in Murchison, ed., *A History of Psychology in Autobiography*, vol. 2, 237–64; also R. J. Richards, *Darwin and the Emergence of Evolutionary Theories of Mind and Behavior*, 375–407.

could be brought into line with the new scientific way of looking at the world. He insisted that there was nothing in his philosophy of emergent evolution to preclude the culmination of progress in one unique individual, but stressed that "mythology—and all that is supernatural in *that* sense—is the chaff to be winnowed from the pure grain of the teaching of Christianity."[69] His papers contain an undated outline of a projected book on "Christianity and Emergent Evolution," and in 1929 he was in correspondence with his publishers about a book on "The God of Philosophy and the God of Religion," neither of which was completed.[70] The manuscript of the latter was cut down into another book, *The Emergence of Novelty,* but the references to religious issues were eliminated in the process. Even so, few of Morgan's readers doubted that one of his main aims was to retain the idea that the human spirit was the product of a divinely instituted evolutionary process. A similar position was proposed in 1930 by the biology lecturer Herbert F. Standing, who acknowledged the support of Morgan and the influence of Henri Bergson's idea of a creative life force, or *élan vital,* struggling to overcome the limitations of the material world.[71]

E. W. MacBride, professor of zoology at Imperial College, London, was long known as an opponent of Darwinian materialism. As a Lamarckian, MacBride was able to argue that mind played an active role in directing evolution, and he occasionally allowed a theistic inference to be drawn from this: "If, therefore, we fall back on the principle that the Creator endowed living matter with something that strives to meet adverse circumstances and can control its own growth, we have reached the most fundamental explanation of adaptation which it is now possible to hold."[72] MacBride wrote an article for the Victoria Institute's *Transactions* opposing the materialistic view of evolution.[73] The Victoria Institute encouraged a discourse between evangelical Christianity and science, and in 1933 the rationalist Joseph McCabe suggested that MacBride had undergone a conversion (presumably to Christianity) decades earlier.[74] There is no evidence of this in MacBride's writings, how-

69.   Morgan, *Emergent Evolution,* 31; *Life, Mind, and Spirit,* xii.

70.   The outline on Christianity is in the Lloyd Morgan papers, Bristol University Library, DM 612. A letter from B. N. Langdon-Davies of Williams and Norgate to Morgan, 24 October 1929, mentions the book on the God of religion, DM 128/444. Letters from Morgan to Langdon-Davies in 1932–1933 record the cutting down of his original manuscript to form his *Emergence of Novelty,* DM 612. The typescript of "Evolution and Creative Purpose" is in the undated material in the Morgan papers.

71.   Standing, *Spirit in Evolution,* 12.

72.   MacBride, "The Oneness and Uniqueness of Life," 158 (this is MacBride's contribution to *The Great Design*).

73.   MacBride, "The Present State of Organic Evolution."

74.   McCabe, *Existence of God,* 110.

ever, and he wrote with very mixed feelings on the creeds of Christianity in the *Hibbert Journal* for 1934.[75] Frederic Wood Jones, who had long argued that the human species was only distantly related to the apes, began to adopt an explicitly antiselectionist position in the 1940s. This position was based partly on the Lamarckian theory as defended by MacBride, but also on the claim that the whole history of life was the unfolding of a great plan aimed at the production of humanity.[76]

The vision of a spiritual progress though evolution was also exploited by Oliver Lodge, who used his belief in spiritualism to link the evolutionary model to the physics of the ether. Lodge was a member of the Cambridge network of physicists with close links to Arthur Balfour, but his religious views, unlike those of Rayleigh and J. J. Thomson, departed substantially from anything that would be acceptable to an orthodox Christian. Lodge's scientific work on electromagnetic radiation convinced him that the ether was the medium that unified the whole of nature. He became involved with spiritualism at an early age—a link deeply intensified by the loss of his son Raymond in World War I—and became committed to the idea of a spirit realm on the ethereal plane. He wrote an attack on Ernst Haeckel's monistic philosophy in 1905, and followed it with a stream of popular books and articles designed to prove the reality of spiritualist phenomena and to explain them in terms of his overall worldview.[77] He was critical of orthodox Christianity because its insistence on the Fall contradicted his faith in progress, but he promoted a more general theism that supposed that all life was ultimately destined to become one with its Creator. Since he believed that superior beings existed on the ethereal plane, Lodge had no difficulty seeing Christ as one such being who chose to become incarnated in order to show us the way forward. The Resurrection was clearly intended to refer to Christ's ethereal body becoming visible to His disciples after the Crucifixion, and Lodge thought that orthodox Christianity's insistence on the reappearance of His physical body was a mistake. Why, he argued, should Christ's resurrection be different from the one we can expect in the spirit world?[78] Perhaps death will

75. MacBride, "The Scientific Atmosphere and the Creeds of the Christian Church." See the response by G. W. Butterworth, "What the Scientists Forget." MacBride's obituaries do record, however, that he built up links with the Anglican church at Alton in Hampshire, his home after retirement.

76. F. W. Jones, *Design and Purpose;* on Lamarckism, see his *Habit and Heritage.*

77. Lodge's attack on Haeckel is his *Life and Matter;* see also his *Man and the Universe, Making of Man, Evolution and Creation,* and finally, his *My Philosophy.* See also Jolly, *Sir Oliver Lodge,* and on the link between physics and psychic research, Oppenheim, *The Other World.* On Lodge and Campbell's New Theology, see chap. 7 below.

78. Lodge, *My Philosophy,* 310.

be conquered in the sense that we shall eventually be able to slough off our physical bodies without pain as soon as we reach the right level of spiritual development. This was the liberal theology that Lodge presented to parents and teachers in his catechism, *The Substance of Faith Allied with Science.*

Lodge's interest in evolution shows how the biological and physical sciences could be joined into a coherent presentation of the new natural theology. By the 1920s, however, it was becoming apparent that the physics of the ether had been undermined by relativity theory. Einstein was well known as a determinist and an opponent of the idea of a personal God, but he was convinced that the universe had a rational structure imposed by a designing intelligence, and was widely quoted as saying: "Science without religion is lame, religion without science is blind."[79] The new physics both reintroduced an apparent role for the conscious observer in the description of nature (thus checkmating materialism) and implied that the sensations observed were all the products of an underlying Intelligence at the heart of reality. This was the position endorsed in Eddington's *Nature of the Physical World.*

The other great popular expositor of the new idealism was the cosmologist James Jeans, whose *Mysterious Universe* of 1930 argued that the universe could be pictured "as constituting pure thought, the thought of what, for want of a better word, we must describe as a mathematical thinker."[80] Jeans had no Christian faith, and readily conceded that his purely rational Creator was hardly the God of traditional religion. Nevertheless, his books were immensely popular and played an important role in the effort to convince the general public that the revolutions in physics, far from undermining the renewed interest in design, were actually reinforcing it. But the wave of enthusiasm for Jeans's and Eddington's books was the last major boost that the proposed reconciliation between science and religion would receive. Few other contemporary physicists took up the theme, and there were no younger biologists following in the footsteps of Haldane, Thomson, and Morgan.

METHOD AND MEANING

The scientists who adopted a liberal Christianity believed that theology itself would benefit from a more scientific approach. To them, science took on a moral dimension through its willingness to admit the need for old ideas to be rethought in the light of experience. They saw the liberalization of re-

79. Einstein, "Science and Religion," 605. See Jammer, *Einstein and Religion* for details of what Einstein actually meant by this.

80. Jeans, *Mysterious Universe*, 136.

ligious belief as a parallel process that was necessary for the reconciliation to go forward. The claim that science offered hypotheses, not facts, could also be used to limit the aggressive power of materialism. The old materialism was presented as a philosophical dogma imposed on science, but a true science that left itself open to correction could never be used as the basis for destroying particular elements of belief. By playing down the dogmatic character of religion and by portraying materialism itself as a dogma, these scientists were able to use a more flexible image of the scientific method to support the proposed alliance between science and religion.

There was another side to this emphasis on the provisional nature of scientific knowledge, however. As science increasingly focused on the role of the observer in interpreting sensory data, it became possible to argue that its ability to pass judgment on the validity of other areas of experience was limited. Some more conservative thinkers began to argue that our understanding of nature—to say nothing of human nature—must take account of intuitions telling us that consciousness is real and the cosmos a purposeful system. All too often, such a policy led to constraints being placed on scientific theorizing: materialistic approaches were deemed inadequate on the grounds of intuition rather than on the grounds of observation and experiment.

An early attempt to link Christianity and the scientific method was Gerald Leighton's *The Greatest Life* of 1908. Leighton was medically trained and wrote popular surveys of natural history. His book was an attempt to bring Christianity into line with scientific thinking—he argued that if the churches did not take a more scientific approach, they would soon find that they had lost all authority among educated people while they were busy fighting among themselves.[81] Christianity made a "scientific offer" of a philosophy to live by—that is, an offer that could be tested by its results. Leighton wrote at length on the process by which organisms gained immunity to harmful influences, arguing by analogy that Christ's teachings were what was needed to make us all immune to evil.[82] Leighton's vision of Christianity seems to have been limited to its ethical teachings, and his views on how religion should modernize itself would have been regarded by most practicing Christians as a recipe for abandoning the true message of salvation. In this respect, Leighton anticipated all the problems encountered by Modernism.

The need for a better understanding of how science worked was stressed by J. G. Adami in his paper presented to the Modern Churchmen's Union in 1926. He argued that the true method of science "was not understood or

---

81. Leighton, *The Greatest Life*, ix–x.
82. Ibid., 262–64 and 275.

taught by the dominant scientists of the mid-Victorian era. Those held that the laws which had been discovered regarding material phenomena had been tested and proved, and were therefore fixed and immutable, representing the exact truth upon which they could proceed to build a sound superstructure."[83] Adami noted that Catholic scientists such as Pasteur had always rejected this position, and that their more restricted model of science's powers had been amply confirmed by the revolutions of the early twentieth century. All laws are based on assumptions, he argued, and the emergence of one contradictory fact requires that a law be rejected. This is what we now call the hypothetico-deductive method, but Adami's purpose was to argue that such a flexible, pragmatic approach to the acquisition of knowledge was exactly parallel to that followed by the religious believer: "We cannot prove that there is a God, but our conviction of His existence may be arrived at by progressive assumptions."[84] Science and religion were both striving to see through a glass darkly, slowly groping their way toward clearer understanding.

The same point was made a year later in another Modernist forum—the volume *Adventure: The Faith of Science and the Science of Faith*, edited by Canon B. H. Streeter—in a chapter by Alexander Smith Russell, an Oxford lecturer in inorganic chemistry. Here the method of conjecture and refutation was presented as an "adventure" toward the gathering of knowledge, exactly paralleling the method of religious thinkers seeking enlightenment.[85] Perhaps the most authoritative expression of this view of science by a Christian scientist was William Bragg's Riddell Memorial Lectures of 1941, published as *Science and Faith*. Stressing the role of hypothesis in science, Bragg insisted that the old dogmatism was gone. He presented Christianity as an "experimental religion" that was also willing to learn from experience, with dogma now being treated in the same way as a scientific hypothesis.[86] The "demand for the absolute acceptance of definite items of faith" was no longer acceptable.[87] For Bragg, as for Adami, recognizing that science worked by hypothesis was valuable because it destroyed the alleged certainties on which materialism was based, but it did not open the way for religious values to dictate what the scientist might hope to discover. On the contrary, it was religion that borrowed from science a method that would transform it from a collection of dogmas to a flexible and progressive view of the purpose of human life.

83.   J. G. Adami, "The Eternal Spirit of Nature," 510–11.

84.   Ibid., 514

85.   A. S. Russell, "The Dynamic of Science."

86.   Bragg, *Science and Faith*, 11 and 15–16.

87.   Ibid., 24

Scientists who adopted a more generalized form of religious belief were in a better position to urge the established faiths to adopt a more flexible approach in line with the scientific method. J. B. S. Haldane, in the 1920s still a pantheist rather than a materialist, contrasted science's willingness to abandon its once-cherished theories with the stagnation of religious knowledge. According to Haldane, the kind of faith that transformed people's lives was not faith in dogma, and most Christians probably accepted that they did not really know the true relationship between God and Jesus—although no church dared admit this. If our ideas about moral conduct have changed since the early Christian era, and if our moral sense really is one of the means by which we perceive the divine, then why not admit that religious faith must change too?[88] At this point, Haldane certainly did not think that science gave any certain knowledge of ultimate reality, and believed that the religious sense remained important. The physicist Herbert Dingle, who shared Haldane's suspicion of the mysticism of Jeans and Eddington, also saw science as a limited activity that left room for a religious view of the world. Far from offering knowledge of reality, science, too, spoke in metaphors, and thus science and religion were merely saying the same things with a different emphasis.[89]

The argument that experience allows a "scientific" test of religion was developed in a Roman Catholic context by the chemist and historian of science F. Sherwood Taylor in the 1940s. Contrasting the way of life of the materialist and the Christian, Taylor thought that the superiority of the latter offered evidence in favor of the church's position. "Science commends the test of experiment, and Jesus Christ said 'Ye shall know them by their fruits': so an examination of the effect of the adoption of one or the other philosophy upon the life of Man, as individual, family, or state, should reveal their respective worth."[90] In the darker atmosphere following the rise of Fascism, it may have been plausible to suggest that the fully traditional view of Christianity offered the only alternative worth testing, but Taylor's position deliberately rejected what the Modernists had tried to achieve. For them, the whole point of subjecting religion to scientific testing was to modify its central tenets to bring them into line with what was now considered acceptable. They believed that the willingness of science to abandon outdated theories when necessary was the model that religion should follow. For Taylor, the resulting

88. J. B. S. Haldane, "The Duty of Doubt," in his *Possible Worlds*, 248–49.

89. H. Dingle, "Physics and God"; for his highly critical opinion of Jeans's *Mysterious Universe*, see his "Physics and Reality" and *Through Science to Philosophy*, 7–8. On science's refusal to offer certainties, see also H. S. Allen, "The Search for Truth."

90. Taylor, *Two Ways of Life*, preface.

modifications had not worked, and the only way out of the dilemma of modern life was a return to traditional Christianity.

The idealism promoted by physicists such as Eddington and Jeans offered another philosophical assault on the residue of Victorian materialism. Here the revolutionary developments of modern science seemed to have directly undermined the assumption that an objective world existed entirely independently of the conscious observer. But Jeans and Eddington were not the only exponents of idealism—indeed, J. S. Haldane openly criticized them for continuing the myth that physics was the more basic science from which such major reinterpretations must flow.[91] Haldane accepted that nineteenth-century materialism was "an insignificant eddy in the stream of human progress," but for him it was biology that had emerged as the central science, confirming personality as the foundation of reality.[92] Such a view did not gain wide acceptance, though, and it was the idealist interpretation of the new physics that religious thinkers found most useful in the campaign to argue that the human mind could again be presented as the centerpiece of reality.

To the Marxist philosopher Maurice Cornforth, the new idealism was the end product of a trend in Western philosophy that had been designed to place limits around science's ability to understand and control the material world, thereby establishing an unassailable sphere for the activities of religious belief.[93] He claimed that the same message could be extracted from any version of the philosophy of science that denied that the method of discovery offered certain knowledge of the real world. There were some religious scientists who merely wanted to ensure that science did not rule out intuitive knowledge of the values and purposes lying behind the universe. So long as those intuitions did not generate insights that placed limitations on the scientists' investigations, the various projects of science, religion, ethics, and aesthetics could each go about their business unhindered. But, as Cornforth realized, such a separation of responsibilities was not the real aim of many exponents of idealism and nonrealist philosophies of science. Rather, their purpose was to encourage the belief that the limited certainty of our objective descriptions of nature left room for intuitions revealing evidence of purposeful design and perhaps even of spiritual interference in the regular flow of phenomena. Far from separating religion and science, the aim was to bring them back together again by showing that science was groping its way toward

91.   J. S. Haldane, *Materialism*, 48.

92.   Haldane, *Mechanism, Life, and Personality*, 73.

93.   Cornforth, *Science versus Idealism*.

a new understanding of nature that would square with a modernized form of religious belief.

Such a vision of science's role was offered by the Cambridge professor of pure mathematics E. W. Hobson in his Gifford Lectures for 1921–1922. Hobson insisted that science offered no knowledge of ultimate reality, only constantly modified descriptions of phenomena. The mechanistic viewpoint was still perceived as a threat by many religious thinkers, but in fact it had been abandoned in the face of both the new physics and neovitalist biology. Amply confirming the fears later expressed by Cornforth, Hobson then proceeded to isolate science from philosophy and religion, which could continue their search for knowledge of reality unaffected by its superficial descriptions.[94] Intuition became once again a legitimate source of knowledge and values, miracles could not be ruled out merely because they contradicted the law of uniformity, and purposeful forces could be legitimately inferred to be at work within evolution.[95] These conclusions, especially the latter, brought out the hollowness of Hobson's claim that science (however much weakened) should be divorced from religion. In the end, he allowed the nature of the world to be predetermined by the assumptions of those describing it. Religious intuitions placed restrictions on the conclusions that scientists could derive from their descriptions.

It was precisely this straightjacket that Joseph Needham sought to eliminate by redefining the relationship between science and religion. He acknowledged the role played by idealism in destroying Victorian materialism,[96] but unlike Haldane, accepted that a more flexible mechanistic approach was essential for progress in biology. He insisted, however, that this did not make mental processes illusory, especially now that the physicists had shown the centrality of the observer's role in the description of phenomena. Drawing on the Italian philosopher Antonio Aliotta, who had surveyed the idealist reaction of the turn of the century, Needham insisted that science and religion were two attitudes of mind, not two theories of the universe—they could coexist and complement each other in the thought of the same individual.[97] Needham thus separated science and religion into two spheres of activity, each of value, but each operating independently of the other. Our religious insights, he argued, should be confined to questions of ultimate pur-

94.   Hobson, *Domain of Natural Science*, 452–53.

95.   Ibid., 466–67, 488–90, 495.

96.   Needham, *Sceptical Biologist*, 131–34.

97.   Ibid., 61. Aliotta wrote the chapter on science and religion in the nineteenth century for Needham's *Science, Religion, and Reality*; see also Aliotta's *Idealistic Reaction against Science*.

pose and value—they should not dictate what kind of a material world the scientist should expect to discover. Like Adami and Bragg, Needham was prepared to see Christianity change in the future, perhaps beyond recognition—a view that became more pronounced after his acceptance of socialism.[98]

SCIENCE AND VALUES

The exponents of neovitalism and teleological evolutionism were more open than Hobson in their expectation that science would discover a world whose activity substantiated the idea of a Creator, or at least made it more plausible. Scientists such as Lodge were essentially realists, and they expected the reality they were uncovering to show direct evidence of the activity of both individual minds and the Creator. Science and religion could thus become part and parcel of the same search for meaning, both willing to transform their ideas by testing them against experience. This position made it legitimate for them to claim that science could play a valuable role in the moral education of young people. Science was not just a collection of facts required only as part of a technical education. The model of science as something more than mere fact-gathering could thus become an important component of the claim that science could promote moral values, although this was a dangerous card to play. To the rationalists, science deserved to replace religion because its dedication to the discovery of the truth offered a moral imperative far more important than anything validated by outdated superstition. Religious scientists were placed in a difficult position: they wanted a more active view of how science sought knowledge, but they wanted science to be seen as a force that could transform rather than replace traditional values. They played a role in the attempt to introduce more science into the school curriculum, but were always at the mercy of more conservative forces who preferred the status quo, in which classics and conventional religious instruction shaped the values of the next generation.

To serve a moral rather than a utilitarian purpose, science teaching would have to give the student a sense of the wonder being uncovered by the new natural theology. Henry E. Armstrong, a chemist from the City and Guilds College, complained about the dryness of much science teaching and insisted that the subject must be presented in a way that allowed students to see the meaning and purpose of the natural world.[99] More actively, Lodge published his "catechism for parents and teachers," which proclaimed the evolutionary

---

98.   See Needham, *Time, the Refreshing River*, 59–60.
99.   Armstrong, "The Chemical Romance of the Green Leaf," 193–94.

vision of humankind's origin: "I am a being alive and conscious . . . a descendant of ancestors who rose by gradual processes from lower forms of animal life, and with struggling and suffering became man."[100] A noble vision indeed, but hardly one to endear him or his science to more traditional teachers. J. J. Thomson, a more orthodox Christian, chaired a government committee of 1916 charged with inquiring into the position of the natural sciences in the educational system. The committee recommended more science teaching on the grounds that it could inspire a wider vision of humanity. But its findings were criticized by William Temple (the future archbishop of Canterbury), who chaired the British Association's education section that year.[101] In the middle of a war in which the nation was threatened by Germany's "efficiency," science could be dismissed as a utilitarian force all too easily manipulated by evil. Better to keep the traditional system of education, which was proven to inculcate moral values.

Fears about science's practical consequences continued to be expressed through the 1920s, and in the following decade the social problems of the depression highlighted the potential for industrial technology to be misused. There had long been a Christian socialist movement, of course, and Needham turned in this direction, eventually creating his own unique synthesis of Christianity and Marxism. Others saw Marxism as a philosophy of this-worldly salvation that required a complete rejection of all traditional religion. Needham, at least, appreciated the dangers of a socialist technocracy that would be as indifferent to the needs of ordinary people as the old elite. But the debate was increasingly being conducted without the overt involvement of religious beliefs, and the story of the Social Relations of Science movement has been told largely as an episode in the changing relationship between science and ethical or political values.[102] Nevertheless, Needham's involvement with the 1935 book *Christianity and the Social Revolution* shows that there was some effort to involve both science and religion in the call for a new social order.[103]

The fact that this debate over the ethical dimension of science was conducted with only limited input from religion symbolizes the changing identity of the scientific community. At the start of the century there were still many active scientists who held strong religious beliefs. It was possible for re-

100. Lodge, *Substance of Faith Allied with Science*, 6. This book was published in 1907 and had reached a ninth edition by the following year.

101. See Mayer, "Moralizing Science."

102. On Needham, see Werskey, *Visible College*, 91–92. McGucken's *Scientists, Society, and State* has little to say about the involvement of those with religious beliefs.

103. Lewis, Polanyi, and Kitchin, eds., *Christianity and the Social Revolution*.

ligious thinkers to argue that science was no longer (if it ever had been) a hotbed of materialism. In the interwar years, these scientists remained active, and their efforts fueled the hope of some clergy that the time was ripe for a reconciliation. Admittedly, many of these scientists were theists rather than active Christians, but their endorsement of some sort of spiritual dimension to human life offered a lifeline, especially to those clergy who felt that Christianity itself was in need of liberalization. More seriously, the rationalists were able to claim that most of the scientists willing to provide active support for religion were members of the older generation—although their eminence often guaranteed them access to means of publication, thereby fueling the impression that the scientific community as a whole was now working for a reconciliation. There is almost certainly some truth in the claim that the younger generation of scientists was increasingly indifferent to religion and, indeed, to many wider issues. This indifference—discussed in the next chapter—reflected the emergence of a new set of priorities within a profession that increasingly found its employment in dealing with down-to-earth matters. Few were even willing to debate the moral issues arising from the application of science, although more would become active in the troubled times of the 1930s and 1940s. Support by scientists for a revival of natural theology simply dried up as the older generation became less active. Those younger scientists who were religious tended to remain silent, like Fisher, or, like Needham, got a discouraging reaction from their colleagues. In this sense, the campaign to enlist science on the side of free thought had triumphed—but as much through a loss of interest in the wider issues as through the active espousal of materialist or rationalist principles.

CHAPTER 2

Scientists against Superstition

The metaphor of science at war with religion had been created in part by the rationalists of the late nineteenth century. To them, all forms of organized religion were little more than superstitions that would be swept away as science gave us a more realistic vision of the universe in which we live. But if the rationalists had hoped to see the divorce between religion and science made final, they were dismayed to find that in the new century, their influence was being challenged more actively than ever before. As calls for a reconciliation with liberalized theology grew louder, the rationalists strove to retain the plausibility of their claim that science would inevitably undermine religious faith. Their numbers now included eminent intellectuals, such as Bertrand Russell, who were anxious to do their bit for the popularization of the cause (see chapter 10). A number of scientists, heirs to the legacy of Huxley and Tyndall, were also still willing to speak out, although with less self-confidence than the previous generation. The conversion of several eminent young scientists to Marxism in the 1930s added a new dimension to this well-established tradition.

Mary Midgley suggests that the scientists of this period were susceptible to a materialistic vision of the future, while artists and writers were more inclined to fall back on the traditions of religion.[1] This may be a correct interpretation of what happened in the 1930s, although it ignores the wave of antimaterialism that affected many scientists at the turn of the century. The hope of using science to transform humankind was not always seen as a rival to religious belief. Nevertheless, there was some substance to Joseph McCabe's claim that the scientists who defended religion in public were mostly members of the previous generation. Younger scientists were less willing to endorse the new natural theology—but this did not necessarily mean that they were actively hostile to it. It was not that the new generation ignored religion; rather, they felt that their commitment to scientific objectivity precluded deep involvement with any of the broader issues raised by the impact of what they were doing. This attitude was noted with disgust by those who became more politically active in the 1930s. Writing in 1941, the biologist C. H. Waddington claimed that many ordinary scientists actually welcomed the excuse to limit the range of their thinking: "Responsible scientists, looking at their colleagues, saw the obvious fact that most specialists were quite unfitted to play an important part in the evolution of general culture; but, far from acknowledging that this was a sign of science's failure, they accepted it almost with glee as an excuse which let them out of the necessity of thinking about wider issues."[2] It may well be that this reluctance to engage in wider debate prevented many of the younger scientists from professing open support for religion—but it would also have discouraged them from campaigning for rationalism (and later for Marxism). Whether their indifference was real or feigned is a crucial question that may be difficult to answer in many individual cases.

In biology, open support for a religious position certainly seems to have become unfashionable among younger researchers, as Needham discovered. It may have been regarded as essential for professional reasons to distance oneself from any kind of metaphysical speculation. Michael Ruse has argued that evolutionary biologists were especially likely to adopt this attitude because of the legacy of Victorian progressionism. With the exception of Julian Huxley, most scientific evolutionists believed that it was necessary to present their field as totally free from this kind of speculation—whatever their private beliefs. Thus Arthur Cain, who trained with E. B. Ford at Oxford and worked in collaboration with him for many years, could write to his mentor thus: "Even now, my dear Henry, I have not the faintest idea of what your religious views (if any) are; never once in all these years have you men-

---

1. Midgley, *Science as Salvation*, 145. There were, as we have seen, a few scientists who returned to religion in the 1930s.

2. Waddington, *Scientific Attitude*, 81.

tioned them, even in social conversation, let alone in scientific."[3] Professionals found Huxley an embarrassment, and he was increasingly treated as a dilettante and popularizer, to be rigidly excluded from the arena of real science.

Ruse's suggestion that the legacy of progressionism put evolutionary biologists into a vulnerable position with respect to their colleagues in other areas of science presupposes that the culture of objectivity had acquired a momentum of its own within those other areas. Ruse himself adopts this position when he argues that the gradual exclusion of overt appeals to cultural values in evolutionism is typical of the process by which science has matured in the twentieth century.[4] Scientists in other areas were thus already encouraged to keep quiet on broader issues. This would explain the lack of engagement in public debate by Christian physicists such as Rayleigh and Thomson (Lodge is an exception more in line with the older generation of biologists). But other physicists were reluctant even to discuss the topic of religion. Ernest Rutherford seems to have abandoned his Presbyterian upbringing completely, apart from its moral code. A colleague wrote of him: "I knew Rutherford rather well and under varied conditions from 1903 onwards, but never heard religion discussed; nor have I found in his papers one line of writing connected with it."[5] This attitude exactly corresponds to that reported for biologist Ford a generation later. It may be that in physics, long before evolutionary biology, the tendency to marginalize cultural values was already well established. This did not mean that such values did not play a role in suggesting hypotheses (ether physics is the clearest example of this), but it was impossible to use insights gained in this way unless they could be presented to other scientists in a completely objective format. In these circumstances, there would have been a temptation to suppress conscious consideration of such broader issues altogether. Those who did retain some form of belief were increasingly reluctant to discuss it in public, while science became an attractive profession for the kind of thinker who prefers to concentrate on immediate problems for which there is some prospect of a solution. Given the reports quoted above, it is difficult to believe that either Rutherford or Ford was deeply religious in private.

There would thus have been a growing indifference to religion among younger scientists, or at least a reluctance to be identified with any kind of metaphysical exercise. In 1934 the *Nature* reviewer of an American book titled

---

3. Arthur Cain to E. B. Ford, 2 April 1985, quoted in Ruse, *Monad to Man*, 358.

4. Ruse, *Mystery of Mysteries*; see especially the epilogue.

5. Eve, ed., *Rutherford*, 402. Curiously, Rutherford was one of the three British members elected to the Pontifical Academy of Sciences.

Science and the Spirit of Man complained about the number of "muddle-headed rebukes" to the alleged materialism of science.[6] How often this attitude translated into actual opposition to theology is another matter, although the results may have been the same as far as the general public was concerned. If once it were perceived that the majority of scientists were indifferent to the proposed reconciliation with religion, the credibility of the link between science and religion would be undermined. In the short term, vociferous support by a small number of scientists could maintain the illusion that the scientific community as a whole was in favor of a reconciliation, but if McCabe's charge that these activists were mostly senior figures from the previous generation is valid, then as the century progressed, we should expect the level of support to drop as these veterans became less active.

At the opposite end of the spectrum were those scientists who still openly endorsed the rationalist position. They tended to oppose not only formal religious beliefs, but also psychic research, spiritualism, and popular interest in the occult, all of which flourished at the turn of the century. They were not homogeneous in their opinions—some were at least willing to grant the importance of the religious emotions, while others saw the emotional appeal of religion as the real basis of the problem. But there was no lack of support for the claim that the old idea of an external supernatural realm should be abandoned. If we cannot accurately chart the opinions of the "silent majority" of scientists, we can at least show that there were some who continued to speak out openly against the idea of a spiritual dimension to life. Eventually, old-style rationalism was replaced by the new ideology of Marxism, but the rhetoric on this one issue, at least, remained unchanged. The most active scientists were once again those who opposed religion.

SCIENCE AND RATIONALISM

In the last decades of the nineteenth century, tensions had erupted within the rationalist camp as T. H. Huxley became increasingly concerned to distance the new, middle-class scientific elite from what he saw as a dangerous attempt to foment the undermining of social values among the masses. McCabe and other writers for the Rationalist Press Association used science to promote the rationalist challenge to formal religion, although their own backgrounds were often evangelical and they tended to promote rationalism as a new creed that would replace Christianity as a source of moral values. Eventually the rationalists, like

6. "Science and Values," a review of Julius W. Friend and James Feibleman, *Science and the Spirit of Man.*

E. Ray Lankester. Cartoon by "Spy"
from *Vanity Fair,* 12 January 1905.

the professional scientists, joined the ranks of middle-class respectability, although Huxley remained suspicious of the attempt to promote agnosticism as a
new form of religion. To him, and to some of his scientist followers, science was
a method of doubt and inquiry, not the foundation for a new faith based on the
alleged progress of the universe toward higher moral values.[7]

The most effective conduit by which Huxley's view of science was transmitted into the new century was the zoologist E. Ray Lankester. Lankester
lost his last professional position in 1907 when he was unwillingly forced to
retire from his post as director of the Natural History Museum, having alienated the powers that be with his demands for reform. But he remained active
into the mid-1920s and exerted considerable influence on popular culture,
both through his friendship with H. G. Wells and with his "Science from an
Easy Chair" articles in the *Daily Telegraph* (which were collected into a number
of books). A disciple whom even Huxley himself regarded as intemperate,
Lankester conducted a lifelong campaign against the class system and against

7.   See Lightman, "Ideology, Evolution, and Late-Victorian Agnostic Popularizers"; see also
Whyte, *Story of the RPA,* and Budd, *Varieties of Unbelief.*

religion, which he saw as both a prop to that system and an obstacle to free thought. The establishment naturally regarded him with a mixture of suspicion and outright hostility, colored partly by his reputation for lax sexual morality (much publicity surrounded his arrest in 1895 for attempting to protect a prostitute from the police in Piccadilly Circus). He was widely regarded as the model for the evil Professor Llewellin in Guy Thorne's 1905 novel *When It Was Dark*, which depicts the chaos resulting from a collapse of Christianity engineered by a fraudulent archaeological discovery.[8] He was an active supporter of the RPA, which reprinted cheap editions of his popular works.

Lankester did his best to stem the tide of popular support for psychic research and the occult. He had gained a good deal of publicity in 1876 by exposing the fraudulent medium Henry Slade, and continued this campaign into the new century by attacking both Oliver Lodge and Arthur Conan Doyle. Along with his physician friend H. B. Donkin, he criticized J. Arthur Thomson for allowing Lodge to write on psychic research in a 1922 collection, *The Outline of Science*. There is a Max Beerbohm cartoon of Lankester confronting Lodge, and it was reported that he would greet the latter with the words: "Well, Lodge, you old charlatan, how are the spooks?" In 1925 Lankester and Donkin again waded in against Arthur Balfour for supporting telepathy, and their letters to the *Times* were widely reported in the national and provincial press. One of Lankester's last published pieces was a critical review of Conan Doyle's *History of Spiritualism* for the *Sunday Times*.[9]

Lankester promoted a materialistic view of life and a Darwinian view of evolution—the Darwinism in H. G. Wells's *Outline of History* that so outraged Hilaire Belloc was derived from him. His materialism was linked to a profound hope that science would be an agent for freeing the human mind from the ancient superstitions that had kept people in chains. He campaigned both to increase the role of science in education and to secularize the educational system. His most widely publicized effort in this direction was his 1905 Romanes Lecture at Oxford, published as "Nature's Insurgent Son." Here he linked a Darwinian view of evolution with the view that humankind's new level of intelligence gave us the power to transcend nature and dominate the natural world. But science was important not just because it gave us control over nature: its real purpose was to free the human mind, and Lankester concluded the lecture with an attack on Oxford University for its refusal to take

8.   The cover of the 1925 edition cited in the bibliography notes that the novel was praised by the bishop of London preaching in Westminster Abbey. On the link to Lankester, see Lester, *E. Ray Lankester*, 188.

9.   For details of these events, see Lester, *E. Ray Lankester*, 213. The Lankester family papers referred to in this biography contain a file of newspaper cuttings about the 1925 telepathy incident.

science seriously.[10] Not surprisingly, the address got a poor reception in the press, although it was widely hailed by other rationalists.

In 1917 Lankester contributed to a symposium published in the *RPA Annual* on the fate of Christianity after the war—predictably, he thought that rationalism would triumph.[11] Five years later, he commented on the "revival of superstition," which he regarded as a real and potentially dangerous phenomenon generated by the trauma of the war and the craving for notoriety in the popular press.[12] He agreed with the view of religion promoted in H. G. Wells's *Mr. Britling Sees It Through*, in which God is merely the personification of humankind's moral sense, although he cautioned against Wells's tendency to treat this abstraction as a real personality.[13] Lankester recommended to Wells a novel by George Moore that sought to undermine Christianity by depicting Christ as having survived the Crucifixion.[14]

Lankester's attack on Oxford in 1905 was particularly heartfelt because he had spent a significant part of his career at the university and had always felt stifled by its clerical atmosphere. Another scientist who had clashed with the Anglican conventions of his university was statistician and Darwinian evolutionist Karl Pearson. As a third-year student at King's College, Cambridge, in 1877, he protested against compulsory attendance at chapel and refused to take a divinity paper, a stand that led eventually to the college abolishing these requirements. By this time he was a freethinker, arguing that the concept of a God violated the laws of thought.[15] In 1888 he published a book titled *The Ethics of Free Thought*. Pearson became known to the scientific world only after his involvement with statistics began in the 1890s. His positivist manifesto, *The Grammar of Science*, was first published in 1892, and here again Pearson made his position clear. He argued that the regularities that we call the laws of nature may be products of the sorting ability of the human mind, not of the unknown source of the sensations we experience. Science thus can tell us nothing about the world of things-in-themselves. "Here science is perfectly definite and clear; natural theology and metaphysics are pseudo-science." To infer that order and beauty are characters of the real world is as ludicrous as the Brahmins' claim that the world is spun from the bowels of a giant spi-

10.   Lankester, "Nature's Insurgent Son," in his *Kingdom of Man*, 1–61. Lankester's notes for the address reveal his thoughts all the more clearly; see Lester, *E. Ray Lankester*, 163–64.

11.   Lankester, "Will Orthodox Christianity Survive the World War?"

12.   Lankester, "Is There a Revival of Superstition?"

13.   Lankester to Wells, 21 September 1916, Wells papers, University of Illinois.

14.   Lankester to Wells, 19 September 1916, Wells papers, University of Illinois; see G. Moore, *The Brook Kerith*.

15.   E. S. Pearson, *Karl Pearson*, 4; see also Magnello, "Pearson, Karl."

der.[16] Science freely confesses its ignorance of those phenomena it has not yet reduced to order, while the metaphysician claims knowledge of those same areas—only to find that science is constantly reducing the available territory. It is science that leaves a genuine mystery behind the world of the senses, because it accepts that it can never provide genuine knowledge of that ultimate reality.[17]

Pearson contributed little to debates on religion after 1900, perhaps because the eugenics movement absorbed all of his efforts in the wider realm of public affairs. But there were other biologists anxious to preserve the rationalist ideal. Lankester's friend, the zoologist Peter Chalmers Mitchell, was particularly active in opposing the renewed link between science and religion. In 1903 he wrote the preface to the English translation of Elie Metchnikoff's *The Nature of Man*, agreeing with the latter's claim that religion offered no real comfort and that science represented the only hope of understanding and limiting the miseries that afflicted humankind.[18] Like Lankester, he shared Huxley's belief that there should be a kind of priesthood of science: "In every country, the new Order of priests of science, in the vigils of the laboratory, is working for the future of humanity."[19] Over the next few decades he emerged as a strong opponent of neovitalist biology and emergent evolutionism, his Herbert Spencer Lecture of 1930 being particularly outspoken.[20]

It was the anatomist and paleoanthropologist Arthur Keith who perhaps most actively took over Lankester's role as scientific advocate for the RPA, although the two men were never close personally. Keith's high-profile work on fossil hominids and human origins (including the Piltdown fraud) brought him a good deal of public attention. But it was his work on the anatomy of the brain that made him a materialist and which—when expressed in his 1927 presidential address to the British Association—exposed him to massive criticism in the popular press. In a 1930 article for the American magazine *Forum*, Keith claimed that he felt uncomfortable when expressing opinions that he knew were distasteful to many of his contemporaries.[21] Yet he had shown little reluctance to express his views in public throughout the 1920s. He was already active for the RPA in 1922, writing an account called "Why I am a Darwinist" for the *RPA Annual*. Here he warned of the danger of scientists

16.   K. Pearson, *Grammar of Science*, 2d ed., 108.

17.   Ibid., 111–12.

18.   P. C. Mitchell, preface to Metchnikoff, *The Nature of Man*, ix.

19.   Ibid.

20.   Mitchell, *Materialism and Vitalism in Biology*.

21.   Keith, "What I Believe," 220.

losing touch with popular culture, as illustrated by the ease with which G. K. Chesterton and Hilaire Belloc could get away with their claims that Darwinism was dead.[22] In 1925 he gave the RPA's Conway Memorial Lecture on "The Religion of a Darwinist," expounding "my personal religion, or perhaps I had better say, my lack of it."[23] C. A. Watts's publishing house issued a number of Keith's works in a popular format: *Concerning Man's Origin* (1927), *Darwinism and What It Implies* (1928), and *Darwinism and Its Critics* (1935).

Keith's evolutionism was by no means characteristic of the new Darwinism emerging at the time, but he was active in defending the general idea of evolution against its critics. His 1927 address, entitled "Darwin's Theory of Man's Origin as It Stands Today," outlined current thinking on human origins as well as his materialistic view of the human mind. Keith records that he thought the address had bored his audience, but that may have been because it was directed to nonspecialists—it was broadcast by the BBC—and on the following day he woke up to find that it was a sensation in the newspapers.[24] The address attracted much negative comment in the form of letters to the editor of the *Times* and other papers, to say nothing of more formal responses over the next year or so.[25] In 1935 his *Darwinism and Its Critics* collected a number of papers written in response to antievolutionary outbursts by Sir Ambrose Fleming, Arnold Lunn, and others. At this point his sympathies were moving even more strongly into line with the position of the RPA.[26] Toward the end of his active career, Keith's writings on human origins began to stress the tensions between the Christian moral code and what he believed to be the driving force of evolution: tribal and national competition.[27]

In his autobiography, Keith claimed that "most of my medical friends . . . thought about religion much as I did."[28] It is possible that many of those with medical training were inclined to take a materialistic view of the soul because they saw daily evidence of the extent to which the mind was controlled

22. Keith, "Why I Am a Darwinist," 11. In 1927 Keith and Belloc engaged in a controversy in the pages of *Nature*—see their articles entitled "Is Darwinism Dead?"

23. Keith, *Autobiography*, 489.

24. Keith, *Concerning Man's Origin*, vi. The address forms the first three chapters of this small book. On the 1927 meeting, see Mayer, "A Combative Sense of Duty."

25. See *Times*, 1 September 1927, 7; *Manchester Guardian*, 8; *Daily Telegraph*, 9; and *Daily Mail*, 9–10. On the following day, the *Telegraph* published a follow-up article with photographs of Keith and a chimpanzee side by side; see 12.

26. Keith, *Autobiography*, 633.

27. Keith, *Essays on Human Evolution*, chaps. 15–18.

28. Keith, *Autobiography*, 489.

by bodily functions. In *The Unpleasantness at the Bellona Club* (1928), Dorothy L. Sayers involves Lord Peter Wimsey with a Dr. Penberthy, who is convinced that the glands determine personality, and who gets into a controversy with a clergyman about original sin.[29] There were certainly real-life medical exponents of the materialist position, perhaps the most eminent of whom was Percy Lockhart-Mummery, senior surgeon at St. Mark's Hospital, London—like Lankester, a friend of H. G. Wells. Lockhart-Mummery believed that science was only just coming to grips with the problems of the human body—his own contribution was the theory that cancer was the result of mutations in the somatic cells. He was a Darwinist who believed that evolution showed no evidence of an overall directive force toward the human species.[30] Like Wells, he published a futuristic account of the world as it might be transformed by science, his *After Us* of 1936. Here he called for limitations on the expansion of the human population and a eugenic policy in which the state would decide who should be allowed to have children. In his world of the future, factory-made food and better town planning would revolutionize living conditions. As people recognized the power of science to transform their lives, more traditional influences on behavior, especially religion, would be abandoned: "Many people are still trying to find a compromise between the prevailing religious beliefs and the truths of modern science. But there can be none. Scientific truth allows no compromise." The old taboos were being abandoned, and few now saw the need to allow a role for the supernatural: "Man has at last begun to digest the Apple that Eve stole from the Tree of Life."[31] In a chapter entitled "Man His Own God," Lockhart-Mummery portrayed the churches as agents of social reaction and suggested that Christ would not be allowed to preach in a Christian church if He reappeared today. The ideal standard of conduct would be the Christian one, but with many modern additions. There was no room for the idea of a personal God: "Man's real God is not God in the attribute of man, but Man himself. Man's God should be the ideal man, not as he is now, but as he should be in the future."[32]

Rationalists argued that, far from providing mere technical information (which might be easily misused), the scientific methodology of critically testing all received opinions was the basis for a modern morality. They claimed that science could thus replace both classics and traditional religion in the

29. Sayers, *Unpleasantness at the Bellona Club*, 143.

30. See Lockhart-Mummery, *Nothing New under the Sun*, chaps. 6–8. For further information, see Palladino, "On the Contradictions of Humanism."

31. Lockhart-Mummery, *After Us*, 23.

32. Ibid., 148.

building of the individual's character in education.[33] One product of this movement was the volume *Science and Civilization* (1923), based on a summer school for the Adult School Union run during the war. J. Arthur Thomson and Julian Huxley contributed to the volume, but its overall thrust was to set science up in opposition to religion and offer the former as the better source of values. F. S. Marvin's introduction stressed the moral problems and opportunities arising from the freedom conferred by greater control of nature.[34] The most active contrast between religion and science came from the historian of science Charles Singer. His account of ancient medicine stressed how the spread of Christianity, with its focus on the other world, had stifled the spirit of Greek science.[35] Significantly, Singer considerably toned down this negative assessment of Christianity for his chapter in Needham's *Science, Religion, and Reality* a few years later.[36]

Sir Richard Gregory, longtime editor of *Nature*, also stressed the moral value of science, its selflessness and love of truth.[37] His article "Religion in Science," published in *Nature* in early 1939, virtually ignored formal religion to concentrate on the moral issues posed by science itself. A book published under a similar title in the following year noted that the interaction of science with human thought and behavior had featured regularly in the pages of *Nature* during his period as editor. But again there was little on religion beyond praise for H. G. Wells's view that the chief difference between science and religion was that the latter professed finality. Gregory argued that the conflict between science and religion was diminishing because the majority of religious thinkers were adopting a more positive attitude toward the assimilation of scientific knowledge.[38] His overall theme was that science promoted a moral view of the universe based on the idea of evolutionary progress.

In their drive to convince the world that science was itself a source of moral values, some rationalist educators were inclined to adopt a progressionist outlook that other rationalists found unconvincing. A more pessimistic view of nature, similar to that of Lockhart-Mummery, was promoted by the lecturer and science columnist John Langdon-Davies. In his *Science and Common Sense* of 1931, Langdon-Davies showed how developments in science were undermining traditional beliefs. Moral values were having to

33. See Mayer, "Moralizing Science."

34. Marvin, "Science and Human Affairs."

35. Singer, "Ancient Science"; see also his "The Dark Ages" in the same volume.

36. Singer, "Historical Relations of Science and Religion."

37. R. Gregory, *Discovery: Or the Spirit and Service of Science.*

38. Gregory, *Religion in Science and Civilization,* viii–ix; see also Gregory, "Religion in Science."

be completely revised in view of the new discoveries of biology and psychology—although there was no reason why this should lead to a breakdown of the social order. The modern view of the universe had destroyed natural theology, and personal religious experience was of no value from a scientific point of view. The idea of God, he argued, is something we impose on reality to help us get through life. Although science tells us nothing about ultimate reality, neither does anything else—in this sense, religion plays the same role as poetry in our lives.[39] He repeated these points in a contribution to a series titled "The Challenge to Religious Orthodoxy" published by *The Spectator*, also in 1931. Here Langdon-Davies insisted that it was a mistake for those who believed in "the good in man" and the idea of progress to personify the moral nature of the universe as God: science taught the dangers of using emotionally loaded words to describe abstract concepts.[40]

RELIGION WITHOUT REVELATION

There were some rationalists who did not agree with Langdon-Davies' claim that all vestiges of religious language should be abandoned. They wanted to transform rather than eliminate religion. While appreciating the need to sweep away the traditional notion of a supernatural realm, they nevertheless thought the religious emotions had value because they expressed a sense that humanity was part of a wider universe that might offer guidance on how we should behave. In one form or another, they wanted to replace religion with a philosophy that continued to put humankind in touch with a wider dimension, reworking the idea of God in a way that would allow it to function in a scientific worldview. Not surprisingly, most orthodox religious thinkers thought this trend even more dangerous than outright rejection of the idea of God.

It was Julian Huxley who most visibly symbolized the scientists' involvement in the move to redirect the religious enterprise. He combined an enthusiasm for rationalism and humanism with an interest in religious experience and a faith in humanity's role within a purposeful universe. Many commentators have noted Huxley's roots in an Edwardian-style worldview that retained the old belief that there is a purpose in nature, even if it is not expressed in the form of a single inevitable goal.[41] His 1912 book, *The Individ-*

---

39. Langdon-Davies, *Science and Common Sense*, 258.

40. Langdon-Davies, "Science and God," 138. In the following week the *Spectator* published a reply by a Presbyterian minister, H. H. Farmer, under the same title.

41. Dival, "From a Victorian to a Modern"; Ruse, *Monad to Man*, 328–38, 349–54. The brief discussion of Huxley's religious views in Dronamraju, *If I Am to Be Remembered*, 149–50, focuses on his influence on some younger biologists.

ual in the Animal Kingdom, was strongly indebted to Henri Bergson, and he retained a belief in the progressive character of evolution. He was friendly with the Anglican cleric (and future archbishop of Canterbury) William Temple and worked with other Anglicans to promote the church's links with eugenics. As an undergraduate, he read deeply in religious works from St. Augustine to Charles Gore, and he always accepted that the religious emotions expressed something important about the individual's relationship to the wider universe. Yet he was also an active member of the RPA—although he recorded his growing dissatisfaction with that movement's rabid materialism and its refusal to credit the value of religious experience.[42] He published in the *RPA Annual* and gave their Conway Memorial Lecture in 1930 with Keith in the chair.[43] He worked with H. G. Wells on *The Science of Life* and knew Wells's own efforts to reinterpret religion from *Mr. Britling Sees It Through* onward. He was involved with the Society for Psychical Research, but concluded that the evidence for survival after death was illusory, although telepathy was real. Two major books, *Religion without Revelation* (1927) and *What Dare I Think?* (1931), along with countless essays and radio broadcasts, promoted his new approach. One of his earliest statements of that approach was an essay titled "Science and Religion" in F. S. Marvin's *Science and Civilization* of 1923. Here he noted that it was from Lord Morley that he derived the hope that the next great task of science was to create a new religion.[44] In Huxley's new scheme, there was no supernatural and no personal God, but the individual needed to identify himself or herself with the purposeful flow of nature that had created mind and now gave it the power to take control of future progress. God was, in effect, humanity's conception of the universe as a whole and our sense of involvement in that whole.

Unlike the more extreme members of the RPA, Huxley would not rule out the religious feelings as an illusion. For him there was a genuine sense of reverence that all human beings felt when trying to come to terms with the cosmos. The sense of the sacred was universal, "a fundamental capacity of man" that demanded expression.[45] In earlier societies this capacity expressed itself through the creation of imaginary supernatural beings. Science now showed that these creations were an illusion, but this did not mean that the great religions had been a waste of effort—they represented an essential

---

42. Huxley, *Memories*, 152; on his links with Temple, see 150.

43. Huxley, *Science, Religion, and Human Nature*, also printed as chaps. 6 and 7 of *What Dare I Think?* His essay "Rationalism and the Idea of God" (reprinted in his *Essays of a Biologist*, 207–31) was originally published in the *RPA Annual*.

44. Huxley, "Science and Religion," 279 (see also *Essays of a Biologist*, 235).

45. Huxley, *Religion without Revelation*, 35.

foundation upon which more mature conceptions could be built.[46] Huxley appealed both to anthropology and psychology to support these claims. In his Conway Memorial Lecture, he presented the current situation as a tension between nineteenth-century rationalism and revived forms of religious Fundamentalism and mysticism—the latter supported even by some scientists, including Eddington. New religions such as spiritualism and Christian Science flourished, while Anglo-Catholic ritualism grew within the Anglican church.[47] His own approach offered a compromise that stripped religion of all those aspects no longer acceptable to science, but steered clear of the sterile rationalism of the previous century. Huxley admitted that liberal Protestants had gone a good way toward purging Christianity of the old anthropomorphisms, focusing on God as immanent within creation rather than a transcendent personality—yet they were still using old forms of worship that were no longer appropriate in the modern world.[48] The basic postulate of theism had to be abandoned, and the religious emotions channeled in new directions determined by science.

Huxley's views paralleled those of the young J. B. S. Haldane. Before his conversion to Marxism in the early 1930s, J. B. S. was still quite close to his father's idealism. He believed that the universe had a structure that resembled the human mind, rather than a machine, and felt that it was important to be in touch with those realms of thought that transcended the material—anyone who was not aware of this wider domain was little more than an animal.[49] Yet from an early age he had adopted a skeptical attitude toward formal religion, seeing science as a force that would render most of the churches' teachings obsolete. A youthful notebook dating from 1912 reveals the outlines of what would become his Daedalus a decade or more later, while another essay on Darwinism shows that he had already come to view that theory as the harbinger of a major revolution in religious thinking.[50] The churches survived, he thought, because they helped people keep in touch with the ideal, but their creeds were full of out-of-date science and their help was unnecessary to anyone who saw science and art as means of transcending the purely material world.[51] Although he rejected personal immortality, he still believed that at death his mind might

46.   Ibid., 172–73.

47.   Huxley, *What Dare I Think?* 182–85.

48.   Ibid., 228–30.

49.   J. B. S. Haldane, "The Duty of Doubt," in his *Possible Worlds,* 217.

50.   Disbound notebook, J. B. S. Haldane papers, National Library of Scotland, MS 20578. The essay "The Future," 26–59, anticipates *Daedalus,* while Darwinism is the topic of an untitled essay, 68–81; on the revolution in religion, see 68.

51.   Haldane, *Inequality of Man,* 213.

J. B. S. Haldane. Photograph by Walter
Stoneman, 1943. By courtesy of the
National Portrait Gallery, London.

merge with the infinite mind that lay behind nature.[52] The real impact of science, he maintained, was through its promotion of a skeptical attitude toward
received opinions, and as such it was massively subversive of existing religious
beliefs, as he proclaimed in the conclusion to *Daedalus* in 1923. Christianity was
less wedded to a rigid moral law than most religions, but Haldane argued that
in the end all religions came into conflict with science on this point. A morally
evolving religion was possible: "That is the only kind of religion that could satisfy the scientific mind, and it is very doubtful whether it could be called a religion at all."[53] Haldane published extensively in the *RPA Annual* and gave the
Conway Memorial Lecture two years before Huxley.[54] He would certainly have
been perceived as a radical on religious matters, yet until Marxism showed him
a plausible way into a materialist philosophy, he allied himself with those who
wanted to transform rather than eliminate religion.

52.   Haldane, *Possible Worlds*, 210.
53.   Haldane, *Daedalus*, 92; see also his *Haldane's Daedalus Revisited*, 49.
54.   Haldane, *Science and Ethics*; his *Science and Life* is a later collection of *RPA Annual* contributions.

A more senior figure who was equally iconoclastic was the neuroscientist Sir Charles Sherrington. In 1928 Sherrington provided a short introduction to Frances Mason's *Creation by Evolution*, stressing the progressive nature of the evolutionary process.[55] But unlike Huxley, Sherrington saw no moral purpose within this process—like T. H. Huxley, he thought that humanity's moral powers transcended the natural order and would now have to be imposed on nature. This point came out more strongly in his Gifford Lectures for 1937–1938, published as *Man on His Nature*. Here he stressed the gradual emergence of mind in the course of evolution, but argued that until the appearance of humanity, mind had done no more than add to the inherently competitive character of the evolutionary process. Nature has no values, he argued, but now that we have developed a level of altruism, we can judge nature and seek to alter its impact on our lives.[56] Moral progress was now at last possible, and thus "Man is Nature's beginning to be self-conscious."[57] Such a vision of humanity's relation to nature offered at best only a highly transformed religion. Much of traditional natural theology had to be abandoned: we might admire the beauty of nature, but not her values. The human situation was much bleaker than the advocates of a morally progressive evolutionary process would admit, but by the same token humanity took on a new dignity and a new responsibility for the future.[58] Sherrington's endorsement of this more pessimistic view of nature links him to the strand of rationalist thought that saw humanity as responsible for constructing its own morality without guidance from any outside source. Yet his invitation to give a series of Gifford Lectures suggests that he was not regarded as an outright opponent of religion. He did at least give humanity an element of cosmic significance.

Julian Huxley's belief that nature, if not her Creator, offers us guidance on how to behave harks back to an older tradition of cosmic teleology whose heirs in the twentieth century were the exponents of emergent evolution. The fact that most supporters of that position—including the scientists—were theists must encourage us to see Huxley as an intermediate figure, torn between the competing philosophies of rationalism and theism. One way out of this dilemma might have been the philosophy of emergence proposed in Samuel Alexander's *Space, Time, and Deity* of 1920, in which the final step in evolution would be the emergence of God. Alexander's views were endorsed by

55.   Charles Sherrington, introduction to Mason, ed., *Creation by Evolution*, xi–xiii. On Sherrington's influence, see R. Smith, "The Embodiment of Value."

56.   Sherrington, *Man on His Nature*, 381, 389, 398–400.

57.   Ibid., 387.

58.   Ibid., 402–4.

the Cambridge physician Sir Walter Langdon-Brown in his *Thus We Are Men* of 1938.[59] Toward the end of his career, Huxley did indeed help to popularize the similar views of Pierre Teilhard de Chardin, writing an introduction to the English translation of *The Phenomenon of Man* in 1959. But in the interwar years he was not prepared to go this far: he would challenge the rationalist position only so far as to retain a role for the religious emotions. He was not as yet prepared to accept that the apparent teleology of the cosmos indicated the need for God.

MARXISTS AND OTHER RADICALS

The 1930s ushered in a period of social instability caused by the depression and the rise of Fascism. Some elements of the British intellectual community, including several eminent scientists, responded by adopting a politically radical position that drove them toward a materialist, and hence an antireligious, position. The most obvious sign of this movement was the adoption of Marxism by J. D. Bernal, J. B. S. Haldane, and others (including Joseph Needham, who managed to combine his Marxism with his Anglican faith). These radicals believed that science could no longer afford to distance itself from politics: scientists themselves must become politically active to ensure the application of science for the benefit of all. But this Social Relations of Science movement was not exclusively a Marxist preserve.[60] Several scientists accepted the need for greater social involvement without becoming Marxists, thus perpetuating the kind of rationalism once popularized by Lankester and the RPA. Both groups were, however, inclined to be suspicious of religion. The Marxists (Needham excepted) followed the essentially materialistic line pioneered by Marx and Engels, in which religion was seen as an agent of oppression. But this was exactly the line that Lankester had urged, and the non-Marxist radicals could thus be equally hostile to religion.

Haldane continued his work for the RPA after his conversion to Marxism, demonstrating the continuity of the rationalist tradition. We have already noted the skeptical attitude toward theological dogma that Haldane expressed in his earlier writings and the contrast he drew with scientists' willingness to abandon a theory when the evidence went against it. This strategy of using science as a weapon against religious belief had been commonplace among the earlier rationalists, and it was to become a central feature of the

59. Langdon-Brown, *Thus We Are Men*, 260.

60. On the Marxists, see Werskey, *Visible College*, and more generally, McGucken, *Scientists, Society, and State*.

arguments used by the Marxists of the 1930s. In an article based on a radio talk broadcast in 1933, the physicist Hyman Levy stressed exactly this point in order to argue for the sterility of religious dogma.[61]

Haldane described his transition to Marxism in an essay titled "Why I Am a Materialist," published in the *RPA Annual* for 1940. He argued that, contrary to popular belief, idealists were more likely to be selfish than materialists. He himself had been a vague idealist outside the laboratory ("a materialist in practice but not in theory") because he was convinced that the mind was a real agent, something more than a mere epiphenomenon derived from the activity of matter. It was his reading of Engels and Lenin that showed him how the traditional duality of matter and mind could be transcended by seeing minds as properties of brains conceived in a holistic fashion.[62] In 1944, in another article, he sought to undermine the logic of William Paley's argument from design. Paley had used the analogy of a watch and a watchmaker to argue that living things must have an intelligent Creator—but a better analogy, according to Haldane, was the machinery on a modern battlefield, and that analogy would lead one to a polytheistic position in which there were a number of competing gods.[63] In 1947, in "The Limitations of Rationalism," Haldane criticized the arguments of the logical positivists who claimed that theological assertions were meaningless. He thought that he could detect logical fallacies more easily in the arguments of the rationalist Joseph McCabe than he could in those of the religious writings of C. S. Lewis, partly because there were fewer of them. We cannot disprove the existence of God, he argued, but one can be sufficiently certain to take atheism as a guide to action, and future generations could hope to attain certainty on such matters. Even so, Haldane still distrusted those who would explain the religious feelings away—they might go on to explain away moral consciousness and the preference for truth over falsehood.[64] In addition to his writing for the RPA, Haldane—like Keith—took up the gauntlet thrown down by the new generation of creationists. In 1935 a series of letters he exchanged with Arnold Lunn was published, while in 1949 he debated two members of the Evolution Protest Movement, Douglas Dewar and L. Merson Davies.[65]

61.   Hyman Levy, "What Is Science?" part 1 of Adams, ed., *Science in the Changing World*; see, for instance, 55.

62.   Haldane, "Why I Am a Materialist," reprinted in his *Science and Life*, 27–35.

63.   Haldane, "The Argument from Design," reprinted in his *Science and Life*, 43–48.

64.   Haldane, "The Limitations of Rationalism," reprinted in his *Science and Life*, 57–64.

65.   Arnold Lunn and J. B. S. Haldane, *Science and the Supernatural*; Dewar, Davies, and Haldane, *Is Evolution a Myth?*

Among the other converts to Marxism, Needham preserved his faith and contributed to the 1935 volume *Christianity and the Social Revolution*.[66] Lancelot Hogben, while not severing his link with the Society of Friends, became known as an outspoken materialist. Hogben was raised as a Fundamentalist Methodist, and recalled his father's withering criticism of any move to liberalize theology—it was his father's "Sunday denunciations of [R. J. Campbell's] *The New Theology*" that "first familiarized me with the devilry of Darwinism."[67] By 1918 he had become a socialist, and later recalled the moment in the late summer or early autumn of that year when he suddenly realized that a humanist viewpoint made God redundant. The universe was a "limitless expanse of unthreatening and impersonal emptiness . . . without purpose or punishment for a lately arrived animal species, free to make or mar its own destiny without help or hindrance from above."[68] Hogben's strongly materialist *The Nature of Living Matter* of 1930 arose from a symposium at the South African meeting of the British Association in 1929, in which he had challenged J. C. Smuts's "half-baked excursion into Teutonic mysticism with the title *Holism*," along with the views of J. S. Haldane and A. S. Eddington.[69] In addition to dismissing neovitalism and holism as blind alleys in science, Hogben insisted that Eddington's emphasis on personal religious experience was an irrelevance to philosophy because it made public discourse impossible.[70]

Perhaps the most outspoken critic of religion was J. D. Bernal, the son of an Irish Catholic father and an American Protestant mother, whose early life at public school had been marked by a fervent Catholicism.[71] Unlike Needham, with whom he became a close friend, Bernal rapidly became involved in radical politics and Marxism upon going up to Cambridge in 1919—he joined the Communist Party shortly after taking his degree. At the same time he began his scientific career as a crystallographer, working for a time under William Bragg in London before moving back to Cambridge. Here he became the center of a brilliant but unorthodox circle. His own radical vision of science's ability to

66. Needham, "Laud, the Levellers, and the Virtuosi" and "Science, Religion, and Socialism," in *Christianity and the Social Revolution*, ed. John Lewis, Karl Polanyi, and Donald K. Kitchen, 180–205 and 416–41.

67. See his "unauthorized autobiography," *Lancelot Hogben: Scientific Humanist*, 11.

68. Ibid., 69

69. Ibid., 113.

70. Hogben, *Nature of Living Matter*, 261.

71. On Bernal's early Catholicism and his loss of faith, see Werskey, *Visible College*, 67–76; see also Synge, "Early Years and Influences," 12–13, and Steward, "Political Formation," 44–45 (both chapters in Swann and Aprahamian, eds., *J. D. Bernal*). These writers draw on Bernal's diaries and his manuscript "Microcosm" in the J. D. Bernal archives, Cambridge University Library.

J. D. Bernal. Photograph by
Wolfgang Suschitzsky, 1949. By
courtesy of the National
Portrait Gallery, London.

shape humankind's future appeared in 1929 as *The World, the Flesh, and the Devil: An Enquiry into the Future of the Three Enemies of the Rational Soul.* This work was a contribution to the series in which Haldane's *Daedalus,* with its vision of the positive values of science, had been challenged by Bertrand Russell's *Icarus,* which stressed the potential inhumanity of a world run by technocrats. Bernal joined Haldane in seeing the human race taking charge of its own destiny, but allowed that the more progressive spirits might colonize space, leaving those who were unable to adapt to stagnate on earth. The progressive wing of humanity, he believed, would eventually devise techniques for minds to come into direct contact with one another, and might fuse to create a single Great Mind.[72]

There was no room for the traditional idea of God here. In the following year, Bernal published an article titled "Irreligion" in *The Spectator's* series, "The Challenge to Religious Orthodoxy." Here he acknowledged that the rejection of faith was an elusive thing, difficult for those who had experienced it to describe. But the currently popular attempt to rebuild the link between science and religion was based on mere wishful thinking. In fact, it was science itself that, when properly understood, not only destroyed faith but cre-

72.   Bernal, *The World, the Flesh, and the Devil,* 57.

ated its alternative: "Science does not destroy religious faith by simple logical contradiction; it undermines it by the continuous advance of its method. The scientific method is incompatible with the religious attitude. It brings to those who use it an intellectual satisfaction deeper than that of faith. It can criticize religion, but religion can find no place for it in its scheme."[73] It was not enough to reject religion: those who were building the skeptical attitude of science must see that attitude as a way of life "for which men can live and work and fight." The result would be (as already expressed in *The World, the Flesh, and the Devil*) a transformation of morality along rational grounds. Scientists might not want to rule the world, but they would no longer tolerate being ruled by greedy and stupid people and by their "ancient and infantile institutions." Here was a bold new humanism that saw the human race defining its own place in a universe that was beautiful and which offered infinite potential, but no moral guidance.

In his *Social Function of Science* of 1939, Bernal included a short section called "Science and Religion," in which he portrayed the recent efforts to reforge a link between the two areas as an indication of science's inherent conformist tendencies. Less than a hundred years earlier, the conflict between science and religion had been at the center of the intellectual world, but that episode had been an anomaly created by religion's efforts to thwart the new discipline's bid for autonomy. Scientists had turned to atheism for tactical reasons, but "the moment that denial [of God] was no longer formally demanded of him, the scientist of the later period was only too willing to return to religion and, with it, to social conformity."[74] This trend had become all the more noticeable after the Russian Revolution had enhanced the significance of religion as a force for reaction. The whole episode merely confirmed the narrowness of the vision imposed by scientific specialization: those who accepted this limitation still needed religion or mysticism to satisfy their emotional needs. It also confirmed the power of the gerontocracy that ruled science. The remedy for this situation was the message preached by Bernal's whole text—the need for scientists to become aware of their ability to transform human life and to demand a role in determining how that transformation should be directed for the good of humankind. The message may have been couched in the terminology of Marxism, but it echoed a theme with which E. Ray Lankester (who had known Marx as a young man) would have felt thoroughly at home. In this sense, the new radicalism of science merely built upon the foundations of the old.

73. Bernal, "Irreligion," 518.
74. Bernal, *Social Function of Science*, 390.

The demand for the reintroduction of an ethical dimension into science was not, however, confined to the Marxists. We have already noted embryologist and geneticist C. H. Waddington's complaints about the refusal of many scientists to get involved in debates outside their own area of specialization. He endorsed Bernal's view that the artificial isolation demanded by the ideal of "pure" science was a product of the discipline's fight for independence: "Science, at the time when organized religion was its enemy, had signed away its rights to have views on the most general questions in return for freedom to put off its swaddling clothes."[75] But Waddington had no interest in encouraging scientists to revive an interest in religion. Although his antimechanistic biological theorizing was deeply influenced by Alfred North Whitehead's philosophy, he did not mention the religious concerns that Whitehead himself recognized in the fight against materialism.[76] In 1941 he published an important paper titled "The Relations between Science and Ethics" in *Nature*, accompanied by comments from a wide range of scientists, church leaders, and philosophers. The paper and much of the resulting comment formed the basis of a book titled *Science and Ethics*, published the following year. Like Bernal, Waddington preached the need for scientists to get involved in the social process, although his approach to ethics was more in line with Huxley's vision of humanity continuing the course of evolutionary progress by more efficient means.[77] At the same time, he criticized Bernal's Marxism, and especially the claim that Communism itself was scientific.[78] His analysis of the relationship between science and ethics thus tended to push aside the claims of both religion and militant rationalism. Both Marxists like Bernal and church leaders were invited to contribute to *Science and Ethics*, but the debate as a whole was organized along lines that relegated religious concerns to the sidelines.

SCIENCE, RELIGION, AND THE HISTORY OF SCIENCE

The interaction of science with religion and ethics also shaped the emergence of the history of science as a discipline. A remarkably large proportion of the figures mentioned above urged the importance of a study of the history of science for an understanding of the development of Western culture.

75. Waddington, *Scientific Attitude*, 75.

76. On Whitehead's influence, see Waddington, "The Practical Consequences of Metaphysical Beliefs on a Biologist's Work," 74–76 and 79.

77. Waddington, "Relations between Science and Ethics," reprinted in Waddington et al., eds., *Science and Ethics*, 9–19; see 19. For an enthusiastic endorsement by Huxley, see 34–36.

78. Ibid., chap. 7.

On both sides of the debate, history was called in—either to demonstrate the importance of science turning its back on the materialist trend with which it had become associated or to argue that the process of fusion with materialist philosophy must be pushed to its logical conclusion.

Bernal's *Social Function of Science* had hailed the power of science in the modern world, but also included a chapter stressing the historical link between the attempt to understand the world and the effort to control it through technology. This theme was developed at length in his monumental *Science in History*, first published by Watts in 1954. Studying the history of science, Bernal argued, would show scientists how their work had transformed the world and would help to convince them that they should play an active role in future developments. Only then would they cease to be the pawns of the commercial elite.[79] In his second volume, Bernal turned on the reactionary scientific philosophies of the turn of the century: Mach and Ostwaldt's neopositivism, Bergson's *élan vital*, and James's pragmatism had all tried to undermine the link between science and materialism. In so doing they minimized science's ability to improve the lot of humankind and improved its acceptability to religion and the conservative industrial establishment.[80] The notion of a "pure" science with no social responsibilities had been developed so that science could become the handmaiden of capitalism. By situating this reactionary episode against the broad sweep of science's involvement in technical and hence social development, Bernal hoped to convince his fellow scientists that it was both philosophically shallow and morally weak.

The role of the history of science in this ideological campaign had been highlighted by the famous visit of the Soviet delegation to the Second International Congress of the History of Science, held in London in the summer of 1931.[81] Because the president of the congress, Charles Singer, was unwilling to grant the unexpectedly augmented Soviet delegation more time to speak, their papers were hurriedly translated and printed under the title *Science at the Cross-roads*. The audience of mostly rather orthodox scientists and historians were bemused by the Marxist interpretation of the development of science presented by Nicholai Bukharin, Boris Hessen, and others. Even Bernal, who was enthused by the Soviets' ideas, thought that their use of Marxist jargon prevented the audience from appreciating the underlying ma-

---

79. Bernal, *Science in History*, vol. 1, 28–29.

80. Ibid., vol. 2, 570. Ernst Mach and Wilhelm Ostwaldt had argued that science merely correlates sense impressions and cannot uncover the underlying reality of nature. William James's pragmatism held that truth is not absolute, but is determined by whether or not the knowledge helps us to deal with the world.

81. This episode is described in Werskey's *Visible College*, 138–49.

terialist theme. For Bernal himself, though, the whole episode spurred an interest in the history of science, which would culminate decades later in his *Science in History.*

One of the organizers of the congress was Joseph Needham, who had already included a rather conventional history of embryology in the first volume of his *Chemical Embryology.*[82] Indeed, Needham's wide reading had encouraged him to comment on historical issues throughout the 1920s: his *Sceptical Biologist* of 1929 included several essays exploring the history of biological materialism, and a volume he edited, titled *Science, Religion, and Reality,* contained essays by Singer and others on historical topics. The 1931 congress changed his whole approach, however, making him realize how limiting it was to consider the development of science as a purely intellectual process. In 1941 he noted the unwillingness of many scientists to get involved with social issues and suggested that studying the history and philosophy of science would help them to realize the extent to which they were in the pay of the bourgeoisie.[83] His 1946 collection of essays, titled *History Is on Our Side,* sought to link cosmic evolutionism and social history into a philosophy that was anti-Fascist and, of course, socialist. Needham was involved in the creation of the history of science program at Cambridge, where he expressed deep suspicions about the professionalization of the discipline. His own later work on the history of Chinese science would turn him into an academic legend (although his influence on history of science at Cambridge was rapidly curtailed).[84]

The negative reaction of historians to the Soviet delegation in 1931 was not necessarily an indication that they were enmeshed by the bourgeois view of "pure" science castigated by Bernal. Charles Singer, then a lecturer in the history of science at University College, London, and hence one of the few professionals in the field, was an advocate of the more moderate view that the growth of scientific knowledge ought to improve the human situation. Singer came from a Jewish background, although this was hardly apparent in most of his writings. His early paper on ancient medicine in Marvin's *Science and Civilization* volume (1923) had depicted Christianity as damaging to science:

---

82. This history was subsequently reworked into his *History of Embryology.*

83. Needham, "Metamorphoses of Scepticism," in *Time, the Refreshing River,* 7–27, see 7.

84. See the introduction to *Background to Modern Science,* edited by Needham and Pagel. For Needham's disquiet about the direction of the Cambridge program, see Joseph Needham papers, file B309 (which also includes comments on reactions to Alistair Crombie's Roman Catholicism). For a commentary on subsequent developments at Cambridge, see Young, "The Historiographical and Ideological Context of the Nineteenth-Century Debate on Man's Place in Nature."

"With the spread of Christianity, interest in phenomena had dwindled. The light of science had become dim, then fluctuated, and finally went out."[85] Subsequent essays, including the one in Needham's *Science, Religion, and Reality*, portrayed the church in a less negative light, but still depicted the Middle Ages as a period of "intellectual degradation."[86] Singer was suspicious of materialist philosophy, although he knew that the mechanistic approach was the only way forward in biology. He looked for a new interaction between science and philosophy, and still felt that science was unable to explain consciousness.[87] Without being overtly critical of religion, Singer's overall approach thus stressed the power of science itself to increase not only knowledge of the world, but also philosophical awareness. This view was paralleled in the survey of the history of science that constitutes the bulk of Richard Gregory's *Religion in Science and Civilization* of 1940.

The events of the 1930s, coupled with Singer's Jewish background, forced him into a radical reappraisal of Christianity's role in history and led to an outburst in his *Christian Failure* of 1943. For Bernal and Needham, the answer to the threat of Fascism was to couple science firmly with Marxist materialism. Those who saw materialism itself as a threat tended to argue that science should be purged of that philosophy so that it could rejoin and reinvigorate the Christian tradition. Faced with the collusion of the Christian churches in the Nazi assault on the Jews, Singer was forced to conclude that the fault lay not in the influence of science, but within Christianity itself. He still felt that science endowed its practitioners with a "mood" or attitude of mind that was both disciplined and, often, humanitarian. It thus naturally allied itself with the "Religion of Humanity" that had existed throughout the ages as an alternative to Christian and other theologies. The challenge of pantheism in the Renaissance and the seventeenth century pinpointed Christianity's inability to cope with new ideas:

> Yet it is a simple fact that to this day theology has not settled its account
> on this point of the distinction of Creator and creature. This is no
> conflict between religion and science. It is a fundamental antithesis be-
> tween two ways of looking at the world. One of these has developed a
> philosophy which has come to terms with science; the other has not. One

85. Singer, "Ancient Medicine," 70.

86. Singer, *A Short History of Science to the Nineteenth Century*, 161; see also his "Historical Relations of Science and Religion," 114, and *Religion and Science*, 34, 40, 43.

87. Singer, *A History of Biology*, viii; *A Short History of Science to the Nineteenth Century*, 341; *Religion and Science*, 76.

is the Religion of Humanity, the other has remained, in all essentials, Christian theology.[88]

Singer himself did not think the Religion of Humanity was enough, but it had provided better moral guidance than Christianity. The problem was not that science or humanism had undermined Christian morality, but that Christian civilization was itself crumbling from within. He quoted Martin Luther's anti-Semitism and Karl Heim's support for a German state based on race as evidence that Christianity had always been shot through with bigotry. The Church of England's protests against Nazi attacks on the Jews had been feeble. He concluded: "What Christians regard as the 'moral decay' of our times appears to those outside Christianity to have arisen as a disease within the Christian body itself."[89] The churches' attitude contrasted with what he had observed in the Jewish community—his own father, a learned and liberal rabbi, had helped to coordinate relief for refugees from the Russian pogroms of the 1880s. In the end, the Prophetic religion had survived better than Christianity, for all the collapse of Jewish civilization (which he did not hope to see revived through Zionism).

Singer himself doubted that anyone would read his little book. But his suspicion of the mechanistic view of nature, except as a practical tool in biology, parallels what was already a major theme in the work of historians who wished to defend religion, who used history as a means of highlighting the dangers of the philosophical baggage that had become associated with science. Charles Sherrington built a significant element of history into his *Man on His Nature*, focusing on the seventeenth-century physiologist Jean Fernel. He thought that Fernel, could he see what was happening in biology today, would accept evolution as a "triumphant strain of purposive causation"— even though it did not lead in a single definite direction.[90] Sherrington subsequently published a whole book on Fernel. His aim was to limit materialism in biology so that life and evolution could retain a sense of purpose.

In this respect, Sherrington's work parallels that of the naturalist-clergyman Charles Raven, in which antagonism to materialism was coupled explicitly with the defense of the Christian tradition. Raven's contribution to history was significant: his biography of the seventeenth-century naturalist John Ray became a classic.[91] As we shall see below (in chapter 8), Raven's in-

---

88.   Singer, *Christian Failure*, 61.

89.   Ibid., 104.

90.   Sherrington, *Man on His Nature*, 169; see also Sherrington, *The Endeavour of Jean Fernel*.

91.   Charles Raven, *John Ray: Naturalist*; see also Raven, *English Naturalists from Neckam to Ray*.

terest in Ray arose from his argument that science must retrace its steps, slough off its recently acquired materialism, and return to the roots established at a time when the study of nature was seen as the study of God's handiwork.

The Ray biography was written in the dark days of World War II, when Raven joined Needham in pointing to the dangers that materialism had introduced. This was an analysis pioneered in Alfred North Whitehead's *Science and the Modern World* of 1926. Whitehead offered a survey of the history of science in which the triumph of the mechanistic philosophy in the late seventeenth century had allowed scientists to stop worrying about philosophy while they got on with the job of reducing nature to order, with disastrous consequences for the subsequent development of both science and Western culture.[92] The same distrust of the legacy of the mechanistic philosophy runs through the pioneering survey of the history of science by the Anglican W. C. D. Dampier-Whetham, which concluded with a chapter stressing the importance of Whitehead and the other thinkers who had challenged the materialistic perspective of modern science.[93]

The historian of chemistry (and later director of the Science Museum of London) F. Sherwood Taylor, who converted to Catholicism, also used his history to spell out a message about the dangers of materialism. In his *Century of Science*, published in 1940, he joined Raven, Needham, and others in seeing the threat posed by Nazi Germany as the culmination of the harmful direction taken by Western thought, in part because of the development of science. Technology had improved not only the material conditions of life but also the weapons of war, while doing nothing to forward the spiritual improvement of humankind. For Taylor, the Catholic Church was the only organization that had preserved its ideals in the face of the materialistic onslaught.[94] He saw his more detailed work on alchemy as important not because it offered insights of value to modern chemistry, but because it could teach scientists that there were aspects of nature that did not appear in technical descriptions and which still demanded the application of human values if a true understanding of the world was to be attained.[95] Another convert to Catholicism was Sir Edmund Whittaker, although his *History of the Theories of Aether and Electricity* was written long before his conversion, and the early pages,

92. Whitehead, *Science and the Modern World*, chaps. 1–3, see esp. 61.

93. W. C. D. Dampier-Whetham, *A History of Science and Its Relations with Philosophy and Religion*, chap. 10.

94. F. S. Taylor, *Century of Science*, 270.

95. Taylor, *The Alchemists*, 237.

even in the later edition, are still very critical of Thomist philosophy for its influence on medieval science.[96]

The emergence of the history of science as an academic discipline was thus in part conditioned by the fact that it served as a battleground for the wider debate over the significance of science in Western culture. Those scientists who still wanted a reconciliation with religion argued that they were merely reestablishing a link lost since the seventeenth century. Their opponents who hailed science's role in the trend toward materialism sought to construct a view of history that vindicated their wider position. The rationalists had always proclaimed science as a force that tested ancient superstitions and found them wanting, and now the Marxist emphasis on social development allowed this emphasis on history to move center stage. The creation of the academic discipline of the history of science was in part an offshoot of a major ideological disagreement over the relationship between science and religion.

96. Whittaker, *A History of Theories of Aether and Electricity,* vol. 1, 3–4. This material was originally published in 1910.

CHAPTER 3

Physics and Cosmology

To outsiders, including clergymen, trying to assess scientists' opinions, biologists were widely perceived to have been more active than physical scientists in promoting materialism. There were, of course, radical scientific naturalists among the Victorian physicists, including John Tyndall, whose Belfast address to the British Association in 1874 had thrown down the gauntlet to religion. But many late Victorian physicists had been deeply religious. They still saw the world as a unified cosmos, not the chaos of billiard balls often (but quite unfairly) attributed to the materialists. Tyndall's opponents sought to undermine materialism by claiming that the nature of matter was not properly understood. For them, matter itself was mysterious, and thus offered no suitable foundation for the kind of materialism that sought to eliminate mind and purpose from nature.

The key to this nonmechanistic physics was the ether, that mysterious substratum that permeated the material universe and served as the vehicle and coordinating agent for all physical action. In the last decade of the nineteenth century, the most innovative studies of the nature of matter were being made by adherents of ether physics, including Lord Rayleigh and J. J. Thomson. Their position was based on a firm belief that the universe was a divine creation and that life and mind were integral parts of God's scheme. In a deliberate rebuff to Tyndall's and Huxley's new professional-ism, these physicists sought to defend a more conservative way of thought and an elitist social agenda. Oliver Lodge became the most articulate spokes-man of this position, writing an endless stream of books and articles defend-ing the idea that the ethereal universe was both the source of the human spirit and an indication of the purposefulness of the whole creation. Al-though it was formulated in the last part of the nineteenth century, Lodge promulgated his position so actively in the early decades of the twentieth that many clerical and nonspecialist readers thought the age of Victorian materialism was over.

Relativity theory eventually destroyed the credibility of the ether, but quantum mechanics offered an equally far-reaching challenge to the old form of materialism. As Arthur Stanley Eddington proclaimed, mind could once again be seen as a fundamental constituent of reality, and if the human mind was real, then why not the old idea of a universal Mind that oversaw the whole system? James Jeans argued that the Creative Mind worked according to mathematical principles. To many religious writers, this argument seemed to offer final proof that science was on their side, although as it turned out, what Joseph McCabe called the "Jeans-Eddington outbreak" represented the last major thrust of the effort to link science and religion in the interwar years.

Both Jeans and Eddington worked in cosmology as well as in physics, and here, too, their ideas had theological implications. Jeans developed and Eddington endorsed a theory of the origin of planetary systems that implied that our own system might be unique, or at best one of a very few in the whole universe. Such a theory had implications for what had once been called the "plurality of worlds" debate, now known as the debate over the possibility of extraterrestrial life. If Jeans was right, then planetary systems are rare, and the human race is probably the only source of intelligence and moral awareness in the universe. Just as physics was demonstrating that mind was an integral part of reality, cosmology showed that the human race was probably the only source of such awareness. Conservative Christians had long held that the hu-man race *is* spiritually unique and had doubted the claims of more liberal thinkers, who believed that God would hardly have created so large a universe in order to leave most of it barren.[1] Jeans's theory thus revived interest in a long-standing debate and provided support for the more conservative theo-logical position. The new cosmologies based on the discovery that the uni-verse was expanding also linked into existing debates over the significance of

---

1.   On the nineteenth-century background to this debate, see Brooke, "Natural Theology and the Plurality of Worlds."

the apparent direction built into the history of the universe by the laws of thermodynamics.

ETHER AND SPIRIT

A number of historians have explored the development of ether physics in the late nineteenth century.[2] Cambridge-based physicists such as Rayleigh and Thomson showed the power of this theoretical initiative as a guide to scientific investigation. The movement was unabashedly conservative in attitude, reflecting the wider efforts by Cambridge dons to resist what they saw as the commercial basis of the Huxley-Tyndall approach to the professionalization of science.[3] Rayleigh's links with the philosopher-politician Arthur Balfour, a long-standing opponent of Huxley's ideology, symbolize this conservative position. Most of the Cambridge group wanted to see science liberated from the apparently materialistic bandwagon driven by the exponents of scientific naturalism. The ether became a vehicle by which the universe could once more be seen as a unified whole with a purposeful structure. At the same time, these physicists were often very interested in psychic phenomena and spiritualism, to the testing of which they devoted their scientific talents. Not all were converted, but the hope that scientific testing of spiritualist phenomena might produce a demonstration that would convince everyone of the hollowness of materialism was too strong for some to resist. William Crookes had taken this line in the 1870s and now came back to it in the early years of the new century. But the most active convert was Lodge, who linked ether physics, spiritualism, and progressionist evolutionism into a religious philosophy that might not have been orthodox Christianity, but offered religion a powerful new way of reintroducing spiritual values into a world dominated by science.

There was a complex relationship between science, religion, and ideas about the paranormal within this group. Rayleigh and Thomson seldom spoke on religious matters in public, but their religion almost certainly upheld their faith in the reality of the ether, and there seems little reason to deny that this vision of nature helped to shape their very real scientific discoveries. It was paradoxical that Thomson should discover what became known as the electron, thereby doing much to precipitate the revolution that would destroy the paradigm within which he worked. The study of the paranormal was pre-

2. D. B. Wilson, "The Thought of Late Victorian Physicists"; Wynne, "Physics and Psychics." On the link with spiritualism, see Oppenheim, *The Other World*, chap. 8. See also Stenger, *Physics and Psychics*, chap. 12.

3. Rothblatt, *Revolution of the Dons*.

sented by enthusiasts such as Lodge as an opportunity to extend the boundaries of science into territories whose existence was suspected on nonscientific grounds. More cautious scientists such as Rayleigh and Thomson clearly felt that the psychic phenomena had not been demonstrated with scientific rigor, although they were anxious that science should not arbitrarily rule them out of bounds. William Barrett accepted that the real source of support for the paranormal came from preexisting religious belief; science, he believed, was not really adapted to the investigation of this area and would never find conclusive evidence. In fact, many Christians (including some scientists) resisted the claim that we can communicate with the spirits of the dead as vigorously as skeptics such as Huxley. Some of the early advocates of the reality of spiritualist phenomena were looking for an alternative to the Christianity that they felt had been undermined by scientific developments in other areas. Other supporters of psychic research retained their Christian faith, although often in a form that deviated significantly toward a kind of pantheism. Virtually all felt that proving the reality of these phenomena would help turn the tide of materialism and usher in a new age of belief, whether or not the new theology would be seen as a continuation of the old.

The scientific paradigm represented by ether physics occupied an uneasy position between idealism and materialism. Idealist philosophy flourished as an antidote to Victorian scientific naturalism, especially at Oxford, but there seems to have been little contact between the Cambridge physicists and the philosophical idealists. The ether still represented a materialistic worldview, but one in which matter itself was a richer substance that could serve as the basis for a holistic and theistic perspective. There is an ambiguity in the term *materialism*. In its conventional form, materialism represents a philosophy that is hostile to the claim that the universe is a purposefully constructed whole. All matter is governed by law, but the entities that are constructed from it are isolated from one another, and any local ordering in the world is merely temporary. But a very different materialism can be built on a holistic vision of matter as the reflection of a continuous universal medium that allows every part of nature to interact with every other. This belief that the material universe was a coherent whole was vastly important to the ether physicists, and it accounts for both their interest in the paranormal and their belief that so unified a cosmos must reflect the designing intelligence of its Creator.

The ether theorists used their vision of nature to support their religious beliefs. This tradition was pioneered in Balfour Stewart and Peter G. Tait's *Unseen Universe* of 1875, which had argued for the existence of a world that could not be investigated directly by science, but was linked to the physical universe by the bonds of energy, especially electrical energy.[4] The sense that the universe was unified by virtue of this unseen world was powerfully conveyed, along with the assumption that psychic events took place on this higher plane. Energy was a relatively new concept at the time, and no one could deny that its manifestations were being detected in an increasingly complex series of phenomena. Stewart (unlike Tait) was convinced of the reality of paranormal effects, although he repudiated the spiritualists' explanation of them. All in all, the book sought to create a worldview that was still in touch with science, but which transcended materialism and allowed the scientists to believe that the universe as a whole was a divine construct.

Sir William Crookes had become deeply involved in the investigation of paranormal phenomena in the early 1870s.[5] Although Crookes was mainly a chemist, and certainly not a member of the Cambridge group, his work in physics was innovative enough to have inspired Thomson in the researches that led ultimately to the discovery of the electron. Crookes was always on the lookout, both in his science and in his studies of spiritualism, for mysterious forces at the limits of detection—in effect, he was driven by the desire to penetrate the unseen universe. He thought that there might be a fourth state of matter that could be detected when gases were rarefied to the limits of the techniques then available. His early work had convinced him of the existence of a psychic force, although he did comparatively little work in this area after 1875. When he gave his presidential address to the British Association in 1898, he sought a means of introducing the paranormal and consulted Lodge on how to proceed. He decided to devote most of the address to science's impact on the economy and introduced the paranormal only at the end. He now thought that the study of telepathy was the best way of convincing scientists that such effects are real. The newly discovered forms of radiation—possibly X rays—might be the physical means by which information could be transferred directly from one mind to another. The brain might contain receivers "whose special function it may be to receive impulses brought from without through the connecting sequence of ether waves of appropriate order of magnitude."[6] Science, through its study of both the physical world and these more mysterious phenomena, was offering a glimpse

4. See Heimann, "The Unseen Universe." The book was, in fact, written largely by Stewart.

5. On Crookes and spiritualism, see Fournier d'Albe, *Life of Sir William Crookes*, chap. 12, and Oppenheim, *The Other World*, 338–54. Fournier d'Albe's biography, published in 1923, offers an even-handed account of Crookes's beliefs, but treats him as the unwitting founder of the whole spiritualist movement in religion.

6. Crookes, "President's Address," 31.

of "a profounder scheme of Cosmic Law" and revealing the true wonder of the universe.[7] Following the death of his wife in 1916, Crookes returned to spiritualism. He obtained a spirit photograph of his wife, which he regarded as a "sacred trust," although he published nothing on the topic and urged those whom he contacted, including Lodge, to respect his wishes.[8] Crookes's ventures into spiritualism were clearly linked to his interest in the possibility of survival after death, but were offered without overt theological comment.

George Gabriel Stokes gave two series of Gifford Lectures in 1891 and 1893, which explored the response of science to the challenge of materialism. The lectures commented more on problems of evolution than on the physical universe as evidence of design, but in the first lectures of the second series Stokes made clear his belief that the growing support of physicists for the ether had wider implications. The fact that a substance of which we had no direct experience was now admitted to exist on the basis of indirect evidence showed the danger of summarily rejecting the existence of other supernormal phenomena.[9] Such phenomena included the possibility that there might be an ethereally based "directing power" in the body that could survive the death of the body itself.[10] Stokes insisted on the evidence of design throughout the physical universe, including the biological realm, although he introduced his specifically Christian beliefs more actively than Lord Gifford's bequest intended.

J. J. Thomson's presidential address to the British Association, given in Winnipeg in 1909, was a panegyric for the ether theory. The ether is not a fantastic speculation, he argued—it is as essential to us as the air we breathe because it conveys the sun's heat to the earth. "The study of this all-pervading substance is perhaps the most fascinating and important duty of the physicist."[11] The new ideas about the nature of matter and about radioactivity clearly did not, in Thomson's view, invalidate the ether theory, because subatomic particles could be seen as bundles of energy concentrated in the ether. The old sense that all the basic foundations of nature were now understood had been broken down by these exciting new developments. There was still much more to learn, but it would all confirm that "Great are the Works of the Lord."[12]

7. Ibid., 32–33.

8. Fournier d'Albe, *Life of Sir William Crookes*, 404–7.

9. Stokes, *Natural Theology* (1893), 26. On Stokes's religious views, see D. B. Wilson, "A Physicist's Alternative to Materialism."

10. Stokes, *Natural Theology* (1893), 56–57.

11. Thomson, "Inaugural Address," 253.

12. Ibid., 257.

Stokes made no mention of spiritualism, despite his evident belief in survival after death, and Thomson served in the Society for Psychical Research for years without being convinced that the evidence for paranormal phenomena was adequate. In the early twentieth century, however, the public mood seems to have become more receptive to spiritualism and related themes, and some scientists participated actively in this movement. No one could match Lodge's prolific output, but another very active writer was William Fletcher Barrett, professor of physics at the Royal College of Science in Dublin. He had begun investigations of spiritualism in the late 1870s and had enjoyed uneasy relationships ever since with both the scientific community and the Society for Psychical Research.[13] Barrett was an Anglican, although he had reservations about the restrictions of religious dogma and hoped that investigations of the paranormal would stimulate a new spiritual awareness in society at large. In an introduction to the writings of the mystic Charles C. Massey, published in 1909, he expressed concern over the conservatism of formal religion and hoped that the widening of horizons promised by mysticism would create "an awakening to the consciousness of the indwelling spirit of God."[14] In a book titled *Psychical Research*, published as part of the popular Home University Library series a few years later, he appealed to the ether as the basis for a belief in a wider range of phenomena that might, at first sight, seem mysterious. The law of continuity tells us that the universe is a cosmos, not a chaos, and "at the centre and throughout every part of this ever expanding and limitless sphere of nature there remains—enshrouded from the gaze of science—the Ineffable and Supreme Thought which alone can be termed Supernatural."[15] For Barrett, only God was supernatural—the events he studied were natural, however far they seemed to take us beyond the normal realm of experience. He used this point to argue against the suspicions of religious people who feared that spiritualism invoked the supernatural; science merely expanded the bounds of the natural into realms that would once have been seen as mysteries.[16] In his conclusion, he noted the ambivalence of the spiritualists' evidence for life after death, since transmissions from the next world often seemed to fade as though the spirit of the departed was also dissolving. This showed that the ethereal spirit was actually in transition from the psychic to the truly spiritual realm. Psy-

13. On Barrett's psychic researches, see Oppenheim, *The Other World*, 355–71.

14. W. F. Barrett, preface to Massey, *Thoughts of a Modern Mystic*, iii–iv. Massey had been trained in the law and defended the medium Henry Slade against E. Ray Lankester's accusations of fraud; see ibid., 2.

15. Barrett, *Psychical Research*, 13.

16. Ibid., 11.

chic research could thus never take the place of religion because it did not address that last transition to the world of the truly supernatural, the spirit's absorption into God.[17]

In the war years Barrett became caught up in the new wave of enthusiasm for spiritualism, publishing *On the Threshold of the Unseen* in 1917. He insisted that it was science's job to investigate the paranormal; it was only because scientists had refused to accept this responsibility that spiritualists had had to raise public interest in such phenomena.[18] Again he returned to the ether as the medium through which the paranormal operated and invoked the idea of a parallel world of evolution in the ethereal world. For Barrett, all evolution was progressive and represented the influence of a divine power: "If the grosser matter we are familiar with is able to be the vehicle of life, and respond to the Divine spirit, the finer and more plastic matter of the ether might more perfectly manifest and more easily respond to the inscrutable Power that lies behind these phenomena."[19] The higher intelligences of the ethereal world might actually have influenced the direction of evolution on our own material plane. In the published version of a 1910 talk, Barrett had challenged Henri Bergson's vision of evolution as a blind upward striving, insisting that there was a Creative Mind behind it all. He thought that the power of this Mind operated through the psychic element in all life, allowing it to strive against obstacles and thus to progress.[20] In an article on this topic, he appealed to the nonmechanistic physiology of J. S. Haldane to support the claim that the soul had now come back into fashion.[21]

In 1918 Barrett gave a talk to the Clerical Society of Birmingham on the deeper implications of psychic research, in which he offered the speculation that all human minds were part of a wider spiritual whole. Perhaps telepathy was the means by which God communicated religious insights to inspired individuals: "And may there not be some telepathic intercommunication between the Creator and all responsive human hearts, to some being given the inner ear, the open vision, and the inspired utterance?"[22] At the same time, this speech seems to contradict Barrett's own claims that the ethereal realm is the vehicle for the paranormal. Arguing that telepathy is not bound by the inverse square law,

17. Ibid., 246.

18. Barrett, *On the Threshold of the Unseen*, 99.

19. Ibid., 112.

20. Barrett, *Creative Thought and the Problem of Evil*, preface and 43–46.

21. Barrett, "The Psychic Factor in Evolution." The role of the mind was a popular theme in anti-Darwinian evolutionism; see Bowler, *Eclipse of Darwinism*, 73–74, 80–81.

22. Barrett, "Deeper Issues of Psychical Research," 174. This article is reprinted in his *Personality Survives Death*, xix–xxxix.

Oliver Lodge. Cartoon by "Spy" from *Vanity Fair*, 4 February 1904.

he insisted that it "cannot be explained by a process of mechanical transmission."[23] Clairvoyance, too, implied that these effects were taking place at a purely psychic level. If the spiritualist and paranormal phenomena could not be explained in terms of the ether, or any other purely physical mechanism, the study of this realm seemed to be moving not just to the margins, but actually beyond the bounds of science. In the end, Barrett believed that religious faith was essential for salvation. Whether part of science or not, the study of the paranormal could at best help to break the habits of materialism—it could not tell us what to do with our lives and could easily be misused for evil purposes.

Although Barrett was still publishing widely in the early twentieth century, it was Oliver Lodge who most visibly kept the worldview of the old physics before the eyes of the general public. Lodge had made major contributions to the study of radiation while he was professor of experimental physics at Liverpool, but his appointment as principal of Birmingham University in 1900 made him a full-time academic administrator and left him

23. Ibid., 173.

with plenty of time for writing. His seniority gave him wide access to editors and publishers, and he had influential contacts, being in regular correspondence with Arthur Balfour. Like Barrett, he linked ether physics to the scientific study of the paranormal and played a major role in the revival of spiritualism during the Great War. His *Raymond: Or Life and Death*, detailing his communications with his son, killed on the Western Front in 1915, was widely read. But Lodge went much further, linking his worldview firmly with the progressionist evolutionism popularized by some biologists and an optimistic view of the future of humankind. His vision extended to a bold plan to revitalize religion, openly challenging those aspects of Christianity that he felt were no longer compatible with the scientific worldview, although he presented his views as a liberalization of the Christian message rather than a rejection of it.

Trained originally in the materialistic tradition of Huxley and Tyndall, Lodge had become involved with the study of spiritualism in the 1880s.[24] By the early twentieth century he was emerging as a prominent opponent of scientific naturalism. His *Life and Matter* of 1905 was a critique of Ernst Haeckel's *Riddle of the Universe*, defending the view that life, mind, and consciousness were not mere by-products of matter. Rather, they were new properties, emerging as matter came into more complex aggregations, but existing apart from matter and capable of influencing it and hence guiding the operations of the material world.[25] At this point, Lodge seemed quite clear that life (to say nothing of mind) was purely immaterial: "Life may be something not only ultra-terrestrial, but even immaterial, something outside our present categories of matter and energy; as real as they are, but different, and utilizing them for its own purpose."[26] Such views were repeated on and off throughout Lodge's later career—as late as 1932 he insisted that life and mind were not functions of matter.[27] He went out of his way to criticize the materialism of Arthur Keith's notorious 1927 presidential address to the British Association.[28] Yet his writings were also increasingly pervaded by a somewhat different way of looking at things derived from his synthesis of ether physics and spiritualism. Lodge became convinced that the material body was ac-

24. See Oppenheim, *The Other World*, 371–89; see also Lodge's autobiography, *Past Years*, chaps. 22–24, and Jolly, *Sir Oliver Lodge*. For a comparison of the views of Lodge, Jeans, and Eddington, see D. B. Wilson, "On the Importance of Eliminating *Science* and *Religion* from the History of Science and Religion."

25. Lodge, *Life and Matter*, 49–50, 136.

26. Ibid., 198.

27. Lodge, "Religion and the New Knowledge," 213.

28. Lodge, *Phantom Walls*, 52–56; see also *My Philosophy*, 74.

companied by an ethereal one that survives after its death, and although sometimes writing as though the ethereal body was an intermediate between the spirit and the flesh, he often seems to imply that the ethereal entity *is* the spirit. Some passages in Lodge's writings thus seem to express a more subtle materialism in which the ether, with its ability to serve as a unifying principle for the whole universe, is the seat of all the activity we call spiritual, survival being continuation of the ethereal body on this more rarefied plane.

Lodge's commitment to the ether was absolute and unchanging, a view that left him increasingly isolated within the scientific community as the impact of Einstein's theory began to sink in. His last work still maintained that the whole universe was based on radiation and the ether, subatomic particles being merely units of locked-up energy. And the whole cosmos formed a unity displaying design: "I claim that the material universe with its variously designed atoms, and the way they have been used in the construction of all the objects, mineral, vegetable and animal, that we see around us, is a sign also of gigantic Design and Purpose, and is a glorious Work of Art."[29] Lodge responded to "Modern Gibes at the Ether" by insisting that the new physics of mathematical abstractions was unsatisfying: there had to be a material substratum for all activities, and in the end a more sophisticated interpretation of the idea that ether waves were the foundation of all energy would be substantiated.[30] Turning to life and mind, Lodge again insisted that they were not functions of matter, but his main development of this topic makes it clear that he was referring only to brute matter. The ether provides us with a position intermediate between idealism and materialism: there is spirit, but it exists on a higher material plane that that of normal matter. Every sensible object has an ethereal counterpart, and the soul will be shown by future research to be an ethereal body.[31] It is this ethereal body that survives the death of the fleshly body and which, by communicating with the ethereal bodies of mediums, allows messages to be transmitted into the everyday world. In the end, Lodge's philosophy seems to reflect the more holistic form of materialism rather than a dualistic position. As he said in the preface to his *Phantom Walls* of 1929, modern physics showed that all mundane experiences were illusory and that matter itself was mysterious enough to serve as the foundation for the spiritual world.

In the later part of his career, Lodge included substantial discussions of evolution in his expositions. Books such as his *Making of Man* (1924) and *Evo-*

29. Lodge, *My Philosophy*, 31.
30. Ibid., chap. 12.
31. Ibid., 234–35.

lution and Creation (1926) applied his general worldview to the question of the origins of the human soul. Lodge joined those biologists and theologians who argued that there need be no conflict between the theory of evolution and the traditional belief that higher things were created by God's design. The study of evolution was merely the study of the method of creation, and the more we learn, the more obvious it is that the evolutionary process reflects "the constant activity of some beneficent Power."[32] Darwin's selection theory involved technicalities on which Lodge felt unable to comment, but it did not explain variation and heredity and thus did not account for progressive evolution.[33] He did not rule out the Lamarckian effect, though he knew that many biologists were suspicious of it.[34] He certainly stressed that effort was important in evolution, because in striving to overcome obstacles animals improve themselves.[35] In Lodge's view, evolution had been guided by an immanent controlling Power from the start, "the Great I AM operating in an eternal present, an eternal Now."[36] The later stages of evolution saw the gradual emergence of the human soul as the moral and spiritual faculties were formed—Lodge seems not to have liked Lloyd Morgan's image of a relatively sudden appearance of the new categories of being. There had always been a risk that evolution could fail to reach its goal, but the process was the best that could be imagined for the creation of beings with free will. Voltaire had thus been wrong to ridicule the view that this was "the best of all possible worlds."[37]

Lodge's progressionist evolutionism was linked to an optimistic faith in future human progress that easily survived the trauma of the war. He saw science as a force that would improve earthly well-being, but linked this vision to a confidence in the moral improvement of the race brought about by an increasing awareness of the divine origin of the universe and its Creator's intentions for the spiritual beings that emerged in it.[38] His was thus an extremely liberal theology that sat rather uneasily with the orthodox Christian

32. Lodge, *Evolution and Creation*, 16.

33. Ibid., 103.

34. Ibid., 109. Here Lodge concedes that natural selection is important.

35. Lodge, "The Effort of Evolution" and *Making of Man*, chap. 2.

36. *Evolution and Creation*, 123.

37. *Making of Man*, chap. 6. In a letter to Sir Martin Conway, Lodge wrote of God taking risks when He allowed a portion of Himself to split off with free will: Lodge to Conway, 21 April 1924, Lodge papers, Birmingham University Library, OJL 1/97/6. Lodge's distrust of some aspects of Morgan's doctrine of emergence was recorded in a letter from Lodge to Morgan, 9 July 1931, Morgan papers, Bristol University Library, DM 128/505.

38. See, for instance, Lodge, *Science and Human Progress*.

tradition. Lodge was quite open in his desire to reform the faith—perhaps out of all recognition, as far as traditionalists were concerned. As early as 1904, his "Suggestions towards the Reinterpretation of Christian Dogma" had insisted on the need to revise our understanding of the Atonement because the idea of original sin was no longer plausible in an evolutionary worldview. The Fall was merely the first awakening of moral awareness and the sense that we were not as perfect as we could become.[39] This position was stated over and over again in the course of his later writings. Humankind was part of a progressive system, and Christ's coming into the world was intended to show us the future toward which we should aspire. The possibility of failure and degeneration was always there in a world with free will, but help and encouragement would come from higher beings in the spirit world.[40]

Lodge wrote on the supposed conflict between science and religion throughout his later career.[41] He disagreed with those who argued for two separate realms: science inevitably offered insights into the ultimate nature and purpose of the world (especially when it addressed phenomena such as spiritualism), and religion had to take account of the new worldview that science presented. Conflict arose only when science took a dogmatically materialistic stance and religion refused to modify its own dogma in accordance with modern knowledge. Lodge's own philosophy was the bridge between the two areas, showing how science itself could deal with matters of spiritual concern.[42] A quotation from *Man and the Universe* sums up his position:

> Theologians have been apt to be too easily satisfied with a pretended foundation that would not stand scientific scrutiny; they have seemed to believe that the religious edifice, with its mighty halls for the human spirit, can ultimately rest upon ancient events and statements, instead of upon man's nature as a whole; and they usually decline to reconsider their formulae in the light of fuller knowledge and development.
>
> Scientific men, on the other hand, have been liable to suppose that no foundation which they have not themselves laid can be of a substantial character, thereby ignoring the possibility of an ancestral accumulation of sound though unorganized experience. And a few of the less con-

39.   Lodge, *The Substance of Faith Allied with Science,* section 2.

40.   *Evolution and Creation,* 143–49. Arthur Burroughs, bishop of Ripon, accused Lodge of veering toward polytheism by postulating these higher beings; see Burroughs to Lodge, 31 July 1928, Lodge papers, Birmingham University Library, OJL 1/67/4, and Lodge's reply, OJL 1/67/5.

41.   See, for instance, his "Outstanding Controversy between Science and Faith," "Reconciliation between Science and Faith," "On the Conflict between Religion and Science."

42.   Lodge, *My Philosophy,* 36.

siderate, about a quarter of a century ago, amused themselves by institut-
ing a kind of jubilant rat-hunt under the venerable theological edifice; a
procedure naturally obnoxious to its occupants. The exploration was un-
pleasant, but its results have been purifying and healthful, and the permanent substitution of fact has already been cleared of much of the refuse
of centuries.

For Lodge, the old scientific naturalism had cleared the way for the creation
of a new liberal theology.

Rationalists such as E. Ray Lankester were particularly skeptical of
Lodge's appeal to spiritualism.[43] There was a vitriolic attack by Charles
Mercier in the *Hibbert Journal* for 1917.[44] Yet, freed of the somewhat dubious
element of the paranormal, Lodge's philosophy fell closely in line with the
more extreme modernizing trends in Christian theology, including R. J.
Campbell's "New Theology," which created such a stir in 1907, and the Modernism of the Anglican Church. Campbell first wrote to Lodge in 1905, asking him to speak at the City Temple and insisting that the congregation had
already heard much of Lodge's philosophy from Campbell's own lips.[45]
Lodge wrote to Bishop E. W. Barnes, praising his efforts to liberalize the
church's teaching, but arguing that the task would be easier if Barnes accepted
spiritualism.[46] At the controversial British Association meeting of 1927,
Lodge attracted a good deal of press attention for his reply to the bishop of
Ripon's call for a moratorium on scientific research, although he wrote to the
bishop assuring him that most scientists understood his concerns about the
effects of technology.[47] Lodge also gained much press coverage with his defense of spiritualism at the same meeting, and with a speech in support of
evolution given to two thousand miners in Salem Congregational Church.

43. See Lester, *E. Ray Lankester,* 213.

44. Mercier, "Sir Oliver Lodge and the Scientific World." For other examples, see Oppenheim, *The Other World,* 371.

45. Campbell to Lodge, 13 June 1905, Lodge papers, Birmingham University Library, OJL
1/74/1. On the New Theology, see chap. 7 below.

46. Draft of a letter from Lodge to Barnes, 1 December 1930, Lodge papers, OJL 1/25/4.
Lodge was also in touch with Bishop Charles Gore, although he did not approve of the latter's Anglo-
Catholic approach. Gore defended Christianity's reliance on miracles in letters to Lodge, 25 February
1897 and 13 May 1905, Lodge papers, OJL 1/156/6 and 11.

47. *Times,* 5 September 1927, 15; *Daily Telegraph,* 5 September 1927, 9; *Manchester Guardian,* 6 September 1927; Lodge to Arthur Burroughs, bishop of Ripon, 9 September 1927, Lodge papers, OJL
1/67/1. In a letter to Arthur Keith, Lodge suggested that the furor surrounding Burroughs's speech
was a product of the Anglo-Saxons' inability to understand his Irish humor: Lodge to Keith, 17 September 1927, OJL 1/209/2. For details of Burroughs's attack, see chap. 9.

The scientific foundations of Lodge's position were, however, by now completely out of date. The ether was no longer credible as the basis for a belief in a parallel "spiritual" world existing on a material plane higher than that of everyday matter. The physicist James Arnold Crowther wrote of radiation as the "central mystery of creation" in Frances Mason's *Great Design* of 1934,[48] but such efforts to reconcile the old physics with natural theology were now irrelevant as far as many scientists were concerned. Several commentators on Alfred North Whitehead's organismic philosophy have noted the influence of J. J. Thomson's idea that events can be seen in terms of transfers of energy, but Whitehead's philosophy could hardly be perceived as an extension of the old physics.[49] Fortunately for those theologians who hoped to sustain the reconciliation with science, the new physics offered as promising an opportunity as the old.

THE NEW PHYSICS

The plausibility of the ether was destroyed by relativity theory, which showed that the hypothetical medium was undetectable by any physical means. It was also undermined, as was the whole philosophy of continuity, by the growing tendency to interpret all phenomena as controlled by discrete entities, whether subatomic particles or quanta of energy. Thomson and Lodge had been quite happy to treat subatomic particles as bundles of energy concentrated in the ether, their apparently discrete nature being dispelled by the wider sense of the ether's distribution through space. But if the whole subatomic world had to be treated in terms of quanta, the old idea of a unified cosmos was threatened. If the new physics was to have a bearing on religion, it would not be through the erection of a natural theology of the old kind. To see a design in the whole, one had to look to the mathematical nature of the relationships among the parts, which alone seemed able to give a workable representation of subatomic reality. This was the position advocated by James Jeans.

The nature of the reality explored by science was no longer the same because the new physics undermined the old idea of a law-bound universe existing entirely independently of any observer. Lodge's physics had never really escaped from the framework of the old determinism; for all that he etherealized matter, he still had to accept that the workings of the spirit world must be consistent with whatever laws govern the transmission of im-

48.   J. A. Crowther, "Radiation."
49.   Emmet, *Whitehead's Philosophy of Organism*, xxiv; Mays, *Philosophy of Whitehead*, chap. 3.

pulses through the ether. He might claim that the spirit world transcended the determinism of brute matter, but he had no argument to show how. The new physics undermined determinism directly by making it difficult to argue that everything was rigidly predictable. As critics of the religious ideas based on indeterminacy never ceased to argue, this was not the same as vindicating the freedom of the will—but the dead weight of billiard-ball materialism was certainly lifted from those who wanted to explore other possibilities. As Eddington pointed out, the new physics also made it clear that events in the physical world were best understood in terms of the sense data experienced by the observer, so the mind was once more a respectable component of reality, not a mere epiphenomenon. The new physics was certainly popularized as the nemesis of materialism and was accepted as such by many religious thinkers, but philosophers and the more thoughtful theologians both urged caution when interpreting the implications of the new theories.

Albert Einstein became a symbol of the new physics following the vindication of his ideas, although few who read about him really understood what his theory of relativity was all about. One of his chief advocates at this superficial level in Britain was the politician Viscount Haldane of Cloan (brother of the physiologist J. S. Haldane), who published a book, *The Reign of Relativity*, in 1921. But Haldane was a philosophical idealist, and this book merely stressed the relativity of knowledge to the self, with little real reference to Einstein's theory. It was Haldane who held a dinner party at which Einstein told the archbishop of Canterbury that relativity had no implications for religion, being just abstract science.[50] Arthur Stanley Eddington, the British scientist most associated with Einstein's theory, was disappointed by Einstein's unwillingness to confront these issues in public, although he sympathized with his desire to evade questions from people who thought the new physics opened the door to mysticism. He believed that Einstein was by no means committed to an absolute separation between science and broader issues, and indeed, Einstein later made it clear that he did believe in an impersonal God who served as the background to reality.[51]

One rather odd spin-off from relativity was the paradoxical defense of the immortality of the soul developed in J. W. Dunne's *The Serial Universe* in 1934. Dunne was already known for his new approach to the phenomenon of

50. Sommer, *Haldane of Cloan*, 382; Bell, *Randall Davidson*, vol. 2, 1052. There is some disagreement about the wording of Archbishop Davidson's question: these two sources claim that he asked about "morale," whereas Eddington and Frank describe him as asking about religion; see Eddington, *Philosophy of Physical Science*, 7; Frank, *Einstein: His Life and Times*, 189–90.

51. Douglas, *Life of Arthur Stanley Eddington*, 115. For the later expression of Einstein's views, see his "Science and Religion," and for a modern analysis, see Jammer, *Einstein and Religion*.

precognition, published in his *An Experiment with Time*, which had been widely commented on. Although he was not a scientist, his ideas were first expounded to the Mathematical and Physical Society of the Royal College of Science, and they attracted the attention of some scientists. Dunne invoked the idea of relativity by arguing that the observer must be included in his or her own description of the universe, but will then become involved in an infinite regress as it becomes necessary to invoke an observer of the original observer in an attempt to gain a complete picture. The result was an argument for immortality in multidimensional time—although Dunne admitted that it was meaningless to talk about life "after" death.[52] This version of immortality was hardly likely to be of much interest to those with conventional religious views.

It was Eddington, however, whose opinions on the implications of the new physics first began to make headlines. Eddington was raised as a member of the Society of Friends and remained a Quaker throughout his life. As one later commentator on his work put it, a reader could guess that he was a Quaker from his writings because his mysticism was characteristic of that movement.[53] The claim that his ideas belong to the tradition of Plato, Pythagoras, and the more mystical side of Newton makes some sense in the light of his later research, in which he sought a mathematical relationship between the universal constants that would allow the whole cosmos to be described as a unity.[54] But his early impact was made through an exploration of the philosophical implications of relativity and quantum mechanics, which led him to a kind of idealism in which our theoretical knowledge of the universe (if not our practical, everyday knowledge) is merely a mathematical construct from the observations we make with our instruments. Whether this was a genuine idealism is open to debate: A. D. Ritchie thought that Eddington's sense of humor let him carry his illustrations too far and led his readers (including the philosophers who attacked his alleged idealism) into misunderstandings. The famous comparison between the real chair we sit upon and the physicist's chair made of insubstantial wave functions was not one of his better jokes.[55]

Whether he was a true idealist or a Kantian, as Ritchie claims, Eddington's writings were widely interpreted as reintroducing a role for mind in our

52.   Dunne, *The Serial Universe*, 37.

53.   Jacks, *Sir Arthur Eddington*, 3.

54.   Ibid., passim; see also Ritchie, *Reflections on the Philosophy of Sir Arthur Eddington*, 28–31.

55.   Ritchie, *Reflections on the Philosophy of Sir Arthur Eddington*, 1, 27. Actually, it is a table, not a chair, that serves as the example at the start of *Nature of the Physical World.*

A. S. Eddington. By Sir William
Rothenstein, 1929. By courtesy
of the National Portrait Gallery,
London.

understanding of reality. The new physics showed that the human mind in
the form of an observer was necessary to the description of nature and that
physical events were not rigidly predictable. Eddington himself made no
claim to the effect that science now demonstrated the reality of free will. He
claimed only that the barrier of the old materialism had been removed, and
appealed to nonscientific intuition for the rest of his philosophy, including
the reality of the human personality and the possibility that the whole universe might be the construct of a designing Mind.

Eddington studied mathematics at Cambridge and was taught by
Whitehead and by E. W. Barnes. He became Plumian Professor of Astronomy at Cambridge in 1913 and was involved in the famous test of Einstein's
theory of relativity during the solar eclipse of 1919. Much of his later theoretical work involved the relationship between relativity and quantum mechanics. He began to explore the implications of the new physics for religion
in a chapter for Joseph Needham's 1925 collection *Science, Religion, and Reality.* At
this point his views were derived largely from relativity—he was only just beginning to realize that the indeterminacy of quantum mechanics might offer
an even greater challenge to the realist view of the external world. Significantly, at one point where he reintroduces the topic of mind or spirit, he takes
the trouble to note that he does not refer to spiritualism, because his provi-

sional position is that of a disbeliever.[56] In many respects, though, the chapter sketches out the position that Eddington would develop in *The Nature of the Physical World*. It begins with the contrast between the commonsense world and the abstractions of the physicist (this time it is the floorboard that serves as the illustration). This contrast leads Eddington to the claim that the really important distinction is not between the material and the spiritual worlds, but between the metrical and the nonmetrical.[57] Science deals only with what can be measured, yet has come to be seen as the source of a totally self-consistent system ruling out the significance of nonmetrical things, such as human values. In fact, this decision to exclude values is a decision of the mind, which seeks to order the world in terms of those aspects of it that do not change—and Eddington recognized that natural selection may have forced us to think in this way because our survival depends on functioning in the material world.[58] But not only values are thus left out; so, too, is a sense of "becoming," which is absent from the four-dimensional world of space-time constructed by the modern physicist. The chapter concludes with the claim that mind can now be seen as a real agent in our descriptions of the world and the hope that "the actuality of the world is not only in these little sparks from the divine mind which flicker for a few years and are gone, but in the Mind, the Logos." Even so, "science does not indicate whether the world-spirit is good or evil; but it does perhaps justify us in applying the adjective 'creative.'"[59]

These themes were all developed further in Eddington's Gifford Lectures for 1927, published the following year as *The Nature of the Physical World*. The book surveys the impact of modern science on philosophy, beginning with relativity, moving on to cosmology (to which we shall return below), and then to the new factor that Eddington had come to appreciate since 1925: quantum mechanics and the wave theory of matter. Quantum indeterminacy now led Eddington to describe the limitations of human knowledge thus: "A quantum action may be the means of revealing to us some fact about Nature, but simultaneously a fresh unknown is planted in the womb of time. An addition to knowledge is won at the expense of an addition to ignorance. It is hard to empty the well of Truth with a leaky bucket."[60] Later on he faces up squarely to the attempt to spell out the philosophical implications of the new

56. Eddington, "Domain of Physical Science," 214.

57. Ibid., 200.

58. Ibid., 213–14.

59. Ibid., 217.

60. Eddington, *Nature of the Physical World*, 229.

physics, aware of his own limitations and plunging in "not because I have confidence in my powers of swimming, but to try to show that the water is really deep." There follows immediately the bald and much-quoted statement: "To put the conclusion crudely—the stuff of the world is mind-stuff."[61] If the old idea of a law-bound material nature existing independently of conscious observers has been destroyed, we must accept that the very nature of reality is built from the experiences we enjoy or suffer. Eddington qualifies his statement immediately by saying that "The mind-stuff of the world is, of course, something more general than our individual conscious minds; but we may think of its nature as not altogether foreign to the feelings of our consciousness."

Eddington notes that his model of theoretical knowledge as an abstraction from "pointer readings" relieves some of the tension between science and religion. The original tension arose between religious thinkers' desire to see God as active in the world and the determinism of the old science. For the scientist to try to include God within his view of the world was to risk reducing the Deity to "a system of differential equations." This can no longer happen because we now see how abstract the scientific scheme is and how much wider is the actual realm of our experiences. Since those wider experiences are nonmetrical, the world of consciousness wherein we may hope to find a greater Power cannot be bound by mathematical equations. For this reason, "the crudest anthropomorphic image of a spiritual deity can scarcely be so wide of the truth as one conceived in terms of metrical equations."[62]

The breakdown of the old rigid determinism by the principle of uncertainty leads Eddington to the problem of free will. Does the denial of determinism open the way for all sorts of supernatural interference, including that of the human will? He rejects the idea that a conscious choice can affect the body via a causal chain initiated by a quantum jump in a single atom, the direction of that jump representing the influence of the will.[63] Nevertheless, "if the unity of a man's consciousness is not an illusion, there must be some corresponding unity in the relations of the mind-stuff which is behind the pointer readings." Thus "it seems plausible that when we consider their [the atoms'] collective behaviour we shall have to take account of the broader unifying trends in the mind-stuff, and not expect the statistical results to agree with those appropriate to structures of haphazard origin." Many critics as-

61. Ibid., 276.
62. Ibid., 282.
63. Ibid., 313–14.

sumed that Eddington was prepared to see mind influencing the world through its effect on quantum transitions.

The penultimate chapter, "Science and Mysticism," develops the most far-reaching implications of Eddington's new philosophy. He argues that mathematical and poetic descriptions of the same phenomenon (the example is waves in the ocean) are equally valid, although the aesthetic experience is what makes us truly human. The old naturalistic worldview allowed human values to be dismissed as an illusion, but this was no longer possible. "It is because the mind, the weaver of illusions, is also the only guarantor of reality that reality is always to be sought at the base of an illusion." Thus "it is reasonable to inquire whether in the mystical illusions of man there is not a reflection of an underlying reality."[64] Subjective elements are real—humor, for instance, cannot be explained away by trying to explain the point of a joke. For some people, the mystical experience of God is as real as anything else in their existence—they regard anyone who lacks the mystical sense the way most of us regard someone without a sense of humor, as lacking something essentially human. The scientist no longer has the right to dismiss either sense as an illusion—indeed, Eddington thinks that the real danger in the future will come from those who dismiss nonpractical experiences as merely a waste of time in the race for material acquisitions and pleasures—that is, from moral rather than from philosophical materialism.[65]

Science can never offer anything that will prove to the atheist that he or she is wrong. One cannot tell someone who lacks the sense of mysticism about God, any more than one can tell a joke successfully to someone who lacks a sense of humor, although in both cases we can hope to show that life has more to offer if one does open oneself to the subjective experience. At best science can only show that the idea of God is not implausible.

> The idea of a universal Mind or Logos would be, I think, a fairly plausible inference from the present state of scientific theory; at least it is in harmony with it. But if so, all that our inquiry justifies us in asserting is a purely colourless pantheism. Science cannot tell us whether the world spirit is good or evil, and its halting argument for the existence of a God might equally well be turned into the argument for the existence of a Devil.[66]

64.   Ibid., 319.
65.   Ibid., 325.
66.   Ibid., 338.

Thus Eddington has no natural theology to offer, no argument from design that tells us something about the nature of God. Knowledge of God must still come from personal experience—mystical experience, perhaps, but in a more everyday sense than the ecstasies of the great religious mystics. His final chapter states this position clearly: "The conclusion to be drawn from these arguments from modern science, is that religion first became possible for a reasonable scientific man about the year 1927."[67]

But if the scientific and the mystical realms are merely to coexist as equal expressions of reality, does this mean that there can be no conflict between them because they form two separate realms? Eddington thinks there are still frontier difficulties that need resolution, and offers as an example the claim that there is a future nonmaterial state of existence waiting for us (more or less what Lodge thought science *had* demonstrated). His last pages note that science is constantly undergoing change, for which reason "the religious reader may well be content that I have not offered him a God revealed by the quantum theory, and therefore liable to be swept away in the next scientific revolution."[68]

Eddington's book was widely and often negatively reviewed by philosophers who saw no reason to approve of an effort to revive idealism coming from outside the ranks of their profession.[69] But the book sold well, and it served as the launching pad for a wide if rather shallow wave of enthusiasm from religious writers anxious to see the renewed links between science and religion preserved despite the revolution in physics. The *Times* described it as a stiff read, but noted its support for the idea that consciousness was close to the nature of reality and its destruction of the old arguments against free will.[70] In the following year, Eddington gave the Society of Friends' Swarthmore Lecture, published as *Science and the Unseen World.* This lecture repeated many of his arguments and introduced a novel illustration of the power of the unseen world. How, asks Eddington, would a visitor from outer space account for the cessation of movement in the streets at 11:00 A.M. on the 11th of November every year? It is certainly not predictable in the way that an eclipse of the sun may be predicted from planetary orbits—it could be explained only by taking into account the memory of the War and its ability to affect people's behavior. But, he asks, "If God is as real as the shadow of the Great

67. Ibid., 350.

68. Ibid., 353.

69. Hicks, "Professor Eddington's Philosophy of Nature"; Joseph, "Professor Eddington on 'The Nature of the Physical World'"; L. Russell, "A. S. Eddington, 'The Nature of the Physical World.'" The philosophers' attacks on scientific idealism are also noted in chap. 10 below.

70. *Times,* 30 November 1928, 28.

War on Armistice Day, need we seek further reason for making a place for God in our thoughts and ideas?"[71] In 1930 he gave a talk in a BBC radio series on science and religion, in which he repeated many of these ideas.[72]

Eddington's Messenger Lectures at Cornell University in 1934 were used in part to respond to his critics, but also gave a brief statement of the theoretical project that was to dominate the last part of his life: the search for a means of reducing all the universal constants of physics to one.[73] In effect, Eddington thought it should be possible to calculate the constants algebraically, rather than having to measure each one individually. Here was a truly Platonic vision of the world as a single rational system, which goes a long way toward vindicating L. P. Jacks's claim that Eddington saw the Creator as a Supreme Artist.[74] Yet although he published on the topic at a technical level, Eddington did not expand upon the implications of his theory for natural theology, and he openly criticized James Jeans's much-publicized claim that God is a mathematician. In the epilogue to his *New Pathways in Science*, Eddington noted that in *The Nature of the Physical World* he had warned against the tendency to assume that the symbols of the mathematician gave a real representation of the world. Yet now Jeans had made exactly that assumption, failing to realize that the mathematician was dealing only with abstractions imposed by the human mind. "I am not sure that the mathematician understands this world better than the poet and the mystic," wrote Eddington. "Perhaps it is only that he is better at sums."[75]

Are we to assume, then, that Eddington's religious beliefs had nothing to do with his search for an underlying unity in the world? Physicists certainly seem to have been reluctant to invoke such a motivation; they account for his obsession in terms of his scientific and philosophical beliefs. Herbert Dingle thought that Eddington's whole theoretical edifice was driven by the fact that he could not fully accept relativity's total destruction of the old idea of an independently existing material world.[76] A recent study of the theory by a historian of physics does not even mention the fact that Eddington was a deeply religious man.[77] Yet is difficult to believe that this search for an under-

---

71. Eddington, *Science and the Unseen World*, 41–42.

72. *Science and Religion: A Symposium*, 117–30.

73. Eddington, *New Pathways in Science*, chap. 13 (on the critics) and chap. 11 (on the new theory).

74. Jacks, *Sir Arthur Eddington*, 31.

75. Eddington, *New Pathways in Science*, 324; see *Nature of the Physical World*, 209. On Jeans, see below. For Eddington's most complete version of his theory, see the posthumously published *Fundamental Theory*.

76. H. Dingle, *The Sources of Eddington's Philosophy*.

77. Kilmister, *Eddington's Search for a Fundamental Theory*.

lying unity in nature was not at least encouraged by his conviction that there is a Mind underlying the whole of reality. This was a theology of nature, but not a natural theology: as a Quaker, Eddington was more interested in forcing people to think about the moral and spiritual experiences that underlie religion, and he did this by showing that science offered no barrier to belief. To stress the abstract image of a God who is rational, but who may not (as far as the world can show) be moral, would not be the most helpful contribution that science could make to religion. Science could place limits on what was plausible, but could not tell the theologian what was true:

> The bearing of physical science on religion is that the scientist has from time to time assumed the duty of a signalman and set up warnings of danger—not always unwisely. If I may interpret the present situation rightly, a main-line signal which had been standing at danger has now been lowered. But nothing much is going to happen unless there is an engine.[78]

The barrier created by Victorian materialism had been removed, but a revival of religion would have to come from the spiritual dimension of human life, not from the inspiration of science.

If Eddington refused to explore the theological implications of the claim that the world is built on mathematical principles, this aspect of the new physics reached the public very openly from the pen of James Jeans. Like Eddington, Jeans built his career working in both physics and cosmology. He worked on the dynamic theory of gases and prepared an important report on the quantum theory before turning to cosmological theory. He was Stokes Lecturer in Applied Mathematics at Cambridge until 1912, after which he retired to work from his home at Guildford. By the late 1920s he was being openly encouraged by S. C. Roberts at Cambridge University Press to write popularizations of the new science, and to some extent he and Eddington were friendly rivals in this area.[79] His *The Universe around Us* of 1929 was a success, paving the way for the advance publicity that launched his Rede Lecture for 1930, published under the title *The Mysterious Universe*. The former book began with a discussion of the pessimistic implications of the new cosmology and invoked the new physics to show that mind was not, after all, an intruder in a hostile universe. It was the final chapter of *The Mysterious Universe* on the

---

78. Eddington, *New Pathways in Science*, 308.

79. See Roberts's memoir, ix–xvii, in Milne, *Sir James Jeans: A Biography*.

implications of the new physics—entitled "Into the Deep Waters"—that attracted the most attention.

In this chapter, Jeans surveyed the picture of our knowledge of the world built up by relativity theory and quantum mechanics. There were different pictures, none of which could be said to give a true knowledge of ultimate reality—indeed, the greatest discovery of the new science was that we are not yet (and may never be) in touch with such a reality. In this respect, Plato's metaphor of the cave, in which the prisoners have to gauge everything from the shadows on the wall, had gained a new relevance.[80] But all the pictures have one thing in common: they are mathematical, and only the mathematician can thus hope to comprehend the world. Jeans notes that his cosmology has served to demonstrate how insignificant and ephemeral human life must be within the universe, so it is thus of great significance that mind can comprehend the structure of a world in which, at first sight, it seems so out of place. The most obvious conclusion is "that the universe appears to have been designed by a pure mathematician."[81] There is a possibility that we are merely imposing our mathematical constructs onto nature, but the system works so well that it is difficult to believe that concepts so far removed from our everyday experience just happen by coincidence to describe reality. "If this is so, then the universe can best be pictured, although still very imperfectly and inadequately, as consisting of pure thought, the thought of what, for want of a wider word, we must describe as a mathematical thinker."[82]

Jeans concedes that this view has brought him close to Bishop George Berkeley's idealism, in which everything exists in the "mind of some Eternal Spirit." He argues that his position is not, however, pure idealism, although it does blur the distinction between realism and idealism. Matter is certainly not hallucinatory—to develop Dr. Johnson's argument against Berkeley (kicking a stone shows it is real), if we kick a hat into which a small boy has placed a brick, the element of surprise shows that we are not making it all up.[83] For Jeans, the real point of the new physics' reintroduction of mind into reality is that the terror of our isolation in the cosmos is shown to be illusory, as is the old mechanistic view that the universe is but a fortuitous concourse of atoms. There can be no discomfort in a picture of the world that could

---

80.   Jeans, *Mysterious Universe*, 127. A second, revised edition was published in 1932. Jeans expanded on the science in his *New Background of Science* in 1933, and on his philosophical idealism in *Physics and Philosophy* in 1942.

81.   *Mysterious Universe*, 132.

82.   Ibid., 136.

83.   Ibid., 138–39.

only have been created by our minds. And if the implication that there is a greater Mind behind it all is accepted, the isolation disappears completely.

> To-day there is a wide measure of agreement, which on the physical side of science approaches almost to unanimity, that the stream of knowledge is heading toward a non-mechanical reality; the universe begins to look more like a great thought than a great machine. Mind no longer appears as an accidental intruder into the realm of matter; we are beginning to suspect that we ought rather to hail it as the creator and governor of the realm of matter—not of course our individual minds, but the mind in which the atoms out of which our individual minds have grown exist as thoughts.[84]

Here was the abstract God that Eddington did not want to invoke. Jeans admits that the designing power shows no evidence of emotion, morality, or aesthetic appreciation. But he is prepared to rest content with the fact that mind as a whole is not a stranger in the universe—perhaps, in the end, he would have supported Eddington's plea that our spiritual perceptions must be given an equal share in reality.

Buoyed by the success of Eddington's *Nature of the Physical World* and Jeans's own popular cosmology texts, Cambridge University Press made sure that *The Mysterious Universe* would hit the headlines. Ten thousand copies were printed even before the lecture was given, and the newspapers all had advance copies to review. The book sold a thousand copies a day during its first month of publication. It was the claim that God was a mathematician that caught the daily papers' imagination: the *News Chronicle*'s review was entitled "Sir James Jeans: God as a Mathematician," and the *Daily Telegraph*'s, "Sir James Jeans and the Universe—Mathematician's Work." The *Daily Mail* also highlighted this theme and drew a comparison with Plato.[85] The *Times*, which had been specially prepared by CUP, carried a long and actually quite critical review. There was a heavy-handed attempt at humor: "Finally he [the plain man] is told by SIR JAMES JEANS to think of himself as a thought. Knowing the poverty of his own thoughts, their inconsequential coming, their unsung departure, he is not likely to feel flattered by this advice. He feels that after further scrutiny the scientist will assure him he is a dream—and a bad

---

84. Ibid., 148.

85. *News Chronicle*, 5 November 1930, 8; *Daily Telegraph*, 5 November 1930, 8; *Daily Mail*, 5 November 1930, 7.

dream at that."[86] The anonymous reviewer then went on to make the serious point that Jeans was engaged in a battle not between science and religion, but between rival theodicies; Jeans's approach was a return to the dictum foisted upon Plato by the happy invention of Plutarch that "GOD is forever geometrizing." In the *Manchester Guardian*, J. G. Crowther reviewed *The Mysterious Universe* as the "Book of the Day," making the point that would be picked up by most philosophers: "One would have thought, perhaps, that mathematics fitted the external world because it was abstracted out of it as one of its features, rather than because a mathematical mind was behind the external world."[87] The *Times* carried a series of letters over the next few days. Francis Younghusband linked Jeans and Whitehead and argued that since all human beings appeared in the world through the physical love of their parents, so the originator of the universe must be a great lover. William Schooling wrote that just because the world could be described in mathematical terms, this did not mean that it should be seen as the thought of a mathematician—no one would want to claim that music was best understood from an analysis of sound waves. This prompted a response from Jeans, who noted that a scientific analysis of a sonata would reveal the harmonies in the chords and would tell one something about the musical mind that had composed it.[88]

S. C. Roberts at CUP thought that Jeans's book was widely referred to by vicars in their sermons, and we shall see below (in part 2) that it was indeed taken up by many religious writers.[89] It was to protest against Eddington and Jeans becoming "the new oracles of the religious world" that Joseph McCabe included so much on their work in his *Existence of God*. Recognizing that Eddington, at least, derived his faith from a source outside science, Mc-Cabe wondered how many of the theologians rushing to welcome this insight from science realized that the new idealism did not merely introduce spirit into the material world—it replaced the material world with a purely mental universe.[90] Chapman Cohen, in a book called *God and the Universe* issued by the Secular Society, also took issue with the new idealism, noting that more thoughtful figures (he mentioned Joseph Needham) were recognizing that

86. *Times*, 5 November 1930, 15.

87. *Manchester Guardian*, 5 November 1930, 5. Curiously, Jeans was upstaged by a longer review of the Rev. F. L. Cross's *Religion and the Reign of Science*, which praised Cross for his attacks on the mechanistic philosophy of the Bolsheviks; see *Manchester Guardian*, 5 November 1930, 16.

88. Francis Younghusband, *Times*, 8 November 1930, 8; William Schooling, *Times*, 11 November 1930, 15; Jeans, *Times*, 13 November 1930, 15. Younghusband's views are discussed in chap. 11.

89. Roberts, in Milne, *Sir James Jeans*, xi.

90. McCabe, *Existence of God*, e.g., 64, 71–72, 143–46.

the popular rush to make matter less material did not really affect the question of God's existence.[91] Even Modernist Anglicans, if they were familiar with the scientific arguments, saw the dangers of assuming that a nonmechanistic view of matter simply opened the way to a more spiritual view of things. Both Bishop E. W. Barnes and Dean W. R. Inge preferred a realist philosophy.[92]

The popular science writer J. W. N. Sullivan took a more positive view in his *Limitations of Science* of 1934, arguing that the position taken by Jeans did not rule out the Deity having other attributes than an interest in mathematics.[93] In general, though, the reaction of working scientists was not positive. Herbert Dingle gave Jeans's book a very negative review in *Nature*. He asked if it signaled the triumph of the Aristotelians over the seed of Galileo (which he would prefer to see blossoming), and warned that Jeans offered a "darkening council, not by words without knowledge, but, much more dangerously, by knowledge without equivalent balance of judgement. Physics has much to say at the present time; there is no need for it to speak for other departments of thought and feeling as well."[94] J. B. S. Haldane—who rebutted Bertrand Russell's assumption that most physicists supported the new idealism—assumed that Eddington wanted the mind of the observer to affect subatomic events, and thought this was incompatible with Jeans's mathematician God: "It would seem that in so far as Eddington is right, Jeans' creator has scamped his work. But in so far as the universe attains a mathematical perfection worthy of that hypothetical being, it leaves no place for free will, and the apparent influence of our minds on it is an illusion."[95]

THE EARTH AND THE UNIVERSE

Eddington and Jeans were both astrophysicists, and their popularizations reflected new developments in cosmology as well as in physics. Here there were also changes in the scientific outlook during the early decades of the century, with implications for long-standing debates that had always had theological overtones. Was the earth, and hence human life, unique even within the vastness of the cosmos now revealed by astronomy? And what did the latest cosmological theories tell us about the origin and ultimate fate of

---

91.   Cohen, *God and the Universe*, 14. This book contains a reply by Eddington, chap. 3.

92.   E. W. Barnes, *Scientific Theory and Religion*, chap. 17; Inge, *God and the Astronomers*, viii.

93.   Sullivan, *Limitations of Science*, 227–28.

94.   H. Dingle, "Physics and Reality," 800.

95.   Haldane, *Inequality of Man*, 264.

the universe? Eddington and Jeans both played important roles in popularizing the debates sparked by new developments in cosmology: CUP published Jeans's *The Universe around Us* to great acclaim, and in 1933 Eddington published *The Expanding Universe*.

It was now taken for granted that the earth was insignificant within the cosmos as a whole, yet at the end of the nineteenth century a compromise had been reached that still placed limits on the size of the physical universe. There was only one galaxy, and it was assumed that our solar system lay more or less at its center. In the early twentieth century this assumption was destroyed—we were not at the center of the galaxy, and our galaxy was only one among many, all rushing apart as the universe expanded. At one level, these new discoveries did not affect beliefs about the spiritual significance of humankind. Conservative religious thinkers had long ago come to terms with the fact that we might be physically insignificant—this was acceptable as long as it was understood that our unique spiritual status made us the focus of the Creator's attention. Opposed to this view stood a liberal theological position, which assumed that God would not have created such an extensive universe only to leave most of it uninhabited. The debate over the plurality of inhabited worlds pitted the old anthropocentrism of the biblical viewpoint against a more outward-looking philosophy that welcomed the possibility of different forms of life and mentality scattered throughout the universe. However large the astronomers assumed the universe to be, these two positions remained available.

A classic product of the anthropocentric viewpoint was the veteran biologist Alfred Russel Wallace's *Man's Place in the Universe*, published in 1903. Wallace challenged the once popular idea, which he attributed to the influence of William Herschel, that there were other galaxies remote from the system of stars to which our sun belongs. He argued that the whole thrust of late-nineteenth-century astronomy had been to show that there was but a single system of stars.[96] By a complex package of arguments, Wallace then tried to convince his readers that the sun lay more or less at the center of this system, and that this central position gave a unique set of physical conditions essential for the origin and evolution of life. Most stars were binaries, and it was most unlikely that planets could enjoy a stable orbit in a system subjected to the gravitational pull of a very close neighbor. Within our own system, he dismissed the possibility that life could have appeared on other planets. In the 1904 edition of his book, Wallace added an appendix in which he stressed that even if life did appear elsewhere, evolution was unlikely to produce any-

96. Wallace, *Man's Place in the Universe*, 100–103; see Dick, *Biological Universe*, 38–53.

thing resembling the human race.[97] In the conclusion to his main text, Wallace made it clear that the outcome of his whole argument is that the universe has been prepared for the appearance of humankind by an omnipotent God.

Wallace's book was not well received, and over the next twenty years many of his cosmological assumptions were overthrown. In 1918 Harlow Shapley showed that the sun was nowhere near the center of the galaxy, and in 1924 Edwin Hubble confirmed the old idea, which Wallace had associated with Herschel, that our galaxy was but one among a vast number scattered across an immense volume of space. Soon Hubble had demonstrated that the galaxies were moving apart from one another, apparently confirming the theories of Einstein and Georges Lemaître in which the whole material universe was expanding.

The universe could now be seen to contain far more stars than Wallace had imagined. But did this mean that it was teeming with planets on which life could evolve? Jeans addressed the question of how planetary systems are formed and concluded that they must be so rare that—even in a universe with a multitude of galaxies—the number of stars with planets must be very small.[98] His estimates were based on his apparent disproof of the old nebular hypothesis, which had allowed an earlier generation of astronomers to argue that the sun and planets condensed naturally from a cloud of rotating dust and gas. On the assumption that gravity was the only force involved, Jeans showed that this was highly unlikely; the normal end product of such a condensation would be a double star system. In a 1916 paper, Jeans developed an alternative mechanism of planetary origins, his "passing star theory." This theory supposed that the material from which the planets were formed was drawn from the sun by the gravitational attraction of another star that passed close by. Given the vast distances between the stars, however, it was clear that such near collisions must be very rare, and hence the number of stars with planetary systems must be very small. Since there is a very narrow range of conditions that would make a planet suitable for life, our earth—and hence the human species—might be unique.

Eddington included a brief account of the implications of the new vision of the universe in *The Nature of the Physical World*. He surveyed the other planets of our solar system, pointing out that the latest evidence was running increasingly against the possibility that there might be life on Mars. There was by now little scientific support for Percival Lowell's claim that the Martian "canals" were built by intelligent beings. On the wider scale, he cited

97. Ibid., 326–36.
98. See Dick, *Biological Universe*, 172–80.

Jeans's calculations showing that planetary systems were very rare: "The so-lar system is not the typical product of development of a star; it is not even a common variety of development; it is a freak."[99] At best, one in a hundred million stars might have planets, and of these, few were likely to have a planet with conditions suitable for life. The earth may not be unique, but if we take account of the vast amount of time it has taken for intelligent life to evolve, it becomes most unlikely that another intelligence exists elsewhere at this point in time.

> I do not think that the whole purpose of the Creation has been staked on the one planet where we live; and in the long run we cannot deem our-selves the only race that has been or will be gifted with the mystery of consciousness. But I feel inclined to claim that *at the present time* our race is supreme; and not one of the profusion of stars in their myriad clusters looks down on scenes comparable to those which are passing beneath the rays of the sun.[100]

Eddington here implies that the purpose of the universe is to produce con-sciousness, but uses Jeans's ideas to preserve the uniqueness of humankind for all practical purposes.

Jeans addressed the wider implications of the new cosmology in the first chapter of *The Mysterious Universe.* He went out of his way to stress the im-mense isolation of the earth in the depths of space and hence the loneliness of the human race. Stars existed in splendid isolation, like ships on an empty ocean, but the near collision that produced our planetary system had oc-curred once, two thousand million years ago. When the human race finally evolved the intelligence needed to survey the cosmos, the first impression was one of terror: we saw a universe of meaningless distances, vast time scales, and above all, indifferent to life. The normal conditions of the material world (empty space and stars) are "actively hostile to life like our own." Into this universe we have stumbled, "if not exactly by mistake, at least as the result of what may properly be described as an accident."[101] Given the size of the uni-verse, such an accident was bound to happen sooner or later, but it still seems difficult to believe that the whole cosmos was designed to produce life. As yet, Jeans argued, the biologists cannot tell us if the origin of life is a natural out-come of normal physical and chemical processes. If it turns out that life is a

99. Eddington, *Nature of the Physical World,* 176.

100. Ibid., 178; Eddington's italics.

101. Jeans, *Mysterious Universe,* 3–4.

natural product of the rare conditions provided by a suitable planet, this will influence our ideas about the possibility of life existing elsewhere in the universe "and produce a greater revolution of thought than Galileo's astronomy or Darwin's biology."[102] Perhaps the special properties of carbon mean that it was designed to make life possible, but we have no right to assume that this is so.

> If . . . we dismiss every trace of anthropomorphism from our minds, there remains no reason for supposing that the present laws [of nature] were specially selected in order to produce life. They are just as likely, for instance, to have been selected in order to produce magnetism or radio-activity—indeed more likely, since to all appearances physics plays an incomparably greater part in the universe than biology. Viewed from a strictly material standpoint, the utter insignificance of life would seem to go far toward dispelling any idea that it forms a special interest of the Great Architect of the universe.[103]

The impression is reinforced when we look into the future, where we see only the prospect of the earth freezing over and a more general "heat-death" of the universe.

> Is this, then, all that life amounts to? To stumble, almost by mistake, into a universe which was clearly not designed for life, and which, to all appearances, is either totally indifferent or definitely hostile to it, to stay clinging on to a fragment of a grain of sand until we are frozen off, to strut our tiny hour on our tiny stage with the knowledge that our aspirations are all doomed to final frustration, and that our achievements must perish with our race, leaving the universe as though we had never been?[104]

Jeans then turns to the new physics, with its emphasis on the role of consciousness as a key element in reality, to counter what he perceives as the pessimistic implications of the new cosmology. Life may look like a cosmic accident, but it cannot be just that alone, because it is actually needed to make sense of the whole situation.

Later in the book, Jeans returns to the topic of the history of the universe, debating whether the red-shift in the light from distant galaxies ob-

---

102.   Ibid., 7. On contemporary debates about the origin of life, see chap. 5 below.
103.   Ibid., 10.
104.   Ibid., 13.

served by Hubble and others represents the expansion of the universe postulated by the latest cosmological theories. He links this expansion to another
unidirectional process, the breakdown of matter to form radiation, which he
argues occurs throughout the universe, not just in stars. The American physicist R. A. Millikan had postulated that there might be a reversal of this process in the depths of space, with matter being recreated out of energy, in
which case, as Jeans puts it, "the creator is still on the job." He doubts that
such a balance can exist: a cyclic universe in which creation and destruction
are exactly matched would violate the second law of thermodynamics. Even
so, Jeans thinks he can understand the lure of the cyclic model: "Most men
find the dissolution of the universe as distasteful a thought as the dissolution
of their own personality, and man's strivings after personal immortality have
their macroscopic counterpart in these more sophisticated strivings after an
imperishable universe."[105] Whatever the plausibility of this analysis in terms
of psychology, it suggests that Jeans himself was not driven to accept the expanding universe out of commitment to the traditional Christian vision of
history.

The new cosmology offered few challenges and much support to the orthodox Christian viewpoint and was thus unlikely to spark debate among religious thinkers. The suggestion that we may be the only intelligent (and
hence presumably moral) beings in the universe merely confirmed the traditional view, although Eddington's belief that other intelligences might evolve
in the course of time suggests a rather more liberal vision of God's purpose.
It was, in fact, the liberal Christians who found it most difficult to acknowledge Jeans's apparent undermining of the hope that life had evolved throughout the cosmos. Bishop E. W. Barnes took exception to Jeans's theory of
planetary origins on the grounds that it limited the possibility of life evolving elsewhere: "My own feeling that the cosmos was created as a basis for the
higher forms of consciousness leads me to speculate that our theory of the
formation of the solar system is incorrect."[106] Barnes was sufficiently enthusiastic about the prospect of life existing elsewhere in the universe to propose
(in a 1931 *Nature* symposium on cosmology) that radio should be used in an
attempt to communicate with the hypothetical extraterrestrials.[107] It was only
in the late 1940s that serious doubts began to be cast upon Jeans's theory, ush-

105.   Ibid., 75. For a report of Millikan's views in *Nature*, see "Energy and Atoms," which is followed by Jeans's critical response, "The Physics of the Universe."

106.   E. W. Barnes, *Scientific Theory and Religion*, 402.

107.   Barnes, "Contributions to a British Association Discussion on the Evolution of the Universe," 719–22.

ering in the late-twentieth-century wave of enthusiasm for the idea that humankind is not alone.

The theory of the expanding universe added another element of unidirectional development to the old worldview of thermodynamics, but did not seriously conflict with the belief that the cosmos has a beginning and an end. The singularity from which the expansion began (the "big bang" of later debates) established an obvious starting point, and the dispersion of the galaxies into the night did not differ much, in its emotional impact, from the prospect of the heat-death of the universe. The nineteenth century, so often portrayed as an era of faith in evolutionary progress, had also seen the rise of thermodynamics as a cultural icon.[108] Christians could thus continue to believe that—whatever life's origin—the universe itself had a point of origin or creation and would end, however gradually, in the destruction of all material achievements.

Many early-twentieth-century religious writers simply incorporated the new cosmology into their traditional thinking (see chapters 8 and 9 below). Dean W. R. Inge, a long-standing opponent of the idea of progress, wrote his *God and the Astronomers* of 1933 to welcome the new cosmology's confirmation of the old's predictions about the fate of the material world. In the 1940s the mathematical physicist Edmund Whittaker, by now a convert to Roman Catholicism, hailed the new cosmology's confirmation of what he took to be St. Thomas's view that the demonstration of a beginning for the universe was a powerful argument for the existence of God. Like Inge, he wanted a God whose existence stood outside the limited time scale of the physical universe, and believed that the simplest way of understanding their relationship was to assume that God created the universe *ex nihilo* at a certain point in time.[109]

The physical sciences and cosmology thus offered a fertile field in which those who hoped to sow the seeds of a reconciliation with theology could work. Here were areas of science undergoing the most momentous changes, yet there was no hint of a return to the self-confident materialism associated with the Victorian era. Lodge's efforts to link the ether theory to spiritualism and the new natural theology may have been undermined on scientific grounds, but the new physics seemed even more amenable to the claim that it made consciousness an integral component of reality. Whether it made the idea of a Supreme Mind more credible was open to debate, but here Jeans's emphasis on the powers of mathematics offered a way of reviving the old claim that the universe had been built by a rational Architect. The great ad-

---

108. See Brush, *Temperature of History.*

109. Whittaker, *The Beginning and the End of the World,* 63; Whittaker, *Space and Spirit,* 116, 121, 131.

vances in cosmology also seemed to uphold the long-established view that the universe did have a beginning, which might be equated with a creation. Not everyone agreed with the implications drawn from these theories, of course, and there were few physicists or philosophers in the later 1930s willing to endorse the idealism of Jeans and Eddington. Their philosophy was treated with suspicion even by some theologians, who thought that it provided no real evidence for a true spirituality. But the new science could at least uphold the challenge to simple materialism, and many religious apologists were eager to use the ammunition it gave them. Through the early 1930s, the door to a reconciliation between the physical sciences and religion was held open by at least an articulate minority within the scientific community.

CHAPTER 4

Evolution and
the New Natural Theology

In the Victorian era the life sciences had provided one of the most visible threats to the traditional Christian viewpoint. The new generation of scientific professionals, led by T. H. Huxley, had been particularly anxious to throw off natural history's reputation as an amateur amusement for country vicars. To do this, they challenged the traditional image of nature and of humankind's relationship to it. Where natural theology had hoped to show that living things are designed by a wise and benevolent God, these scientists now used Darwin's theory to argue that all species had evolved by a purely natural process that showed no evidence of ultimate purpose. Where physiologists had once insisted that life was a unique entity that could not be explained in terms of the laws of chemistry and physics, they now adopted a mechanistic approach. And where once the soul had been treated as something existing in a purely spiritual realm, the human species was now situated firmly within material nature.

The sciences of the 1860s and 1870s had thus become identified with materialism and the assault on religion. Yet in the last decades of the century, the pendulum had swung the other way. Oliver Lodge's adoption of the view that life was engaged in a process of spiritual evolution matched the rise of non-Darwinian ideas in biology and the emergence of a renewed interest in vitalism, the claim that life transcended the laws of physics and chemistry. Evolutionism had gained the day, only for the Darwinists to find that it had been subverted by those who were determined to re-create natural theology by showing that evolution was the unfolding of a divine plan. The "new" natural theology hailed by many early-twentieth-century writers was really nothing more than the teleological evolutionism proposed by many late Victorian biologists, only now emerging into the public consciousness as the threat from Huxley's scientific naturalism abated. The claim that the human species was the preordained goal of evolution was replaced by a more diffuse progressionism, which merely proclaimed that mind, in some form, was its goal. Henri Bergson's *Creative Evolution,* translated in 1911, was highly influential in promoting this view. Bergson's vision of an unstructured yet unstoppable progress arising from an original impulse, the *élan vital,* caught the attention of many biologists. Coupled with the existing tradition of neo-Lamarckism, in which evolution was supposed to be guided in a purposeful direction by organisms' behavioral innovations, it helped to keep alive a vision of life and mind as something rising above brute matter.

These moves did not go unchallenged by those biologists who preferred to defend the original version of scientific naturalism. E. Ray Lankester and Peter Chalmers Mitchell railed against Bergson and against neovitalism, as they railed against the new wave of interest in spiritualism and telepathy. But those who stayed loyal to Huxley's vision were all too well aware that the public no longer shared their view of science's implications. Evolutionists had always seen humanity as the pinnacle of life's development, and the implication that the process represented the unfolding of a divine plan was now more openly expressed. Anti-Darwinian theories of evolution that had already become popular in the late nineteenth century were now being hailed as the latest developments in science.

Within science itself, however, the pendulum had already begun to swing back in the opposite direction. In the 1920s and 1930s, Darwinism emerged at last as the dominant theory of evolution, thanks to its synthesis with genetics—although the expectation that natural selection would ultimately generate progress remained important for key figures such as Julian Huxley and R. A. Fisher. There was thus a permanent legacy in biology stemming from the temporary eclipse of materialism at the turn of the century. But Fisher's ability to link his Darwinism to his Christianity (in part through his enthusiasm for eugenics) was overshadowed by the far more public attacks on teleology and anthropocentrism by J. B. S. Haldane and others. By the 1920s the influence of the "new" natural theology lay primarily in the public perception of science, rather than in science itself.

SCIENCE AND CREATION

By the end of the nineteenth century, virtually all scientists were committed to the theory of evolution, even though there were still many who disliked the Darwinian explanation of how it worked. Yet a small but vociferous antievolution movement emerged in the 1920s, paralleling the far more active crusade in America. Miraculous creation was still seen as an option by these conservative thinkers, although there was as yet no support for what was to become the most popular modern form of creationism, in which the whole fossil record is supposed to have been deposited during Noah's flood. A popular book on the flood legend by the archaeologist Harold Peake noted that few ministers of the church now took the Genesis story literally, although many ordinary believers were still worried about the implications of science for its veracity.[1] Peake argued that there had been a real flood, although it was confined to Mesopotamia. But if flood geology was not an option, there were some who still wanted to retain the traditional view that all species, and especially the human species, were divinely created. A few scientists and writers with a scientific background were anxious to participate in this creationist movement, although they were seldom active researchers in the areas most directly concerned with evolutionism. Most British scientists looked on with amazement at the popular opposition to evolutionism in America, symbolized by the trial in 1925 of John Thomas Scopes for teaching evolution in the high school at Dayton, Tennessee, which was widely reported in the British newspapers.[2] *Nature* carried a supplement on the topic in which a variety of scientists and some religious leaders reflected on the spectacle of a whole society apparently turning its back on science. Lankester warned against state interference in science, while the Lamarckian E. W. MacBride argued that the mechanistic atmosphere of the Darwinian selection theory fueled the fears of the antievolutionists.[3] Most of the comments were self-congratulatory, based on the assumption that such madness could never occur in the more mature climate prevailing on the other side of the Atlantic. In 1926 *Nature* again contrasted the "propaganda" of American Fun-

---

1. Peake, *The Flood*, preface.

2. The *Times*, for instance, noted antievolution legislation in Kentucky, 18 March 1922, 13, and then carried a whole series of reports on the Scopes trial: 16 May 1925, 13; 13 June, 12; 11 July, 14; 16 July, 10; 17 July, 13; 20 July, 12; 21 July, 13; 22 July, 25; 23 July, 13; 18 August, 11.

3. Lankester, "Evolution and Intellectual Freedom"; MacBride, "Evolution and Intellectual Freedom." The whole supplement occupies pages 70–84 of vol. 116 (1925) and includes an anonymous introduction. For recent accounts of the Scopes trial, see Larsen, *Summer for the Gods*, and Numbers, *Darwinism Comes to America*.

damentalism with the calmer approach taken by British churchmen such as the Rev. W. R. Matthews.[4]

Not everyone shared this optimism, however. Even before the Scopes trial hit the headlines, the geneticist William Bateson had warned that the British scientists' feeling of security might be an illusion, because the opinions of the uneducated masses remained untouched by evolutionism.[5] Bateson himself had been accused of lending support to the antievolutionists in his 1922 American Association for the Advancement of Science address, in which he chided biologists for their overconfidence on the topic.[6] But even he feared the outright rejection of evolution in favor of supernatural explanations for the origin of life.

Where Bateson worried that ordinary people remained untouched by science, some conservative Christians thought that they had been browbeaten into uncritical acceptance of evolution. In the early 1920s the American George McCready Price—later to emerge as one of the founding fathers of mid-twentieth-century creationism—was in London, where he struggled to find an audience for his antievolutionary views.[7] Price's creationism was based on the notion of a "young earth" only a few thousand years old, with all the fossil-bearing rocks having been laid down in Noah's flood. The one center he could rely on in London was the Victoria Institute, founded in 1865 to promote opposition to what was seen as the unnecessarily materialistic tone of science. By the 1920s the Institute's membership had declined from its heyday in the late nineteenth century, but it was still active, and its journal published a number of items by Price. In 1925 the Rationalist Press Association lured Price into a public debate with Joseph McCabe, in which the American suffered frequent heckling from the audience.[8] Price was able to make some capital out of the evident uncertainty then prevailing about the causes of evolution (which had been Bateson's point), but he was unable to convince his hearers that this was a good enough reason to abandon evolution altogether.

In 1927 the eminent physicist and electrical engineer Sir Ambrose Fleming was installed as president of the Victoria Institute. Fleming had been nominated for a Nobel Prize by Marconi and other pioneers of radio com-

4. "Propaganda and Philosophy," *Nature* 118 (1926): 543–45.

5. Bateson, "The Revolt against the Teaching of Evolution in the United States."

6. Bateson, "Evolutionary Faith and Modern Doubts." This was reported quite fairly in the *Times*, 14 February 1922, 8, although Bateson subsequently published a letter complaining about the "comical misinterpretations which have followed," *Times*, 13 April 1922, 8.

7. For an account of Price's visit and a description of British creationism at the time, see Numbers, *The Creationists*, chap. 8.

8. The text of the debate in reproduced in Price and McCabe, *Is Evolution True?*

munication for his invention of the thermionic valve. As a member of the evangelical wing of the Anglican Church, he felt that it was his duty to speak out against the theory of evolution, especially as it applied to the origin of humanity. He was not a complete antievolutionist, although he attacked the theory of natural selection in order to defend the claim that the whole development of life must be seen as the unfolding of a divine plan.[9] Like many opponents of Darwinism, Fleming believed that Huxley and the other early promoters of the theory had been so successful that to challenge it was to invite ridicule.[10] The pervasiveness of this view in the early twentieth century helps to explain the modern misperception that Darwinism was completely triumphant in the nineteenth century—a myth that is amply contradicted by the evidence that the vast majority of early evolutionists (including Huxley) were certainly not Darwinians in the modern sense.[11] It suited the twentieth-century opponents of Darwinism to claim that they were fighting against an entrenched dogma, and their interpretation has all too often been accepted by historians trying to assess the impact of the Darwinian revolution. In fact, by the time Fleming was writing, the selection theory was only just beginning to achieve its modern dominance within biology, and many biologists were still almost as skeptical of it as was Fleming himself.

Fleming's real target was the evolutionists' attempt to show that the human mind or soul could have evolved from the mental powers of animals. In 1927 he responded to Arthur Keith's notorious British Association address on this theme and to Bishop E. W. Barnes's efforts to force the Anglican Church to face up to the implications of evolutionism.[12] In 1935 he gave his presidential address to the Victoria Institute on this topic, making it clear that in this case miraculous creation was essential to explain the appearance of spiritual powers.[13] The address was widely reported in the press under headlines such as "Scientist Attacks Evolution—Product of Imagination—Man the

9.   Fleming, "Evolution and Revelation"; see also his booklet under the same title. Fleming's most substantial attack is his *Evolution or Creation?* Fleming's autobiography, *Memories of a Scientific Life*, says little about his work in this area; on his other interests, see MacGregor-Morris, *Inventor of the Valve*.

10.   Fleming, *Evolution or Creation?* 33.

11.   Bowler, *Non-Darwinian Revolution* and *Eclipse of Darwinism*.

12.   Fleming, "Truth and Error in the Doctrine of Evolution." Barnes's "gorilla sermons" are discussed below in chap. 8. Keith responded to this article with his "Evolution and Its Modern Critics," which elicited a response from G. H. Bonner, "Evolution: A Reply to Sir Arthur Keith" and from Fleming under the same title.

13.   Fleming, *Modern Anthropology versus Biblical Statements on Human Origins*. This was immediately expanded into a full-length book, *The Origin of Mankind*. Fleming specifically repudiated the concept of emergent evolution as an explanation of how a new quality such as mind could appear without divine intervention; see *Origin of Mankind*, 60.

Work of a Conscious Creator," and it attracted a direct response from Keith in the *Literary Gazette* and from W. D. Lang in *Nature*.[14] What is perhaps most striking about Fleming's most substantial account of his views is that in addition to repeating fairly conventional objections based on the paucity of hominid fossils and the impossibility of the spirit having a natural origin, he also promoted the idea of multiple human creations to explain the diversity of living races. He saw "no serious obstacle to the belief that *there were other ethically inferior races of human beings on this earth at the time of the creation of Adam and Eve.*"[15] Dismissing the Mongoloid and Negroid races as having contributed nothing to the work of civilization, he identified them with these pre-Adamic creations: "These broad and distinct types of mankind can only be explained by attributing them to different Creative Acts of Divine Power and ... the unquestionably superior Caucasian branch is alone the derivative by normal generation from the Adamic man, namely the God-worshipping members of the Adamic race which survived the Flood—Noah and his sons and daughters."[16] For Fleming, the flood was a real event that had wiped out bastard races produced by interbreeding and had thus purified the three main races.

On 12 February 1935 Fleming and other key antievolutionists organized a meeting at the Essex Hall in London, which led to the founding of the Evolution Protest Movement.[17] Of those who spearheaded the movement, none shared Fleming's high scientific credentials, but several had scientific training or interests. The most active was Douglas Dewar, who had published widely on ornithology while serving in the Indian Civil Service. Back in England, Dewar became a frequent speaker at the Victoria Institute. His *Difficulties of the Evolution Theory* of 1931 was issued at his own expense, but became one of the standard texts cited by the increasingly active antievolution lobby. Dewar cited the work of the French biologist Louis Vialleton, who had opposed the theory of continuous evolution linking the animal types.[18] Perhaps his most original contribution was a series of calculations based on the distribution of fossil and living species of Indian mammals, suggesting that the fossil record was much more complete than the evolutionists claimed, thus giving clear

14. The headline is from the *Manchester Guardian*, 15 January 1935, 17; see also *Daily Mail*, 15 January 1935, 5; *News Chronicle*, 15 January 1935, 8; *Daily Telegraph*, 15 January 1935, 10, 11–12. Keith's response is reprinted with substantial additions in his *Darwinism and Its Critics*; see also Lang, "Human Origins and Christian Doctrine."

15. *Origin of Mankind*, 109, Fleming's italics.

16. Ibid., 137.

17. See *Times*, 13 February 1935, 10; Numbers, *The Creationists*, chap. 8.

18. There is a quotation from Vialleton facing the title page. Vialleton was also cited as an authority in Hilaire Belloc's attacks on evolutionism, discussed in chap. 11 below.

evidence of discontinuities in the development of life on earth.[19] In 1937 Dewar published another book, *More Difficulties of the Evolution Theory*, but in neither of these works did he develop a clearly defined alternative, and he was careful not to link his ideas openly to the text of Genesis. His position emerged more clearly in his *Man: A Special Creation*, which shares Fleming's conviction that the human spirit must be the product of divine creation. Unfortunately, it also reveals an even deeper right-wing paranoia. Noting that Darwinism was essential to the atheists and materialists in their assault on traditional values, he quoted the (otherwise discredited) *Protocols of the Elders of Zion*, in which Darwinism, Marxism, and Nietscheism are presented as parts of a Jewish plot to undermine Western civilization.[20]

If Dewar was mostly self-taught, the same cannot be said for Lewis Merson Davies, who had trained as a paleontologist at Edinburgh and became an expert on fossil Foraminifera while serving with the Indian Army (the fact that Dewar and Davies both spent much of their careers outside Britain may explain their willingness to challenge home-based scientific orthodoxy). Davies published a series of articles attacking evolution in the *Indian Chronicle*, which were then expanded into a book, *The Bible and Modern Science*, first published in 1925. As the title suggests, he did not share Dewar's reluctance to link his ideas directly to the truth of Genesis. Davies attacked evolutionism as an expression of a more general uniformitarianism that had permeated science. For him, divine creations were part of a generally catastrophist view of earth history, although he specifically distanced himself from Price's flood geology. There must have been multiple creations, he argued, because the geologic record could not be explained as the result of a single flood, and this could be squared with the Genesis account by allowing for a gap between the events described in the first two verses.[21]

It was Dewar who eventually had the greatest impact. The meetings he addressed were attended by large numbers, and reputable scientists went out of their way to defend evolutionism against his attacks. The most substantial response came from the paleontologist A. Morley Davies, who published his *Evolution and Its Modern Critics* in 1937. Like Keith in his response to Fleming, Davies tried to deal with the objections from popular writers such as Hilaire Belloc and G. K. Chesterton, but he noted in his preface that Dewar was the foremost champion of the new antievolutionary movement. He regretted that Dewar's *Difficulties of the Evolution Theory* had not been answered sooner, his

19. Dewar, *Difficulties of the Evolution Theory*, 137.

20. Dewar, *Man: A Special Creation*, 98 n.

21. L. M. Davies, *The Bible and Modern Science*, on Price, see 11 n, 89.

own book having been delayed by work on another project.[22] Davies' tactic was to challenge Dewar's competence in analyzing the fossil record in order to minimize the significance of his allegations of sudden "leaps" or saltations in the history of life. In particular, Davies pointed to the latest work on the origin of mammals from reptiles, in which the gulf between the two classes had now been bridged. The same tactic was being used with hominid fossils in the thorny area of human origins, and here Davies referred to some of Fleming's more bizarre interpretations of Genesis.[23] In his conclusion he admitted that the causes of evolution were still uncertain, although he favored Darwinism because the Lamarckian hypothesis had now been discredited. He also conceded that demonstrating the validity of evolution in the organic world "leaves the psychic side of life untouched"—a point that must have given some comfort to his opponents.[24] Dewar's *More Difficulties of the Evolution Theory* was a prompt response to Davies' defense. In the 1940s J. B. S. Haldane joined in the fray, debating both Dewar and L. Merson Davies in a public exchange of letters organized by the New Zealand branch of the Evolution Protest Movement.[25]

The scientific defenders of evolution had to combat both the technical arguments of Dewar and Davies and the more popular expositions of writers such as Belloc and Chesterton. But where the Evolution Protest Movement reflected a largely evangelical Protestant form of Christianity, Belloc and (after 1923) Chesterton were Roman Catholics. By the 1920s the church had reluctantly accepted at least the possibility of organic evolution, although it held that the theory was still highly speculative and that the human spirit must be excepted from the process (see chapter 9). The most active biologist writing from a Catholic perspective was the anatomist Sir Bertram Windle, a convert from a Protestant background, who moved from Cork to Toronto in the later part of his career. Windle wrote widely to defend the church's positions on affairs connected with science. His *The Evolutionary Problem as It Is Today* (1927) was by no means an antievolutionary text, but it did stress the uncertainties over the mechanism of evolution and the evidence for saltations (again, Vialleton was presented as an authority). Windle insisted that there was no actual proof of evolution, although there was nothing to stop a believer seeing organic transformations as God's method of creation. The possibility of a natural origin for the human soul was simply not open

22. A. M. Davies, *Evolution and Its Modern Critics*, v.

23. Ibid., 233, 235, 240, and 244 on the miracles of the New Testament.

24. Ibid., 248.

25. Dewar, Davies, and Haldane, *Is Evolution a Myth?*

to discussion.[26] Significantly, Windle warned against the biblical literalism of Protestant Fundamentalists.[27]

EVOLUTION AND PROGRESS

Outright opposition to evolutionism was rare among those with scientific training, but many biologists accepted a teleological, and hence non-Darwinian, view of the evolutionary process. Darwinists and non-Darwinists alike saw evolution as inherently progressive, but the supporters of non-Darwinian theories such as Lamarckism (the inheritance of acquired characters) were more inclined to see progress as illustrating a divine purpose in nature. By the start of the new century Lamarckism was on the defensive, although by no means eliminated yet from scientific discourse. But the expectation that the evolutionary process should demonstrate a more purposeful ascent toward the human species than could anything driven by mere trial and error continued.

Michael Ruse has argued that faith in progress has underpinned much of the scientific support for evolutionism.[28] In the early twentieth century that faith was still articulated quite explicitly, and indeed, it has never really disappeared, although Ruse notes that as the century progressed it became much less fashionable for it to be expressed in public. He argues that evolutionary biologists were increasingly forced to adopt a rigid posture of objectivity, which ruled out all appeals to wider concerns, in order to protect themselves from the accusations of other scientists who saw the field as tainted by the legacy of Victorian progressionism. There were still enthusiasts for progress among the ranks of the new Darwinists of the 1930s, but they seldom dared to express their feelings except in private. Julian Huxley broke this rule and was marginalized within the scientific community as a consequence, becoming a popularizer who was at best tolerated by "real" scientists for his skill at public relations. R. A. Fisher, who was able to link his Darwinism to his liberal Christianity, kept quiet about his faith until the later part of his career. Many Darwinists found it difficult to square the much more diffuse progressionism permitted by their worldview with anything resembling a traditional

26.   Windle, *The Evolutionary Problem as It Is Today*, 58. On Windle's career, see the biographical memoir edited by Monica Taylor, *Sir Bertram Windle.*

27.   *The Evolutionary Problem as It Is Today*, 63.

28.   Ruse, *Monad to Man* and *Mystery of Mysteries.* I disagree, however, with Ruse's contention that the phylogenetic research of the late nineteenth and early twentieth centuries was just bad science; for an alternative interpretation, see Bowler, *Life's Splendid Drama.*

Christian faith. A few, including J. B. S. Haldane, openly attacked religion, but most simply ignored the whole question of an ultimate purpose.

Two points of significance emerge from Ruse's interpretation. First, it supports the accusation of rationalists such as McCabe that those evolutionary biologists who did express open support for the "new" natural theology were behaving in a manner characteristic of the older generation—which was in fact the generation to which they belonged. The younger biologists simply did not speak out in this way, whatever their beliefs. The second point reflects the effect that the new air of professional objectivity had on the relationship between scientific innovation and the wider—although now private—beliefs of the innovators. The favorite theoretical perspective of the old evolutionary natural theology was antimechanistic and anti-Darwinian, but the scientific developments of the twentieth century were almost universally made by those who approached biology from the opposite direction. It was mechanism and Darwinism that generated practical research insights, while the old teleology seemed increasingly like empty mysticism. This transition certainly did not involve a loss of faith in progress, although that faith was driven underground in most cases. But the faith had to be structured in a way that allowed it to cohere with a Darwinian view of evolution, and that was a view that was traditionally seen as incompatible with the Christian view of human nature. Those such as Fisher who did remain Christians had to adopt a liberal faith that could reconcile itself with the theory of natural selection. We thus end up with the apparently paradoxical situation in which the demands of scientific objectivity forced scientists with diametrically opposed value systems to work within the same theoretical paradigm. Fisher, the Christian, and Haldane, the Marxist, are both founding fathers of the modern genetical theory of natural selection.

A small number of evolutionists from the older generation, including Alfred Russel Wallace and the paleontologist Robert Broom, still adopted an anthropocentric form of teleology in which the human race was portrayed as the intended end product of a divinely planned ascent of life. Most evolutionists, however, were now suspicious of so explicit a continuation of the argument from design. The idea of progress was reformulated in less rigid terms to depict it as a background force that would enhance mental and moral powers, but was not drawn toward a single goal, such as the human form. Lamarckians had always claimed that mind was the driving force of evolution and the expansion of mind one of its chief goals. Several mutually contradictory philosophies combined to reinforce this model of evolution in the early twentieth century, ranging from Haeckel's monism (very much a product of the last century) to Bergson's *élan vital* (widely perceived as a challenge to the old materialism). Twentieth-century biologists were deeply in-fluenced by these ideas, especially those of Bergson, and thus continued to support theories in which evolution is pushed forward by a general progres-sive force somehow linked to the expansion of mental powers. There was dis-agreement, however, on whether mind was an integral part of all living things or a property that emerged at a key point in the ascent of life.

A few biologists still preferred an anthropocentric form of theistic evo-lutionism in which progress was explained as the unfolding of the divine plan toward its goal, humankind. This was the kind of teleological evolution-ism once advocated by the anti-Darwinian (and Roman Catholic) anatomist St. George Jackson Mivart.[29] An unexpected convert to this position was the aging Wallace, once a rigid Darwinian as far as animal evolution was con-cerned, who published his *World of Life* in 1910 to expound a highly structured version of the argument from design. Wallace had long held that the emer-gence of the human mind required a supernatural impact on an otherwise natural evolutionary process. He now insisted that a directive Life Principle must have been active throughout, implying the existence of a Creative Power whose ultimate purpose was "the development of Man as the crowning prod-uct of the whole cosmic process of life-development."[30] He believed that ear-lier mechanists such as Huxley had denied the existence of a life force only because they saw no way of applying the concept in their scientific work. But a survey of the history of life on earth proved that such a directing power must be at work. The amount of cruelty in nature implied by the Darwinian theory had been exaggerated, although a certain amount of suffering was es-sential to stimulate organisms to develop.[31] Although life exhibited infinite variety, the whole complex was designed to make the appearance of the hu-man race possible, and to emphasize the preordained nature of the process, Wallace invoked both the influence of angels from the supernatural realm and directed lines of organic variation.[32]

Wallace thus combined an enthusiasm for vital forces as agents of the creative power of God with a very traditional anthropocentrism that still assumed that the human race was the preordained outcome of the process. Another scientist who insisted on a rigid predetermination of the evolution-

---

29.   Mivart's *Genesis of Species* of 1870 was an influential anti-Darwinian work. Gregory Elder credits Mivart with playing an important role in converting many Anglicans to a belief in providential evolution, see his *Chronic Vigour*, chap. 3.

30.   Wallace, *World of Life*, vi. Wallace's position here coincided with the anthropocentric cos-mology of his *Man's Place in the Universe*, discussed in chap. 3 above.

31.   Ibid., chap. 19.

32.   Ibid., 392–95.

ary process was the paleontologist Robert Broom. Broom spent most of his life in South Africa, where he made his reputation first by describing the mammal-like reptiles of the Karoo formation and later by discovering fossil hominids. He was a staunch Nonconformist and was also deeply interested in spiritualism. His views bear a strong resemblance to those of Wallace, although he confessed that he had not seen *The World of Life*.[33] After years of publishing technical papers on fossil reptiles, he startled the British Association at its 1931 meeting by offering the opinion that, since all other animals were too overspecialized to evolve further, the human race must be the final goal of the evolutionary process. The *Daily Herald* report of the meeting portrayed Broom as a "malicious sprite . . . who stuck another pin into the profound scientists."[34] In the following year Broom let his views be known in the concluding chapter of his *Mammal-like Reptiles of South Africa*. Stressing once again that all the branches of the tree of life had now become too specialized to be capable of any profound modification, he argued: "Apart from minor modifications evolution is finished. From which we may perhaps conclude that man is the final product; and that amid all the thousands of apparently useless types of animals that have been formed some intelligent controlling power has specially guided one line to result in man."[35]

In the following year Broom laid out his whole philosophy in *The Coming of Man: Was It Accident or Design?* Here he gave guarded support to the Lamarckian theory, which allowed new habits to shape the evolution of the species, but insisted that there were many aspects of evolution that could be explained by neither Lamarckism nor Darwinism.[36] He thought that intelligent forces had guided evolution, although evidently there were at least two such forces that did not always work in concert. The lower-level force could only shape species in response to current pressures—it could not plan for the future.[37] It was represented by a psychic agency—perhaps the entelechy of Driesch or the organizer of Spemann—which could be seen at work by the physician who studied cases of faith healing.[38] The more important trend worked on a longer time scale and was responsible for the overall advance of life toward humankind; its aim was "the production of human personalities, and the

---

33. Broom, *Coming of Man*, 225.

34. Quoted in Findlay, *Dr. Robert Broom*, 59. The *Times* report of the debate merely notes Broom's objections to Darwinism; see 25 September 1931, 7.

35. Broom, *Mammal-like Reptiles of South Africa*, 333; see Bowler, *Theories of Human Evolution*, 118–19, 219–22, and *Life's Splendid Drama*, 310–17.

36. *Coming of Man*, 21–22, 30.

37. Ibid., 66–67.

38. Ibid., 197–99.

personality is evidently a new spiritual being that will probably survive the death of the body."[39] Since biological evolution was now finished, this force had probably transferred its efforts to promoting human spiritual development, perhaps through the appearance of exceptionally gifted individuals.

Broom knew that his views were anathema to most scientists—he quoted a negative review of the philosophical sections of his *Mammal-like Reptiles* at the start of *The Coming of Man*. He described his ideas in letters to Bishop Barnes, who had been reading R. A. Fisher's *Genetical Theory of Natural Selection*, and who found Broom's teleological approach to evolutionism implausible.[40] One scientist who did lend support, at least in private, was the Oxford professor of geology W. J. Sollas. In the conclusion of his archaeological work *Ancient Hunters and Their Modern Representatives*, Sollas had openly criticized natural selection as an "idol of the Victorian era" and expressed the view that "the fundamental cause in the whole process of evolution is in reality an affair of the mind."[41] He and Broom engaged in an extensive correspondence, and in response to the latter's book, Sollas wrote:

> I read with great pleasure all that you say, about Man and some great
> power—a mind—behind it all. Like you I cannot get away from it, and
> further I feel confident that this power is not unmindful of us, and not
> inaccessible but always at hand to help. And I see nothing inconsistent
> with the scientific attitude in this. I don't think it is "scientific heresy"
> but I admit that it is so regarded by the general run of scientific workers.
> There is an "odium scientificum" quite as virulent as the "odium theo-
> logicum"—not fiercely persecuting, but contemptuous and disdainful.[42]

Sollas was another member of the older generation of scientists who preserved a wish to link science and religion, but he chose to hold his tongue because he realized that his views were now perceived as outdated by the scientific community. Broom's ideas about the lack of further evolution beyond humankind did, however, influence Julian Huxley, who incorporated a faith in the progressive character of evolution into his own nontheistic religion.[43]

---

39.   Ibid., 221.

40.   Broom to Barnes, 26 November 1930, 12 and 21 January 1931 (E. W. Barnes papers, EWB 11/2/49, 51, 52); Barnes to Broom, 10 April 1931 (EWB 10/4/51); see also the newspaper cuttings Broom sent to Barnes (EWB 10/4/50 and 52).

41.   Sollas, *Ancient Hunters*, 570–71.

42.   Sollas to Broom, undated letter quoted in Findlay, *Dr. Robert Broom*, 44–45.

43.   See below and Swetlitz, "Julian Huxley and the End of Evolution."

Outside science, Broom's views were cited with approval in Gerald Heard's *The Source of Civilization,* part of a popular series on the progress of life and of human society.[44]

If most scientists were now unwilling to explain evolution as the unfolding of a divine plan, many were still anxious to see it as a more generally progressive force. The newer model of progress as a broadly based trend had its origins in a variety of philosophies with very different origins. Paradoxically, even the old-fashioned kind of Darwinism could be made to yield this implication. At the turn of the century, the German Darwinist Ernst Haeckel had issued a defiant statement of the monistic philosophy underlying his vision of nature, translated into English as *The Riddle of the Universe at the Close of the Nineteenth Century* in 1900. Haeckel attacked a whole swath of traditional religious beliefs: There was no separate spiritual realm because mind and matter were different aspects of the same fundamental substance. The only viable religion was a pantheism that identified God with nature. There could be no revelation, no survival of bodily death, and no transcendental source of moral values. Like Huxley and Tyndall, Haeckel went out of his way to ridicule the claims of the spiritualists.[45] His most explicit target was the Roman Catholic Church, which he portrayed as the inveterate enemy of science. Protestantism was more open to innovation, but liberal Protestants had ended up with a kind of pantheism very similar to Haeckel's own position.[46]

Haeckel's book was rightly perceived as an attack on traditional religion, and it elicited critical responses from both Lodge and Wallace, both determined to defend the existence of a distinct spiritual realm.[47] Yet there were elements in Haeckel's thought that distanced him from mechanistic materialism. Although he denied any transcendental goal toward which evolution was aimed, his whole evolutionary philosophy had been based on the assumption that the laws of nature combine to force life steadily to ascend the scale of psychic development. He insisted that not only the most primitive forms of life, but even the particles of matter itself, had a rudimentary and unconscious mentality. The conscious minds of higher animals and humans were created by the synthesis of ever more complex material structures, and in turn played an active role in directing the future course of evolution.[48] There was a strong element of romanticism in Haeckel's philosophy and a

44. Heard, *Source of Civilization,* 66–67.
45. Haeckel, *Riddle of the Universe,* 313–14.
46. Ibid., 331.
47. Lodge, *Life and Matter;* Wallace, *World of Life,* 4–6.
48. Haeckel, *Riddle of the Universe,* 183–91.

strong Lamarckian component in his evolutionism.[49] Increasingly, the idea that mind was an active agent in directing the world's upward ascent would be incorporated into efforts to forge a version of evolutionism in which teleology would be ushered back into science, or at least into philosophies that could claim a consistency with science.

The new direction of thought is most visible through the impact of another Continental thinker, Henri Bergson. Widely hailed as one of the driving forces of the revolt against materialism that characterized European thought in the years before World War I, Bergson has also been identified as a supporter of an essentially conservative ideology in France.[50] Although in many respects no more teleological than Haeckel's, his evolutionism was perceived as an alternative to Darwinism and as a vehicle that would reintroduce mind as the driving force of nature. Bergson's *Creative Evolution* was translated in 1911, but the French original had already attracted much comment in Britain. Its vision of an irregular but ultimately unstoppable upward ascent of life driven by the *élan vital*'s efforts to overcome the resistance of brute matter allowed a reconfiguration of the argument from design (see chapter 11). Life was pushed upward in many different directions by the *élan*, not pulled toward a single goal. The life force was the motor of progressive evolution, and God could be seen as the Creator of this force, who allowed the details to be worked out in the course of history in order to leave room for free will in the highest products of the trend.

There were certain tensions within this approach, however. Did the emergence of mind or consciousness represent merely an enhancement of the most basic life force to be found in even primitive organisms, or did it mark the appearance of a new level of nonphysical reality, transcending life itself and capable of exerting even greater control over the evolutionary process? This was the crucial difference between Bergson's creative evolutionism and the emergent evolutionism of Lloyd Morgan. As the leading Bergsonian H. Wildon Carr put it in a letter to Morgan, the difference between them was that to him every moment in creative evolution was equally critical—one should not try to identify particular turning points at which new properties emerge.[51] Another problem was the question of how life or mind controls the direction of evolution if, as seemed increasingly likely, the Lamarckian mech-

49. See Bowler, *Non-Darwinian Revolution*, 82–90, and Di Gregorio, "Entre Mephistophele et Luther." For a controversial assessment of the impact of Haeckel's thought, see Gasman, *Haeckel's Monism and the Birth of Fascist Ideology.*

50. Grogin, *Bergsonian Controversy in France, 1900–1914.*

51. H. W. Carr to Lloyd Morgan, 20 January 1920, Morgan papers, Bristol University Library, DM 128/366.

anism of use-inheritance had to be rejected as incompatible with the new genetics.

These tensions can be seen in the writings of the biologist who became most closely identified with the new natural theology, J. Arthur Thomson, professor of natural history at the University of Aberdeen. Having given up plans to become a minister in the Free Church of Scotland as a young man, Thomson now advocated a theism in which science would be reconciled with belief in the existence of a Creator via an antimechanistic view of life. He went even further than Lodge in "modernising" Christianity to effect this reconciliation: his popular book *The Gospel of Evolution* replaces the good news of Christ with an appeal to the purposeful character of the created world.

Thomson's impact as a scientist was limited, partly because he spent much of his time in teaching and in lecturing and writing for a broader audience. In 1904 he contributed a joint essay with the sociologist Patrick Geddes to a collected volume edited by an Anglican clergyman. Thomson and Geddes insisted that life was something more than a mechanical phenomenon and that evolution had a moral purpose, as explained by writers such as Henry Drummond and Peter Kropotkin.[52] Drummond and Kropotkin had both argued that—contrary to the principles of Darwinism—evolution generated increasing levels of moral awareness (Kropotkin's *Mutual Aid* of 1902 offered evidence of altruism in the higher animals). Thomson saw Bergson's philosophy as a contribution to this approach, and wrote a very positive review of *Creative Evolution* for *Nature*, in which he declared that "the time is ripening for a closer cooperation of philosophy and science, and the man of the time is Henri Bergson." He conceded that "while nature-poetry is in no sense biology, it may be a very important complement."[53] Over the next two decades he conducted a campaign—often in conjunction with Geddes—against mechanistic biology.[54] Their chosen tactic was not direct participation in scientific research, but the production of nonspecialist literature and textbooks aimed at college students.

Thomson's rejection of mechanism was linked to a teleological interpretation of evolutionism that allowed it to be presented as evidence that there was a moral purpose to creation. His Gifford Lectures for 1915 and 1916, published as *The System of Animate Nature*, offered his first major exposition of this

52. Thomson and Geddes, "A Biological Approach."

53. Thomson, "Biological Philosophy," 475, 477.

54. See, for instance, *System of Animate Nature*, chap. 5; Geddes and Thomson, *Biology*, chap. 9; Thomson, *Purpose in Evolution*, 18; Thomson and Geddes, *Life: Outlines of General Biology*, 448, 649, 656, 665. The Geddes papers in the National Library of Scotland contain correspondence related to this collaboration.

philosophy. Given that these were the war years, he set out to qualify the Darwinian vision of evolution through struggle. The tendency of a previous generation of scientists to portray the world as "a dismal cockpit" and a "chapter of accidents" had "engendered what may be called a natural irreligion, and it is the object of this course [of lectures] to show that such views are scientifically untenable."[55] The struggle for existence was often less severe than Darwin supposed, much of it being between groups, where it selected for the instincts that promote social cohesion and ultimately morality. Drummond and Kropotkin were cited to support the claim that evolution had a moral purpose.[56] Evolution exhibited a general trend toward progress, although of course there were side branches where stagnation occurred. The human species was "the summit of the whole" because we exhibit the greatest ability to control our own behavior: we are the outcome of a persistent trend toward freedom of mind.[57] Thomson quoted Goethe in praise of nature, but insisted that we could not worship her—so should we not seek to worship Him who nature reveals?[58]

In 1923 Thomson contributed a chapter on Darwin's influence to F. S. Marvin's *Science and Civilization*, again stressing "the age-long man-ward adventure that had crowned the evolutionary process upon the earth."[59] Two years later he published a substantial book titled *Science and Religion*, based on lectures given at the New York Theological Seminary. Noting the continued reprinting of A. D. White's book on the conflict between the two areas, he set out to show that this was a false antithesis. There was need for a "frontier commission" to police the boundary between them, however, because although they addressed different questions, they had to interact over the significance of the material universe.[60] Noting that the new physics had made matter itself less concrete, Thomson warned against assuming that this automatically guaranteed a role for the spirit: the real problem was that the minds of living things had all too often been ignored by science. His own view, which he admitted as a "personal heresy," was that of panpsychism: *"there is nothing inanimate."* Although Thomson had elsewhere praised the concept of emergent evolution, here he insisted that for mind to emerge from matter, it must have

---

55.   *System of Animate Nature*, v.

56.   Ibid., chap. 9; see also chap. 18, which dismisses most diseases as the result of human interference with nature.

57.   Ibid., 397, 565–66. The claim that there was a central "trunk" to the tree of life was not uncommon at the time; see Bowler, *Life's Splendid Drama*, 424–35.

58.   Ibid., 651–52.

59.   "The Influence of Darwinism on Thought and Life," 217.

60.   *Science and Religion*, 2, and chap. 1 generally.

been there all the time—and behind it the will of God.[61] The unity of nature, the self-directing activity of life, and the progress of evolution do not
actually prove the existence of God, but they are all suggestive—can it really
be a coincidence that nature favors what we call clear-headedness, beauty, and
altruism?[62] The book ends with an indictment of writers such as John Stuart
Mill, T. H. Huxley, and William James who had depicted nature as an amoral
system indifferent to human values.

In 1928 Thomson wrote a piece entitled "Why We Must Be Evolutionists" for Frances Mason's *Creation by Evolution,* contending that the world of
the modern evolutionist was one "in which the religious man can breathe
freely."[63] He provided the introduction to the follow-up volume, *The Great Design,* proclaiming: "There is something very grand in the conception of a Creator who originated Nature in such a way that it worked out His purpose: an
orderly, beautiful, progressive world of life with its climax, so far, in Man,
who echoes the creative joy in finding the world 'good.'"[64] Thomson repeated
these ideas in a BBC broadcast in 1931 and in his *Gospel of Evolution,* which again
insisted that nature is a "materialized ethical process" and that *"Nature became
articulate and self-conscious in Man."*[65]

Mason's books were an Anglo-American effort to promote the new natural theology, and several other British scientists contributed to them, including Lodge. The neurophysiologist C. S. Sherrington wrote an introduction to *Creation by Evolution,* stressing that there had been a great ascent up to
humankind, driven by forces that were still shaping us today.[66] The same
theme reemerged in Sherrington's Gifford Lectures, although here he stressed
that nature herself had no values, so that it was only after mind had appeared
in the higher animals, and especially in humans, that morality came into the
world. He did, however, echo Thomson's view that "Man is Nature's beginning to be self-conscious."[67]

61.   Ibid., 43–44, Thomson's italics. Thomson told Lloyd Morgan that he found *Emergent Evolution* difficult reading; see his letter of 26 July 1925, Morgan papers, Bristol University Library, DM
128/354.

62.   Ibid., 177.

63.   "Why We Must Be Evolutionists," 22. Thomson and Lloyd Morgan corresponded about
the efforts of "the dear and energetic Mrs Mason" to promote the new natural theology; this comment is from Thomson's letter to Lloyd Morgan, 4 May 1930, which also notes the success of her
books: Morgan papers, Bristol University Library, DM 128/432.

64.   "Introduction," in Mason, *The Great Design,* 14.

65.   Thomson, in *Science and Religion: A Symposium,* 23–36; *The Gospel of Evolution,* 211 and 203,
Thomson's italics.

66.   Sherrington, introduction to Mason, ed., *The Great Design,* xi–xiii.

67.   Sherrington, *Man on His Nature,* 387.

Another key contributor to Mason's books was the psychologist C. Lloyd Morgan, whose concept of emergent evolution was widely appealed to by writers anxious to stress the appearance of higher qualities in the course of evolution. Morgan wrote a chapter called "The Ascent of Mind" for *The Great Design* and one called "Mind in Evolution" for *Creation by Evolution*, stressing in both that the ascent of life toward mind and spirit suggested that the whole process reflected "the direct presence of God."[68] Morgan's psychological researches had suggested that it was illegitimate to attribute higher mental powers to animals, forcing him to accept that those powers had appeared without antecedent at a key point in human evolution. He was impressed by the psychological insights in Bergson's theory of creative evolution.[69] His own concept of emergent evolution, proclaimed in the Gifford Lectures for 1922, was meant to explain how such new qualities could appear by a natural process. In effect, Morgan believed that as life ascended the scale of organic development, its increasing complexity allowed entirely new properties to manifest themselves from time to time. Life itself was the first emergent property, followed by mind and at last by the human spirit. He saw Bergson as an early advocate of this position, although Samuel Alexander had now become the chief philosopher of the movement.[70] Morgan had begun his career as a monist, convinced that there were mental (but not conscious) properties even in matter itself. He now wanted to argue that the higher qualities transcended these primitive manifestations in a way that might almost suggest supernatural intervention at key points in evolution. They did *not* require a miracle, although in a sense each new property was itself supernatural when contrasted with the previous level.[71] Morgan was anxious to stress that his philosophy did not postulate supernatural forces at work guiding nature. It was a strictly naturalistic position fully compatible with science[72]—but the fact that the world was set up in a way that promoted the successive emergence of higher qualities was clearly suggestive to anyone who thought about its ultimate purpose. There was a single main line of evolutionary advance,

---

68. Morgan, "Mind in Evolution," 345.

69. Morgan lectured on Bergson in 1912, and his papers contain several sets of notes on Bergson's philosophy: Morgan papers, Bristol University Library, file DM 612. See also his correspondence with H. Wildon Carr, Morgan papers, DM 128/294, 305, 307, 308, 366. There is a single letter from Bergson to Morgan acknowledging receipt of his *Instinct and Experience*, DM 128/278.

70. Morgan, *Emergent Evolution*, 3; see Blitz, *Emergent Evolution*, and McLaughlin, "The Rise and Fall of British Emergentism."

71. Morgan, *Life, Mind, and Spirit*, x.

72. *Emergent Evolution*, 2.

and its goal was the production of human personality.[73] God was necessary to complete the philosophy, and he was ever active within the human spirit.[74]

Morgan's teaching was often coupled with that of the South African soldier and statesman Jan Christiaan Smuts, whose interest in science was strong enough for him to be elected president of the British Association in 1931. Smuts's *Holism and Evolution* of 1926 had stressed the organizing power of evolution and the ability of more organized structures to manifest properties that could never have been predicted from a knowledge of their physical components. Smuts believed that the main steps in evolution's advance were the appearance of life, mind, and personality—a position very similar to that adopted by Morgan, although Smuts objected to the latter's willingness to see some traces of unconscious mind even in matter itself. For Smuts, the new properties exhibited by wholes were entirely novel, and there was no need to trace mind all the way back down the scale.[75] In his 1931 presidential address, Smuts drew together the teachings of the new physics, the new cosmology, and the new organismic biology to present an optimistic vision of the world as a purposeful cosmos: "In this holistic universe man is in very truth the offspring of the stars. The world consists not only of electrons and radiations, but also of souls and aspirations."[76]

Thomson, Morgan, and Smuts were widely quoted by religious writers to confirm that teleology was now respectable in evolutionary theory, but there were several minor scientists who also took up the theme. Herbert F. Standing, a lecturer in biology at Woodbridge College, sent the manuscript of his *Spirit in Evolution* (1930) to Morgan, who read it with approval despite minor differences of interpretation between them.[77] His thesis was "that the whole evolutionary process is fundamentally a manifestation of divine purpose and activity, and that man's spiritual experiences, involving the realization of his highest ideals of Beauty, Truth, Goodness, and Love, are in line with the upward trend of Organic Evolution, and alone give an intelligible meaning to the whole World-Process."[78] The driving force of the progression was the mental powers of living things themselves, which transcended mere mechanical activity and directed evolution increasingly along a path de-

73. Ibid., 31.
74. Ibid., 300–301; *Life, Mind, and Spirit*, lecture 10.
75. Smuts, *Holism and Evolution*, 321 n.
76. Smuts, "The Scientific World-Picture of Today," 17–18.
77. Standing, *Spirit in Evolution*, 12.
78. Ibid., 13 (the same words are quoted on the book's dust jacket).

termined by their desires and values. The same position was expressed in the entomologist Charles J. Grist's *Science and the Bible*, which presented the creation story as a metaphor. Evolution was a progressive trend driven by the power of mind, which reflected God's spirituality pervading the cosmos. It led inevitably toward the human mind and beyond—Grist saw Christ as evidence that the whole human race would eventually move into a more spiritual phase, and speculated that such higher races already existed elsewhere in the universe (perhaps they were the angels referred to in the Bible).[79] A wholesale attack on the materialistic view of evolution came from Albert Eagle, who had been an assistant to J. J. Thomson and taught at Imperial College, London, and the University of Manchester. In a privately printed book of 1935, Eagle insisted that evolution was driven by spiritual forces along the lines suggested by Bergson, while natural selection was an absurdity.[80]

THE ROLE OF LAMARCKISM

The logic of teleological evolutionism depended on the inevitability of progress, even if the ascent to higher levels was irregular and unpredictable. This philosophy assumed that life and mind were themselves creative forces—expressions of the divine will—that steered evolution in directions that would not have been possible in a world governed solely by purely physical forces. But how was mind supposed to direct evolution, assuming it did not directly interfere with the laws of physics?

In the late nineteenth century, neo-Lamarckians such as Samuel Butler had rejected Darwinian natural selection because it seemed too mechanistic, arguing that it was the inheritance of acquired characteristics that allowed mind to serve a creative role in evolution. Animals responded to environmental challenges through behavioral innovation, and the new habits generated new structures because the bodily modifications produced by the new activities were transmitted to the next generation. At the same time, new mental characters acquired by learning would become part of the species' inbuilt behavior—and this would include the social instincts that are the foundation of morality.[81]

For all their complaints about the materialistic Darwinism of the late nineteenth century, the exponents of the "new" natural theology were con-

79.   Grist, *Science and the Bible*, 124, 150, 163, and on higher races, chap. 17.

80.   Eagle, *The Philosophy of Religion versus the Philosophy of Science*; see chaps. 9 and 10. In chaps. 7 and 8, Eagle attacked Eddington's philosophy and dismissed both relativity and quantum mechanics.

81.   See Bowler, *Eclipse of Darwinism*, chaps. 4 and 6; R. J. Richards, *Darwin and the Emergence of Evolutionary Theories of Mind and Behavior*, chaps. 8 and 9.

tinuing along a path already marked by the anti-Darwinians of that era. Their task was made more difficult, however, by the increasingly strong campaign mounted against Lamarckism by the geneticists. Was there another way in which behavioral innovation could direct evolution, if acquired characters could not be incorporated into the genes? An alternative had indeed been suggested by Lloyd Morgan, along with the American psychologist James Mark Baldwin. Organic selection, more frequently known as the "Baldwin effect," was proposed in the 1890s as a means of allowing Darwinism to steal the Lamarckians' best argument. Acquired characters might not be inherited directly, but they allowed the species to adjust to a new lifestyle while random variation came up with the equivalent hereditary modifications—which would then be favored by natural selection. Yet organic selection was seldom mentioned in the early twentieth century, even by Lloyd Morgan himself. The discussions of the evolutionary process by Morgan, Thomson, and others were often remarkably vague, amounting to little more than an often-repeated assumption that the purposeful activities of living things must somehow be reflected in the evolutionary process. There had been some technical objections to organic selection at the time, but the reticence of later writers to discuss the topic may have been prompted by the fact that Baldwin was discredited in 1909, having been detained by the police in a brothel. Whatever the reasons, the idea that behavior might control the direction of natural selection was seldom discussed in detail. Instead, various holistic and organismic concepts were exploited to give the impression that the innovations prompted by the exercise of mind could not be completely excluded from the hereditary impulse. It was argued that the mechanistic theory of the unit-character gene could not be the whole story. Without ever stating a belief in Lamarckism, Morgan and Thomson were thus able to borrow that theory's teleological model of evolution.

A few scientists remained committed to Lamarckism whatever the evidence from genetics. In the 1890s one of the most active writers in defense of Lamarckism had been the clergyman-botanist George Henslow. In the new century he turned his attention to spiritualism, publishing several books defending the authenticity of the evidence from seances. This objective evidence from scientists, he claimed, was essential for the very survival of the Christian church.[82] His two interests were combined in his *Religion of the Spirit World* of 1920, in which he (or rather, the spirits who talked to him) used the idea of self-adaptation in organic evolution to argue that life itself was a di-

---

82.   Henslow, *Proofs of the Truth of Spiritualism*, 3. On Henslow's botanical Lamarckism, see Bowler, *Eclipse of Darwinism*, 85–86.

recting force imparted by the Creator, and that each of us must strive to adapt ourselves to the requirements of the next world so that we are fully prepared to "pass over."[83]

Support for Lamarckism was by no means driven solely by religious concerns; as with all opposition to materialism, philosophical considerations played a major role. This was certainly the case for the psychologist William McDougall, who defended Lamarckism openly in his *Body and Mind,* first published in 1911. McDougall thought that the Darwinians rejected Lamarckism precisely because they wanted to eliminate mind from nature. He felt that the refutation of the inheritance of acquired characters was by no means definitive, but that even if the evidence against the theory were to become stronger, organic selection would still allow mental activities to play a role.[84] In a later attack on the theory of emergent evolution—which, like many critics, he saw as little more than an evasion of the real problem—he openly criticized Lloyd Morgan for not discussing the actual mechanism of evolution that allowed mind to influence the process.[85] He now claimed that the evidence for Lamarckism was getting stronger, citing his own paper on the alleged inheritance of learned behavior in rats.[86]

The leading British biologist defending Lamarckism was the embryologist Ernest William MacBride. It was MacBride who led the defense of the Austrian biologist Paul Kammerer, whose experiments on the inheritance of acquired characters in the "midwife toad" were attacked by Bateson and the geneticists.[87] Like MacDougall, MacBride was strongly opposed to materialism and was thus willing to lend his support to the new natural theology. He contributed a paper on his favorite line of defense for Lamarckism, the recapitulation theory, to Mason's *Creation by Evolution* and a more general piece to *The Great Design.* Here he defended the view that all living things exhibit mental properties and the role of habit in guiding evolution. Could anyone seriously suggest, he asked, that the regulating and directing powers of life could arise from the chance encounters of atoms? He concluded: "If, therefore, we fall back on the principle that the Creator endowed living matter with something that strives to meet adverse circumstances and can control its own

83. Henslow, *Religion of the Spirit World,* 31–33.

84. McDougall, *Body and Mind,* chap. 18.

85. McDougall, *Modern Materialism and Emergent Evolution,* 151–52.

86. Ibid., 154; see McDougall, "An Experiment for Testing the Hypothesis of Lamarck," and Bowler, *Eclipse of Darwinism,* 102–3.

87. See Koestler, *Case of the Midwife Toad;* Bowler, *Eclipse of Darwinism,* 99–102. Like Fleming, MacBride had very right-wing views on the race question; see Bowler, "E. W. MacBride's Lamarckian Eugenics."

growth, we have reached the most fundamental explanation of adaptation which it is now possible to hold."[88] MacBride shared Lodge's view on the overall purpose of evolution and wrote to Lodge praising his ability to arouse the interest of young people in religious issues. The two agreed that the mental character of animals was vital for evolution, although Lodge was suspicious of MacBride's efforts to argue that the effects of habit were laid down in some physical change in the germ plasm.[89]

The pathologist J. George Adami, an Anglican who campaigned actively for the reconciliation of science and religion, also defended Lamarckism. His Croonian Lecture to the Royal College of Physicians in 1917, titled "Adaptation and Disease," challenged E. Ray Lankester on this topic and claimed that most workers in pathogenic bacteriology accepted the direct adaptation of bacteria to changes in the environment as an explanation for the rapid appearance of new diseases.[90] Adami made no explicit reference to this topic in his general writings on science and religion, although he certainly believed that "Nature move[es] towards an ever increasing perfection, directed by a beneficent spirit."[91]

A more explicit attack on Darwinism came from the pen of Hermann Reinheimer, whose books *Evolution by Cooperation* (1913) and *Symbiogenesis* (1915) played an important role in creating the modern concept of symbiosis. Reinheimer presented his vision of organisms coming into increasingly close cooperation with one another in the course of evolution as an alternative to the Darwinian worldview based on struggle. He wrote on the topic for the *Hibbert Journal,* stressing the implications of his views for human affairs. Symbiosis and parasitism represented two incompatible ways of life, the one equivalent to industry, the other to crime.[92] Another article on science and religion stressed the damaging effect of the Darwinian worldview on religious belief and morality.[93]

Lamarckism also figured prominently in the thinking of the anatomist Frederic Wood Jones, who was identified with the "tarsioid theory" of human origins, according to which the human species was derived

88.   MacBride, "The Oneness and Uniqueness of Life," 158.

89.   MacBride to Lodge, 17 March 1927 and 5 July 1931; Lodge to MacBride, 29 June and 6 July 1931; Lodge papers, Birmingham University Library, OJL 252/2 and 6–8.

90.   The lecture is reprinted in J. G. Adami, *Medical Contributions to the Study of Evolution,* part 1; see 9–11, and on the controversy with Lankester, appendix 2.

91.   Adami, "Eternal Spirit of Nature," 516.

92.   Reinheimer, "Cooperation among Species," 175. On Reinheimer's contributions to biology, see Sapp, *Evolution by Association,* 60–66.

93.   Reinheimer, "Science and Religion."

not from the great apes, but from a lower primate such as the spectral tar-sier. His earliest account of this theory had been published by the Society for the Promotion of Christian Knowledge in 1918, and concluded with a clear statement of what Jones took to be the wider implications of this view:

> Man is no new begot child of the ape, bred of a struggle for existence upon brutish lines—nor should the belief that such is his origin, oft dinned into his ears by scientists, influence his conduct. Were he to re-gard himself as an extremely ancient type, distinguished chiefly by the qualities of his mind, and to looks on the existing Primates as the fail-ures of his line, as his misguided and brutish collaterals, rather than as his ancestors, I think it would be something gained for the ethical out-look of Homo—and it would also be consistent with present knowl-edge.[94]

A similar theory was used in America by Henry Fairfield Osborn to dis-tance evolutionism from the link between humans and apes, which many people found distasteful and which thus played into the hands of the cre-ationists. The theory demanded a great deal of parallelism in evolution to account for the similarities between humans and apes. Lamarckism had tra-ditionally been associated with the idea of parallel evolution because it was thought to generate more active directing trends than would be possible with natural selection. Jones explored these implications in detail in a se-ries of books published toward the end of his career. *Habit and Heritage* (1943) proclaimed open support for Lamarckism, while *Trends of Life* (1953) promoted vitalism and the claim that evolution could be directed along predetermined channels. The religious implications of this position were explored in *Design and Purpose* (1942), in which Jones attacked Darwinism for its destruction of belief in a wise Creator, praised Smuts's holism, and sup-ported the transformation of the argument from design in a way that al-lowed evolution to be seen as a preordained spiritual progress instituted by a Cosmic Mind. He attacked church leaders such as Bishop Barnes and Dean Inge who had fallen in with Huxley's old idea that nature itself was amoral.[95]

94. F. W. Jones, "The Origin of Man," 131, also published separately as a pamphlet by the SPCK. On Jones's theory, see Bowler, *Theories of Human Evolution*, chap. 5.

95. Jones, *Design and Purpose*, 71.

DARWINISM REVIVED

While Lamarckism continued to receive some support, it was obvious to most scientists that its incompatibility with genetics left it with a serious credibility problem. By the 1920s it was clear that genetics—although initially seen as incompatible with Darwinism too—was now being synthesized with it via the genetical theory of natural selection developed by R. A. Fisher and J. B. S. Haldane. Those commentators who wanted to stay abreast of the latest scientific thinking would have to take this development into account. But the implications of the new Darwinian synthesis were not as clear-cut as might have been expected. Darwinism had long been associated with materialism, and to scientists such as E. Ray Lankester and Arthur Keith, this was exactly where it belonged. Anyone wanting to explore the broader implications of evolutionism would have to accept that the advance of life was not aimed as directly at the production of the human race as the exponents of the new natural theology assumed, nor was the driving force of the evolutionary process so obviously the product of a Creator who shared the moral values to which we aspire. Yet Keith himself was not a Darwinist in the modern sense of the term, and still saw evolution as being under the influence of purposeful forces internal to the organism. Julian Huxley, who would eventually give the new theory its title in his book *Evolution: The Modern Synthesis*, strove openly to interpret it in a manner that would make it compatible with his own nontheistic religion. And of the two British founders of the genetical selection theory, one (Haldane) became a Marxist while the other (Fisher) remained a lifelong Anglican. The modern synthesis may not have seemed as friendly to natural theology as the non-Darwinian theories popular at the turn of the century, but many of its supporters wished to preserve a role for the idea of progress. The transition from the old evolutionism to the new was thus not as dramatic as we might at first suppose.

Those scientists who supported the rationalist program were appalled at the new directions being taken in the early twentieth century. Lankester wrote a vitriolic preface to Hugh Elliot's *Modern Science and the Illusions of Professor Bergson*, dismissing "those illusions [as] worthless and unprofitable matter, causing waste of time and confusion of thought to many of those who are induced to read them."[96] Science, he insisted, had succeeded by interpreting nature as a vast mechanism, and did not seek to inquire about the ultimate nature of reality. Whatever the true nature of things—and Lankester con-

96. Lankester, preface to Hugh Elliot, *Modern Science and the Illusions of Professor Bergson*, vii.

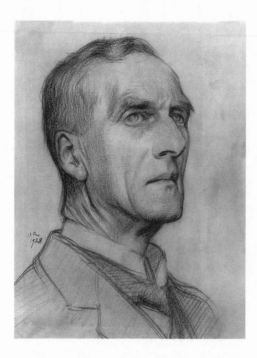

Arthur Keith. By Sir William
Rothenstein, 1928. By courtesy of the
National Portrait Gallery, London.

fessed himself not uninterested in the question—the mechanistic order of
the world we live in remained unassailable. Bergson's efforts to challenge that
order represented the worst kind of speculative metaphysics, but it was his
use of half-understood scientific information as a means of giving his system
a spurious credibility that was really dangerous. *Nature*'s review of Morgan's
Emergent Evolution betrays the same kind of impatience. The author, J. Johnstone, insisted that concepts such as "emergence" and "organicism" should
be critically and perhaps unsympathetically examined in the interests of
sound thinking.[97]

   In private, Lankester was equally scathing about Arthur Keith's dalliance
with holism[98]—yet in many respects, it was Keith who took over Lankester's
role as the RPA's chief scientific hatchet man. This fact illustrates how careful we need to be in our analysis both of what "Darwinism" meant in the
1920s and of the ideological labels attached to it. For Keith, Darwinism was
no longer the purely "trial and error" process caricatured by its critics; the
theory had developed considerably since Darwin's time and was now far more
sophisticated. It was easy to discredit a theory in which selection could pick

---

97.   J. Johnstone, "Emergent Evolution."
98.   Lankester to H. G. Wells, postcard, 3 October 1926, Wells papers, University of Illinois.

only from purely random variation, but Keith had no interest in genetics and was not prepared to see random mutation as the sole source of new characters. He insisted that the mechanisms of individual development would automatically coordinate any new character with the rest of the body's structure, and stressed the role of hormones in bringing this balance about.[99] A holistic view of heredity and development thus allowed Keith to propose a version of Darwinism that was significantly out of step with the ideas being developed by the population geneticists.

Despite this venture into organismic biology, neither Keith nor his critics thought that he had abandoned those aspects of Darwinism that made the theory so distasteful to religious thinkers. In his controversial 1927 address to the British Association, Keith went out of his way to stress that evolution was not a simple, predetermined upward sweep toward humankind. The multitude of different hominid fossils (including the Piltdown fraud, which Keith took very seriously) showed that the emergence of modern humanity had to be represented not as the ascent of a ladder, but as the bushing out of a tree, most of whose branches had to be pruned. For Keith, the winnowing out of whole species was at least as important as the selection of individual differences. Nor was the path of development along our own line of ancestry a smooth one—for all that he stressed the coordinating power of the body, Keith saw evolution as a process made up of a mosaic of individual modifications, with each branch of the hominid family tree experiencing its own unique combination of changes.[100] Where the exponents of the new natural theology looked to the creative power of life to bring coherence and purpose to the whole sweep of evolution, Keith saw a purely materialistic interaction within the body, which had no power to anticipate the future or to impose a degree of unity and coherence on the tree of life. A similar image of evolution as a haphazard process was stressed by J. P. Lockhart-Mummery, who insisted that the human species is by no means perfect—indeed, we are as yet poorly adapted to our upright posture.[101]

Keith was not prepared to see life or spirit as a force that transcended material nature. In his 1927 address he insisted that the evolution of humanity's special qualities was purely a matter of the expansion of the brain. A follow-up lecture in 1928 reiterated this point and argued that virtually all medical experts saw the mind as merely a by-product of the functioning of the

99.   Keith, "Why I Am a Darwinist"; "Is Darwinism at the Dusk or the Dawn?"; "Is Darwinism Dead?"; *Concerning Man's Origin*, chaps. 1–3 (the latter being the substance of his 1927 British Association address).

100.   *Concerning Man's Origin*, 8–9. On Keith's views, see Bowler, *Theories of Human Evolution*, 91–97.

101.   Lockhart-Mummery, *Nothing New under the Sun*, chaps. 6–8.

brain.[102] Whatever Keith's interest in holism, he had no intention of portraying life as a creative force that could direct evolution along new channels, nor of allowing any credence to the belief that the human soul could survive the death of the body. It was this element of materialism that attracted the attention of the press to his 1927 address, and which so incensed his religious critics. Keith records that he was attacked by "Daytonists" (what we would now call creationists—a reference to the Scopes trial in Dayton, Tennessee) and by "Daytonian Darwinists" (who accepted animal evolution but insisted that a miracle was needed for the creation of the human soul). He listed other attacks, including one by Ambrose Fleming, to which he felt obliged to reply in detail.[103] At the same time, he became embroiled in a controversy with Hilaire Belloc in the pages of *Nature*, in which each accused the other of being out of touch with the latest scientific work (in Belloc's case, this was Vialleton's saltationism).[104]

Keith insisted that Darwinism was certainly not dead, and indeed, went on to develop a full-scale theory of human evolution that focused on the role of intertribal competition as the key to our emergence from the apes. He argued that the instinct to join with our own social group in attacking rivals was deeply ingrained in human nature, and that without it, progress would not have occurred. Any attempt to stifle that instinct in the modern world would dehumanize us and lead to cultural stagnation. He stressed that this Darwinian vision of human nature was at variance with the traditional ethics taught by Christianity, and argued that the churches were losing ground because they ignored this vital part of human nature.[105]

Here Keith's interpretation of Darwinism ran counter to that being articulated by one of the leading advocates of the synthesis of natural selection and genetics. Julian Huxley regarded himself as a rationalist, yet he could not bring himself to accept that the universe itself was without purpose. He also felt that humanity was an important part of that purpose—and that true religion rested in the sense of awe and reverence that this prospect engendered in the human spirit. A key facet of this new humanistic religion was the idea of progress: Huxley remained convinced throughout his life that evolution was progressive and that the human race represented its most important outcome. The fact that we do not represent a goal willed by a per-

102.   Keith, *Darwinism and What It Implies*, chap. 2.

103.   Keith, *Concerning Man's Origin*, introduction, and *Darwinism and What It Implies*, chap. 3. On Fleming, see above; other attacks included "Keith versus Moses" in the *New Statesman* and Bonner, "The Case Against 'Evolution.'"

104.   Keith, "Is Darwinism Dead?"; Belloc, "Is Darwinism Dead?"

105.   Keith, *Essays on Human Evolution*, chaps. 15–17.

Julian Huxley. Photograph by
Howard Coster, 1939. By courtesy
of the National Portrait Gallery,
London.

sonal God was irrelevant because science itself was able to provide evidence
of this broader purpose for human life.

Historians have agreed that Huxley was a creature of the optimistic turn-
of-the-century culture who found it difficult to adjust to the darker tones of
twentieth-century thought.[106] His willingness to see the cosmos as a unified
whole has been linked to the idealism popular among Oxford philosophers
in his younger days, to the new initiatives of Jeans, Eddington, and Whitehead, and also to Morgan's emergent evolutionism. In fact, Huxley was critical of the physicist-idealists, dismissing Jeans's mathematician-God as yet another example of anthropomorphism in religious thought and insisting that
this whole mystical trend was potentially dangerous because it encouraged a
rejection of rational thought on the issue.[107] His appeal to progress in evolution was well established long before the idea of emergence was popularized. He was certainly influenced by Bergson: his first book, *The Individual in
the Animal Kingdom* (1912), openly proclaimed this debt,[108] and although he
concealed his vitalist leanings in his later career, he continued to mention

106.   See Dival, "From a Victorian to a Modern" and other contributions to Waters and Van
Helden, eds., *Julian Huxley*; J. C. Greene, "The Interaction of Science and World View"; Ruse, *Monad to
Man*, 328–38.

107.   Huxley, "Man and Reality," 195–96; *What Dare I Think?* 182–83.

108.   Huxley, *Individual in the Animal Kingdom*, vii–viii; see Ruse, *Monad to Man*, 336–37, and Smocovitis, *Unifying Biology*, 111.

Bergson. Even *Evolution: The Modern Synthesis* praises Bergson's vision, while admitting that it does not represent science. But where Bergson saw the life force as a creative spark in conflict with brute matter, Huxley's vision of a unified cosmos extended to panpsychism: like Thomson and Morgan, he was prepared to believe that every particle of matter was in some sense alive.[109]

To Huxley, progress was essential to give this unified cosmos—and human life within it—a moral purpose. As early as 1921 he wrote on this topic in the *Hibbert Journal*, arguing that if one did not believe in God, it was still necessary to believe in something molding the world. Science provided a verifiable doctrine of progress that would allow us to believe in a universe whose natural forces conspired to produce the moral values we cherish.[110] His chapter "Science and Religion" in F. S. Marvin's *Science and Civilization* tried to strike a balance between the Bergsonian view that the course of progress could not be predicted and the claim that the human species does represent the outcome of a main trend in evolution. The universe, he argued, exhibits a blind movement in a direction that we, with hindsight, see as conferring value, and "the main direction gives us cause for optimism."[111] The theme of progress was central to Huxley's contribution to Mason's *Creation by Evolution*; here he argued that the attribution of human values to the evolutionary process did not involve a circularity because we could also see episodes of retrogressive development. For this reason we could be sure that we are "in the main stream of biological progress, not in an eddy or backwater."[112] Huxley also began to develop the theme that was to become central to his humanistic philosophy of later years: the claim that human mental powers now allowed us to take charge of the evolutionary process, thus endowing us with immense responsibility.

By the time he wrote the conclusion of *Evolution: The Modern Synthesis*, Huxley had refined his definition of progress to highlight independence of and control of the environment, by which criteria humanity was clearly at the cutting edge, thanks to its ability to manipulate the world. He had now begun to focus even more strongly on the existence of a main line of evolution leading toward humankind. Following Broom's idea, he now saw evolutionary specialization as a steady whittling down of the potential for further change:

---

109.   Huxley, *Religion without Revelation*, 75; see also his "Science and Religion," 284–85, in which this belief is associated both with emergence and with the old monistic philosophy.

110.   Huxley, "Progress: Biological and Other," reprinted in his *Essays of a Biologist*, 3–64.

111.   Huxley, "Science and Religion," 299, reprinted in his *Essays of a Biologist*, 235–304.

112.   Huxley, "Progress Shown in Evolution," 337.

One curious fact emerges from a survey of biological progress as culminating for the evolutionary moment in the dominance of *Homo sapiens.* It
could apparently have pursued no other general course than that which it
has historically followed: or, if it be impossible to uphold such a sweeping and universal negative, we may at least say that among the actual inhabitants of the earth, past and present, no other lines could have been
taken which would have produced speech and conceptual thought, the
features which form the basis for man's biological dominance.[113]

The fact that all other branches of evolution were blind alleys identified our
own as the main stem, and our responsibility in the future was to maintain
the course of progress. Huxley remained convinced that although progress
was not inevitable, it *was* possible if we took our position as trustees seriously.
His later thought may have been presented more under the banner of humanism than of "religion without revelation," but the teleology of his cosmic progressionism was as obvious as ever. His support for Teilhard de Chardin's mystical evolutionism in the 1950s can be seen as a natural outcome of
this trend in his thought.

In this respect Huxley moved in exactly the opposite direction to that
conventionally associated with the new Darwinism. Most of the new generation of evolutionists retained an interest in the idea of progress, but they
were increasingly less likely to see its advance as predetermined. One of the
leading architects of the genetical theory of natural selection was R. A.
Fisher, who adopted a Darwinian perspective at an early stage in his career,
yet was determined to show that the selection theory was compatible with the
Modernist Anglicanism to which he was equally committed. At first sight,
this might have seemed an impossible task—the rhetoric of the new natural
theology had been based in part upon the need to transcend the alleged materialism of the selection theory. But as James Moore points out in his study
of the nineteenth-century debates, it was the more liberal theologians who
had wanted a non-Darwinian evolutionism to preserve cosmic teleology. A
small number of orthodox Christians had welcomed Darwin's mechanism
because it made evolution historically contingent and because its harshness
fitted their vision of a world based on suffering.[114] Provided one deflected attention away from the specific problem of human origins, it was possible to

---

113.   *Evolution: The Modern Synthesis,* 569; Huxley notes that he had first suggested this point in a
British Association address in 1936. See Swetlitz, "Julian Huxley and the End of Evolution."

114.   J. R. Moore, *Post-Darwinian Controversies,* chap. 12.

argue that natural selection was more compatible with Christianity than were the more overtly teleological alternatives. This position had been forgotten in the rush to depict Darwinism as a materialistic bogeyman, but now Fisher was able to revive it, in part because he accepted the harsh view of the human situation characteristic of genetic determinism.

Fisher wrote little on his religious beliefs until the 1950s, but historians studying his contributions to selection theory are convinced that his faith played an important role in sustaining the worldview that made his work possible.[115] He had been taught mathematics by both Jeans and Whitehead, and he visualized the world in terms of the theory of probability, which governed the behavior of large masses of entities—whether gas molecules or organisms. In physics, the laws of nature led to degradation (the second law of thermodynamics), while in biology, natural selection showed how it was possible for nature to move in the opposite direction, toward progress. For Fisher, there could be no long-range trends, however: God had instituted a much less direct method of creation based on adaptation to the local environment by natural selection. Yet this was still a process of creation, of the production of novelty, and it was important that progress was achieved in this indirect manner because only thus was a rigid determinism circumvented, leaving the resulting creatures (including ourselves) with a degree of free will.

In his Eddington Memorial Lecture of 1950, Fisher showed that he, like Huxley, had been inspired by Bergson. His purpose was, however, to argue that the French philosopher was mistaken in thinking that an almost supernaturally creative life force was needed to challenge the deterministic world of the physical universe. Once the revolutionary nature of nineteenth-century probabilistic thinking was absorbed, it was possible to see how the behavior of large numbers of individuals could be law-bound and yet allow for the production of novelty (for Fisher, the indeterminism of quantum mechanics merely drove home the message of this earlier revolution). Fisher also praised Smuts's holism for its effort to make the evolutionary process seem creative. But it was Darwinism, not Bergson's vitalism nor a Lamarckian emphasis on effort, that explained the creative process of evolution in terms that made sense to the Christian:

> There is indeed a strand of moral philosophy, which appeals to me as pure gain, which arises in comparing Natural Selection with the Lamarckian group of evolutionary theories. In both of these contrasting hypothe-

---

115.   See especially Hodge, "Biology and Philosophy (including Ideology): A Study of Fisher and Wright"; Ruse, *Monad to Man*, 301–3.

ses living things themselves are the chief instruments of the Creative activity. On the Lamarckian view, however, they work their effect by willing and striving only; but, on the Darwinian view, it is by doing or dying. It is not the mere will, but its actual sequel in the real word, its success or failure, that is alone effective.

We come here to a close parallelism with Christian discussions on the merits of Faith and Works. Faith, in the form of right intentions and resolution, is assuredly necessary, but there has, I believe, never been lacking through the centuries the parallel, or complementary, conviction that the service of God requires of us also effective action. . . . Good intentions, and pious observances are no sufficient substitute, and are noxious if accepted as substitute.[116]

Fisher was anxious to defend the traditional view that human beings have free will. In 1934 he had conceded that even if we live in a determined universe, we have evolved the illusion of free will for a purpose—to make us believe that our choices really do matter.[117] He wrestled with the same problem in letters to the neurophysiologist C. S. Sherrington, insisting that somehow (perhaps due to quantum fluctuations in the brain) there was an indeterminate factor in choice, which a religious person could ascribe to guidance by a higher purpose.[118] The element of choice was important to Fisher because it allowed him to argue that we ourselves can contribute to the good of the world, and combat the evil in it, by our actions. This in turn allows us to participate in and extend the evolutionary process, by taking control of selective breeding in the human population through a eugenics program.

Having presented natural selection as the mechanism that God had chosen to use for the creation of humanity, Fisher had to accept that a winnowing out of maladaptive characters was always necessary. If the human race had been formed by such a process, it was self-evident that its gene pool must still contain many characters that needed weeding out by further selection—but it was now possible to do this by conscious choice. Those with vision and ability must use their influence to prevent the reproduction of less fit individuals. Much of Fisher's work on heredity and selection was driven by the hope of illustrating how the human race might be improved in this way.

116.   R. A. Fisher, *Creative Aspects of Natural Law*, 19–20. Without the link to a specifically Christian doctrine, this distinction between the two evolutionary mechanisms is also made in his 1934 article "Indeterminism and Natural Selection," 111.

117.   "Indeterminism and Natural Selection," 108.

118.   Fisher to Sherrington, 22 January 1947, quoted in Bennett, ed., *Natural Selection, Heredity, and Eugenics*, 261–62.

Fisher's work in this area was appreciated by Modernists in the Anglican Church such as Bishop Barnes; in fact, Fisher, Huxley, and Barnes were all involved in an attempt to interest the Anglican Church in a proposal to pay clergy family allowances in order to encourage them to have more children.[119] In this area, at least, Fisher and Huxley agreed on a program for the practical improvement of the race:

> To the traditionally religious man, the essential novelty introduced by the theory of evolution of organic life, is that creation was not all finished a long while ago, but is still in progress, in the midst of its incredible duration. In the language of Genesis we are living in the sixth day, probably rather early in the morning, and the Divine Artist has not yet stood back from his work and declared it to be "very good." Perhaps that can be only when God's very imperfect image has become more competent to manage the affairs of the planet of which he is in control.[120]

Most importantly, it was our duty to improve the human race itself. Stripped of its reference to God, this was exactly what Huxley taught, and the idea was by no means unacceptable to a certain type of Christian thinker.

Any attempt to correlate religious beliefs with particular theoretical innovations is vitiated by the fact that the other British founder of the genetical theory of natural selection, J. B. S. Haldane, was skeptical of formal religion and converted to Marxism in the early 1930s. Although originally inclined to an idealism similar to his father's, Haldane was always anxious to challenge conventions, and from an early age had been an enthusiastic Darwinian. He did not actually work on the theory in the early part of his career, but was induced to do so in the 1920s by the skepticism of popular writers such as Hilaire Belloc. The first chapter of his *Causes of Evolution* (1932) is introduced by the epigram "'Darwinism is dead'—*Any sermon*," but an earlier article is more specific: "'Darwinism is dead'—Mr H. BELLOC." It has been argued that it was Belloc's attack on H. G. Wells's Darwinism that persuaded Haldane that he should do what he could to put the selection theory on a firmer foundation.[121] He also played an important role in public debates

119.   Minutes of a conference intended to prepare a presentation to the Convocation of Canterbury, E. W. Barnes papers, EWB 9/16/29.

120.   Fisher, "Renaissance of Darwinism."

121.   McOuat and Winsor, "J. B. S. Haldane's Darwinism in Its Religious Context." See Haldane, *The Causes of Evolution*, 1, and "Darwinism To-Day" in Haldane, *Possible Worlds*, 27–44, 27. Haldane also criticized G. K. Chesterton and George Bernard Shaw; see his "Possibilities of Human Evolution" in Haldane, *Inequality of Man*, 78–96.

defending the theory, including a series of open letters exchanged with the Catholic Arnold Lunn, and another series in which he took on Douglas Dewar and L. Merson Davies.[122]

Although Haldane shared Huxley's and Fisher's view that natural selection was an agent of progress, he had a far more realistic understanding of the vast amount of evolution that has gone in nonprogressive directions. An unpublished essay written as early as 1912 stresses Darwinism's potential for upsetting religious beliefs and argues that a world governed by natural selection "presents itself as an arena of immense struggle, very largely futile, but occasionally leading to the production of creatures like ourselves." Purpose comes in through the behavior of individual organisms, "but it is impossible to trace any general plan running through the whole."[123] His article "Darwinism To-Day" also insists that evolution is not an "intelligent" process with a directing agency resembling mind, most obviously because it does not learn from its mistakes (in the form of species that have gone extinct from overspecialization).[124] The conclusion of *The Causes of Evolution* addresses the same issues, contrasting the conventional image of progress through to humankind with the prevalence of degenerative evolution throughout the animal kingdom. Haldane thought that the widespread use of the term "progress" represented "rather a tendency of man to pat himself on the back than any clear scientific thinking."[125] After stating his preference for a monistic philosophy that allows mental properties to be attributed to what are often called material entities, he attacks the doctrine of emergence as unscientific when used to suggest that mind suddenly appears at a key point in evolution. Mental capacity has increased gradually, he argues, and it is only through individual organisms that mind can influence evolution. The claim that evolution has been guided by a divine power is falsified by the vast number of blind alleys in the history of life and by the appearance of parasites—Haldane asks if we really want to make God responsible for the tapeworm.[126] Noting that philosophers such as Bergson and populist Lamarckians such as George Bernard Shaw professed to be horrified by the possibility that mind plays only a limited role in evolution, Haldane proposes a different reaction:

122. Lunn and Haldane, *Science and the Supernatural;* Dewar, Davies, and Haldane, *Is Evolution a Myth?*

123. Haldane, essay on Darwinism, disbound notebook, National Library of Scotland, MS 20578, 68–81, see 88 and 85.

124. In *Possible Worlds*, 29–30.

125. *Causes of Evolution*, 153.

126. Ibid., 159–60.

If evolution, guided by mind for a thousand million years, had only got as far as man, the outlook for the future would not be very bright. We could expect very slow progress at best. But if now for the first time the possibility has arisen of mind taking charge of the process, things are more hopeful. . . . There is at least a hope that in the next few thousand years the speed of evolution may be vastly increased, and its methods made less brutal.[127]

Haldane shared some of Huxley's and Fisher's hopes for the human race taking charge of evolution, but he did not see this as the continuation of an existing purposeful process. Nor did he support eugenics as the best way forward. Although firmly convinced that individual character is partly determined by heredity, his left-wing sympathies made him far more alert to the extent to which the effects of heredity in the general population were masked by the environmental differences between rich and poor. We must have "equality of opportunity," he argued, before selection on eugenic grounds is justified. Haldane attacked Dean Inge, a prominent Anglican Modernist, who advocated a stringent eugenic policy to stave off the degeneration of the human race.[128] Haldane thus revealed a very different side of the new Darwinism. While not denying progress in evolution, he cautioned against optimism on these grounds because he saw no evidence that the overall process could be seen as directed toward the production of the human mind. And while well aware of the potential of the new genetics to influence the future of humanity, he drew no lessons from the history of life about the choices we must make to shape that future.

The revival of Darwinism was a useful weapon against the teleological image of evolution exploited by the new natural theology, although lingering respect for the idea of progress allowed figures such as Huxley to formulate a hybrid philosophy. Fisher was able to reconcile Darwinism with his own Christian belief, but that belief was of a decidedly liberal character. In any case, he did not publicize his views in the interwar years, and most religious thinkers would have remained unaware of his vision of how faith could be reconciled with the new biology. Haldane's more materialistic interpretation of Darwinism reflected a long-standing tradition espoused by figures such as Lankester and Wells and was still taken by nonspecialists as a reflection of the theory's true character. Most nonscientific commentators also seem to have

127. Ibid., 164.

128. Haldane, "Inequality of Man," in *Inequality of Man*, 12–26, and on Inge, *Science and Ethics* (also reprinted in *Inequality of Man*, 97–118).

remained unaware that the situation in science had changed. The anti-Darwinian campaign by figures such as Belloc continued unabated, for all the efforts of Keith and Haldane to checkmate the claim that the selection theory was dead. Many religious thinkers, as well as many popular writers, continued to believe that the teleological interpretation of evolution promoted by the exponents of the modernized natural theology represented a consistent new direction taken by the biological sciences. The efforts of scientists such as Thomson and Morgan to promote the anti-Darwinian viewpoint popular in the early decades of the century were immensely successful outside science, although their views were being abandoned by most active researchers. Non-Darwinian evolutionary theories had been a major factor in late-nineteenth-century biology, and while their influence within science declined in the new century, their impact on popular culture actually increased. The progressionism of the new natural theology formed a major component of what most nonspecialists took to be the implications of current science—even though it was being promoted mostly by an older generation of biologists out of touch with the latest thinking.

CHAPTER 5

Matter, Life, and Mind

The debates over the mechanism and course of evolution meshed with another set of issues concerning the nature of life itself and the role of mind in directing behavior. Materialists asserted that living organisms—including their brains—were no more than complex machines, with mind only an epiphenomenon, a ghostly illusion incapable of having any effect on a real world of matter. Materialism also destroyed the sense of individual responsibility for moral decisions that was a vital part of Christian belief. The opponents of materialism saw life and mind as active agents, capable of taking decisions and actions that had a real effect on the world—actions that could not have been predicted from the laws of chemistry and physics and were thus in some sense "free." Those who preserved some form of religious belief often wanted to see this innovative behavior as a manifestation of the creative activity of God Himself. Henri Bergson's *élan vital* was a creative force that challenged the laws of physics, and according to his philosophy, the universe was dead until this almost supernatural urge was introduced into it. Lloyd Morgan's philosophy of emergence did not go this far: it allowed for life to appear as a consequence of the normal activity of the physical world, but insisted that once it had appeared, it took charge of organisms' physical structure in the same creative manner as Bergson's *élan*.

Those who wanted life to be seen as a creative force were tempted by the old philosophy of vitalism, the claim that life is a distinct nonmaterial entity. Vitalism, although the deliberate target of the mechanists, had been revived in biology by Hans Driesch in the late nineteenth century. His "neovitalism" in its purest form had a very limited influence, yet it encouraged a whole generation of biologists to probe the limits of the mechanistic philosophy. They sought a workable theory in which the activity of a living body, while not actually defying the laws of physics, was nevertheless capable of integrative activities that could not meaningfully have been predicted from those laws. The question of the origin of life also had a bearing on these issues: If life was a distinct nonmaterial force, its introduction into the material world would presumably have to be supernatural. If it could be understood as a physical activity (however much transformed by the integrative behavior of complex systems), it might be possible to explain its origin in purely natural terms. At the opposite end of the scale, some psychologists tried to develop means of understanding the behavior of complex living things in terms that could be defined only by mental qualities unpredictable on the basis of any lower level of explanation.

There were debates on these topics within various specialized areas of science, but the very nature of the questions sometimes allowed different specializations to attempt an answer. The various areas of science came at these problems with different expectations and prejudices, but their claims were likely to be taken up by outsiders wanting to see a message coming from science as a whole. Such outsiders were equally likely to seize upon the writings of a particular group of scientists whose work appealed to them and hail their views as indications of a new direction in thought, even when the majority of scientists in the same or related fields were indifferent or even hostile to those views.

The wider significance of these debates was obvious to all. Anne Harrington has written of holism as the vehicle by which German culture reacted against the "disenchantment" of nature by nineteenth-century materialism.[1] The British reaction was similar, but was complicated by the interaction of religious and philosophical concerns. The opponents of materialism were certainly articulate, especially when addressing a nonscientific readership, and their views were widely hailed by religious thinkers anxious to see a new generation of scientists emerging from the slough of Victorian materialism. In previous centuries Christian thinkers had sometimes opposed vitalism on the grounds that to invoke a natural "creative" force would demean God's sover-

1. Harrington, *Reenchanted Science.*

eignty. Now that a powerful philosophy of materialism had emerged in science, they were more likely to endorse any challenge to that more serious threat. Not all antimechanists were religious, however, and their work was thus often applied by religious thinkers to ends that they would not necessarily have endorsed. Some biologists and psychologists opposed materialism on largely philosophical grounds; they may have been aware that their work was being used by apologists for religion, but it was not their primary intention to provide support for theism, let alone for Christianity. Some non-mechanistic philosophies—J. S. Haldane's idealism is a case in point—did have genuine religious implications, but were difficult to use in support of any conventional theology.

At the opposite end of the ideological spectrum, rationalists and materialists opposed all efforts to create a role for what they regarded as supernatural entities in the world. They rejected the alleged higher powers of life and mind because they wanted to portray the universe as an essentially amoral and purposeless system that gave no support to the claim that it was created by a Power able to offer us a transcendental source of values and beliefs.

THE ORIGIN OF LIFE

The question of how life first appeared on earth had been widely debated in Victorian times, with the consensus going against those scientists who claimed that a purely natural origin was demonstrable, or even plausible. Although mechanists such as Huxley and Tyndall had treated life as an activity of a material substance, protoplasm, the efforts of scientists such as Henry Charlton Bastian to demonstrate the appearance of life from nonliving matter were greeted with much skepticism, especially when Louis Pasteur seemed to have proved that living cells came only from other living cells.[2] Although some evolutionists insisted that a thoroughgoing naturalism required the origin of life from nonliving matter, Bastian came to be regarded as a loose cannon by the inner circle of Darwinists, including Huxley and Tyndall. Huxley himself shifted the emphasis to the distant past, arguing that scientists could legitimately consider the hypothesis that life first appeared on earth by natural causes, perhaps under conditions quite different from those that have obtained though its later history.[3] Huxley's suggestion eliminated

2.   See Farley, *Spontaneous Generation Controversy*, chap. 7 and Strick, "Darwinism and the Origin of Life."

3.   T. H. Huxley, "Biogenesis and Abiogenesis," in his *Discourses Biological and Geological*, 229–71; see 256.

any necessity for an appeal to the supernatural to explain the first "breathing in" of life, a terminology that even Darwin had used in the conclusion to his *Origin of Species*.[4] In private, Darwin had regretted truckling to public opinion by adopting this biblical metaphor, but he sensed the importance of not linking evolutionism too closely to highly controversial claims about the ultimate origin of life.

One way of blocking the search for a natural origin without appealing to miracles was to sideline the whole question by arguing that life was first brought to the earth from elsewhere in the universe. In 1871 the physicist Lord Kelvin—who was certainly opposed to the materialistic view of life—suggested that living matter from colliding planets might be shot into space, where it could ultimately be transported to new worlds. The Swedish scientist Svante Arrhenius modified this view by reviving the old hypothesis of panspermia—the idea that minute "seeds" of life abound throughout the universe, capable of surviving endless journeys through space, and ready to start up the process of evolution on any new world upon which they land.[5] Arrhenius implied that the origin of the first living seeds was beyond scientific explanation, and might thus be linked with the creation of the universe itself.

This evasive approach appealed neither to the rationalists nor to those whose religious beliefs predisposed them to accept the supernatural creation of life. For the rationalists, an important addition to Huxley's position was made by E. Ray Lankester in his article on the Protozoa for the *Encyclopedia Britannica*. At this point Lankester did not challenge the conventional view that life was a property of the chemically complex "protoplasm," but he suggested that the earliest forms of this substance could not metabolize carbon dioxide and instead consumed the less complex organic chemicals that had been the stepping-stones to their formation.

A conceivable state of things is that a vast amount of albuminoids and other such compounds had been brought into existence by those processes which culminated in the development of the first protoplasm; and it seems therefore likely enough that the first protoplasm fed upon these antecedent steps in its own evolution, just as animals feed upon organic compounds at the present day.[6]

4.   Darwin, *Origin of Species* (6th ed.), 429. The regret at "truckling" is expressed in a letter to J. D. Hooker, 29 March 1863, reprinted in F. Darwin, ed., *Life and Letters of Charles Darwin*, vol. 3, 17–18—but he did not remove the term.

5.   Arrhenius, *Worlds in the Making*, chap. 8; on Kelvin's ideas, see 218–19.

6.   Lankester, *Zoological Articles*, 3.

All trace of the original chemical evolution had thus been destroyed, and there could be no equivalent process going on in the world today because the intermediate steps could not accumulate. Lankester paved the way for an important step forward in the debate by showing that experiments disproving spontaneous generation in the modern world were irrelevant to the question of the ultimate origin of life by chemical evolution.

Lankester's position was quoted with approval by J. Arthur Thomson in his *Science and Religion,* which he wrote to defend the vision of a purposeful universe without the need for any recourse to the supernatural.[7] For Thomson, the increasingly popular view among scientists that the origin of life would ultimately be explained in natural terms was not a threat to religion because, whatever its origin, life was self-evidently a purposeful entity. His position fitted in well with Lloyd Morgan's emergent evolutionism, although Morgan himself was a psychologist more interested in the emergence of mind, and said little about the actual process by which life itself might have been formed.

The rationalists continued to argue that the origin of life was a purely natural process. J. B. S. Haldane's "The Origin of Life," originally published in the *Rationalist Annual* for 1929, came to be regarded as a pioneering step toward a scientific program in this area. Haldane was not yet a Marxist, so his initiative cannot be explained in terms of that philosophy—it was an attempt to explore the latest developments in science, which might throw light on this intractable problem. In his typically iconoclastic style, Haldane began by exploring the wider implications of the issue. Many conservative evolutionists, he claimed, had welcomed Pasteur's work because it allowed them to retain a special status at least for life itself. Materialists had tried to bridge the gap between the living and nonliving, without much success, but a few idealists had been happy to support the project too, because they believed that mind could be associated with matter itself. Pasteur's work appealed to those who wanted to separate mind from matter, including the churches (although Haldane noted that the early Christians were themselves materialists, believing in the resurrection of the body, not the immortality of the soul).[8]

Haldane's argument started from the growing difficulty encountered by scientists trying to define a boundary between living and nonliving matter. Viruses and bacteriophages seemed to blur the distinction—depending on which definition of life one chose, they could be regarded as merely complex

7. J. A. Thomson, *Science and Religion,* 95–96.

8. J. B. S. Haldane, "The Origin of Life," in his *Inequality of Man,* 148–60, see 148–49. John Farley makes the point that Haldane's paper cannot have been inspired by Marxism; see *Spontaneous Generation Controversy,* 163–65.

chemicals or very primitive living things. Haldane thought they were, in effect, naked genes. Below this level were active agents such as enzymes, obviously not alive, but capable of stimulating great levels of organic activity. From these clues one could reconstruct the probable steps in the origin of life. In the oxygen-free atmosphere of the early earth, complex organic chemicals, including proteins, had formed. Such proteins are destroyed by microorganisms if formed today (Lankester's point), but would then have accumulated to "the consistency of hot dilute soup." Large molecules capable of reproducing themselves in this rich environment would be formed, and after a vast period of time, oily membranes enclosing collections of such molecules would form the precursors of the first cells. The fact that all modern creatures utilize chemicals that polarize light in the same way suggests that they are all descended from the same ancestral cells. Eventually the cells became self-supporting and began to synthesize their own food from sunlight. Haldane did not think that he would live to see his conclusions vindicated by a laboratory reproduction, but he insisted that the various steps would eventually become susceptible to experimental tests. He concluded by returning to the philosophical issues. He admitted that his views were materialistic, but argued that they could not be dismissed on those grounds alone, because they were also compatible with the view that mind can associate itself with certain types of chemical substances. In fact, all attempts to describe the relation between mind and matter were no more than clumsy metaphors.[9]

Haldane's brief essay played an important role in spurring later scientific activity in this area, but it also illustrates that the origin of life was an issue the rationalists could not afford to ignore without risking their credibility in the public debate. If religious writers were anxious to show that life could not have a natural origin, the apparent inability of scientists to tackle the question would support their claims. Haldane felt that it was necessary to meet this challenge head on by showing that there were at least indirect arguments that favored the removal of the traditional barrier between dead and living matter. Yet his initiative was not necessarily recognized by religious writers. W. Osborne Greenwood's *Biology and Christian Belief* of 1938 dismissed the idea of life from outer space and insisted that the choice was between a natural or a supernatural origin on earth. But Haldane's work was not mentioned; Greenwood claimed that Bastian's experiments had never been repeated, leaving most biologists indifferent to the whole issue. Christianity could accommodate itself to the idea that life had a natural origin, if that were eventually proved, but in the absence of such a demonstration, it was still legitimate to

9. Haldane, "Origin of Life," 160.

believe in some form of creation. This did not tie the Christian to the most literal model of creation, because there might be more subtle ways in which the Creator could influence the material world.[10] Conservative religious thinkers continued to draw hope from what they perceived as science's inability to come to grips with this issue (see chapter 9).

VITALISM AND ORGANICISM

Whatever T. H. Huxley's doubts about the spontaneous generation of life, his essay "On the Physical Basis of Life" was a clear statement of the claim that science reserved the right to explain the life-giving activities of protoplasm in materialistic terms. Huxley specifically warned that this was a methodological prescription for the scientist, not an endorsement of radical materialism, but his essay was widely perceived as a contribution to the latter cause.[11] In the reprinted version of this essay (1893), he notes that it came in for more than its fair share of criticism, and recommends the reader to Michael Foster's *Textbook of Physiology* for information on the extent to which the progress of science had vindicated his pronouncement. But the mechanists had not had it all their own way, and—as in other fields—the late nineteenth century had seen a reaction against them. In this case, the reaction produced a temporary resurgence of vitalism, the belief that life consists of a specific nonphysical force. Outright neovitalism did not survive long in science, but the early twentieth century saw an ongoing battle between those who still strove to reduce life to no more than physicochemical activity and those who felt that the attack on mechanism had been in part justified. There was no vital force, the latter argued, but neither did it make sense to try to explain all the properties of a living organism in reductionist terms. This organicist approach paralleled the philosophy of emergence to the extent that it saw complex living systems as exhibiting purposeful behavior that required biological categories for its proper description.

Evidence that vitalism was seen as a weapon in the battle to defend theism is provided by the invitation to the leading vitalist, Hans Driesch, to give the Gifford Lectures in 1907–1908. In the 1890s Driesch had performed studies of sea urchin eggs whose results seemed to suggest that the developmental process had the capacity to overcome even the most drastic interference by the experimenter. He became convinced that the life force, which he termed the "entelechy," was an active, purposeful entity that could never be explained in phys-

---

10.   Greenwood, *Biology and Christian Belief,* chap. 6.

11.   Huxley, "On the Physical Basis of Life," in his *Methods and Results,* 130–65; see 164–65.

ical terms. His *Science and Philosophy of the Organism*—the published version of his Gifford Lectures—was an extended critique of mechanism and a statement of the need for life to be seen as something more akin to a constructive force at work in the world. The first volume presented the development of the embryo as a goal-driven process tailored to the production of coordinated structures. Many examples were offered as evidence that a purely mechanistic approach was incapable of explaining its coordinating power, including the ability of many lower animals to regenerate lost parts. Driesch also attacked the Darwinian theory of evolution, favoring Lamarckism and the possibility of preordained evolutionary trends directing life toward certain goals.[12] In the second volume, he explored other activities of living things which, he claimed, confirmed their status as purposeful entities. A long philosophical section then defended the concept of the entelechy, concluding with its moral implications. Vitalism was the "high road to morality" because "How could I feel 'morally' toward other individuals, if I *knew* that they were machines and nothing more?—machines which someday I *myself* might be able to construct like a steam engine!"[13] Vitalism also pointed to the existence of "an original primary entelechy" that made the universe, thus accounting for the overall harmony of nature. Science itself was engaged in the search for God, and a natural theology was still possible.[14] In a chapter entitled "The Breakdown of Materialism" written for Frances Mason's *Great Design* in 1934, Driesch was still arguing for the entelechy as a spiritual agent directing the activities of the living body.[15]

Driesch's defection to philosophy was an indication of how rapidly his neovitalism was ejected from scientific biology. One of the few biologists to defend an outright vitalism was the Catholic anatomist Bertram Windle, who linked scientific arguments based on the regeneration of lost organs in lower animals to the Scholastic philosophy of the church.[16] But many biologists continued to share Driesch's doubts about the full-scale mechanist program, and this led them to explore less drastic ways of trying to defend a role for purely biological categories of explanation. The philosophies of holism and organicism were intended to show that even a purely material system, if sufficiently complex and coordinated, could exhibit properties that could never have been predicted on the basis of physics and chemistry. In some cases, the

12.   Driesch, *Science and Philosophy of the Organism*, vol. 1, 260–305. On developments in early-twentieth-century biology, see G. Allen, *Life Science in the Twentieth Century*.

13.   *Science and Philosophy of the Organism*, vol. 2, 358, Driesch's italics.

14.   Ibid., 370–71.

15.   Driesch, "Breakdown of Materialism," 301.

16.   Windle, *Vitalism and Scholasticism*.

techniques used to expound these philosophies led to a good deal of confusion and a widespread belief outside science that vitalism was still holding its own. Even the supporters of organicism sometimes referred to it as "neovitalism." Religious thinkers were thus able to continue writing about the new nonmechanistic slant in science as though it provided support for the claim that nature was pervaded with almost supernatural activity—much to the disgust of scientists such as J. S. Haldane, whose views were hardly supportive of religious orthodoxy. The scientists who argued for organicism included some with strong religious feelings and others whose distaste for mechanism was based almost solely on philosophical grounds.

J. S. Haldane adopted an antimechanistic approach to biology and an idealist philosophy at an early stage in his career. When he moved from Edinburgh to Oxford in 1887, he began the long series of researches on the regulation of the body that brought him fame, beginning with a study of the way the carbon dioxide content of the air controls the rate of breathing. Since regulation ensured that the body responded appropriately to changes in its environment, Haldane saw his work as a vindication of his belief that

J. S. Haldane. Photograph by Benjamin Stone, 1902. By courtesy of the National Portrait Gallery, London.

teleology was an essential concept in biology. His views were expounded in lectures to colleagues and in articles from the 1880s onward, but became more widely known when he published a series of books, beginning with his *Mechanism, Life, and Personality* in 1913.

Haldane was quite happy to borrow the arguments against the mechanistic view of life proposed by the vitalists, but insisted that Driesch's entelechy was unacceptable because it was inconsistent with the law of conservation of energy.[17] Instead, he treated the organism as fundamental to biology: we perceive the organism as a self-regulating entity, he argued, and every effort to analyze it into components that can be reduced to a mechanical explanation violates this central experience.[18] Although Haldane always insisted that he was not a vitalist, his idealist philosophy led him to defend his organismic viewpoint in ways that were open to misapprehension. Instead of stressing the interactive properties of complex systems, he insisted that life was a primary reality of our perception, not an artifact created from physical components.[19] These views were repeated in his Gifford Lectures for 1927–1928, in which they were linked to his broader idealist philosophy, with its acceptance of the universe as a great Mind, along with a complete repudiation of the supernatural.[20] In his last collection of essays, however, he attacked Jeans and Eddington for their continued endorsement of the claim that physics was primary—like Whitehead, he saw the universe itself in biological terms.[21] Although personality was not his chief concern as a scientist, he supported those psychologists who saw it as irreducible. He was also anxious to use his idealism and his organicism to bolster a social philosophy in which the state had a significant role to play in coordinating the actions of individuals. He believed that the liberal philosophy of extreme individualism could be challenged by appealing to an analogy between society and the living organism: both functioned as coordinated wholes structured to ensure the well-being of all their components.

Many of the religious writers who quoted Haldane in support of the claim that biology now endorsed the existence of nonphysical forces did not realize that his opposition to mechanism rested on a philosophy in which the world was seen in terms defined by the human mind. The more conventional approach was to defend holism by arguing that the complexity and coordi-

17.   J. S. Haldane, *Mechanism, Life, and Personality*, 26–29; on the links between Haldane's philosophical position and his biology, see Sturdy, "Biology as Social Theory."

18.   *Mechanism, Life, and Personality*, 79–83.

19.   Haldane, *Organism and Environment as Illustrated by the Physiology of Breathing*, 100.

20.   Haldane, *Sciences and Philosophy*.

21.   Haldane, *Materialism*, chap. 2.

nated nature of bodily interactions produced effects that, while governed by the laws of physics and chemistry, could not be understood at that level because they worked to maintain the integrity of the whole. Even the rationalist Arthur Keith accepted that the role played by hormones in individual development helped to coordinate the process in an apparently purposeful way. E. Ray Lankester was disgusted by this turn in Keith's thought—in a postcard to H. G. Wells in 1926, he complained, "We have all been there—forty years and more ago. It is a barren blind alley."[22] But others thought that the holistic approach offered a valuable compromise between mechanism and vitalism, especially after Smuts popularized the term "holism" in the mid-1920s.

One powerful advocate of this approach was J. Arthur Thomson, often working in collaboration with Patrick Geddes. In his Gifford Lectures, Thomson noted both Driesch's and Haldane's critiques of mechanism, stressing the need for a methodological vitalism that used nonphysical characteristics to describe living things without implying that they were driven by vital forces.[23] His *Science and Religion* also stressed that life had to be seen as a creative force, a "victorious insurgence" that could not be accounted for by the laws of the physical world.[24] Similar views were expressed in textbooks intended to introduce students to the latest thinking in biology. In 1925 Geddes and Thomson's text on biology for the Home University Library series stressed the role of nonphysical explanatory systems so heavily that they conceded that some readers might think they hankered after Driesch's entelechy or Bergson's *élan vital*. Their real purpose, though, was to argue that the psychological characters of living things gave them a degree of spontaneity—this was the real life force.[25] A few years later another of their textbooks openly proclaimed itself the vehicle for projecting a neovitalism that showed "the untenability of a Biology which denies that Mind counts."[26] They cited the Bergsonian philosopher Wildon Carr as an exponent of "positive vitalism" linked to Smuts's holism—according to Carr, Smuts was an "entelechist" without realizing it because his system allowed for the emergence of wholes with new properties.[27] This was immediately qualified, however, with the observation that the most tenable position was methodological vital-

22. Lankester to Wells, 3 October 1926, Wells papers, University of Illinois.

23. J. A. Thomson, *System of Animate Nature*, chaps. 4 and 5.

24. Thomson, *Science and Religion*, 100–102.

25. Geddes and Thomson, *Biology*, 172–74, 273.

26. Thomson and Geddes, *Life: Outlines of General Biology*, vol. 1, vii.

27. Ibid., 448.

ism—science dreaded the introduction of vital forces in disguise, but biology had to use concepts of its own that could not be reduced to those of physics. Other scientific opponents of mechanism included J. Gray and A. Daniel Hall.[28]

J. S. Haldane had insisted that the idea of life was closer to reality than the ideas of matter and energy used by the physicists.[29] If one area of science was going to swallow up the other, it would be biology swallowing physics, not the other way around as the mechanists predicted. Most biologists found these views difficult to take seriously, but they were prophetic in view of the growing enthusiasm for Alfred North Whitehead's philosophy. One biologist who did quote Haldane (and Whitehead) with enthusiasm was Joseph Needham, whose career encompassed a move from mechanism to organicism that paralleled his adoption of Marxism, both transformations being mediated by a continuing commitment to Christianity. Needham's aim as a scientist was to synthesize biochemistry and morphogenesis (the development of structure in the organism), and he was thus addressing exactly those issues that had prompted Driesch to declare that a mechanistic explanation of life was impossible. But for the young Needham this was a counsel of despair: for science to be possible, the development of complexity must have a causal explanation, and in his early papers he freely used mechanical analogies.

At this point Needham had to wrestle with the conflict between the methodological necessity of adopting a mechanistic approach to life and his religious beliefs, which told him that human life was a far richer activity. His essay "The Sceptical Biologist" reveals the tensions within his methodological commitment. Mechanism, "the well-established firm of Democritus, Holbach and Huxley Ltd.," was now a Victorian ruin, brought down by the efforts of Bergson, Whitehead, and others, "and finally Eddington ploughed it over and sprinkled salt on it."[30] Yet the alternative school of neovitalism promoted by Driesch and Haldane was of no use because it erected barriers to science labeled NE PLUS ULTRA. Science could not abandon mechanism, yet mechanism dehumanized the world, so it had to be treated as a method without metaphysical foundations. The mechanistic and nonmechanistic viewpoints were each only attitudes of mind without counterparts in

28. Gray, "Mechanistic View of Life" (this critique was given as the presidential address to the zoology section of the British Association in 1933); Hall, "The Faith of a Man of Science."

29. Haldane, *Mechanism, Life, and Personality*, 102–4.

30. Needham, "Sceptical Biologist," in his *Sceptical Biologist*, 11–39, see 13–14. On vitalism, see also "Hunting of the Phoenix," in *Sceptical Biologist*, 89–130. On the development of Needham's views in biology, see Haraway, *Crystals, Fabrics, and Fields*, chap. 4, and Abir-Am, "The Philosophical Background of Joseph Needham's Work in Chemical Embryology."

Joseph Needham. Photograph by Ramsay and Muspratt, 1937. By courtesy of the National Portrait Gallery, London.

reality, and the scientific habit of mind must hold the mechanistic tendency in check so that other activities relying on intuition could still play a role in human life. Another essay extended this argument explicitly into the relationship between science and religion. The modern idealists had robbed the material universe of its terrors, so that science and religion could themselves be seen as two differing attitudes of mind, each with a valuable role to play and neither able to tell the other that it was barking up the wrong tree.[31] It was pointless, however, to think in terms of a "reconciliation" between two such different approaches to human experience.[32]

Needham's 1928 essay "Organicism in Biology" seems at first sight to move in a new direction. Here he adopts a sympathetic tone toward Haldane's prophetic remarks on biology swallowing up physics and points to Whitehead's philosophy of the organism as the vindication of this position, especially as refined by Morgan.[33] According to this new philosophy, the whole universe consisted of wholes, each level of organization revealing new prop-

---

31. Needham, "The Limitations of Optick Glasses: Some Observations on Science and Religion," reprinted in *Sceptical Biologist*, 43–66. See also "Materialism and Religion," in *Sceptical Biologist*, 233–69.

32. Needham, "Religion and the Scientific Mind."

33. Needham, "Organicism in Biology," 32–33 (he notes that Lloyd Morgan had published a perceptive review of Whitehead's *Science and the Modern World*); this essay is reprinted in *Sceptical Biologist*, 69–86. Needham wrote to Morgan saying how helpful the concept of emergence was: Needham to Morgan, 18 November 1925, Lloyd Morgan papers, Bristol University Library, file DM 612.

erties. By the end of the article, however, Needham had dismissed the concept of the organism as something valuable in philosophy but of no use to science, which must keep up its mechanistic, analytical approach for progress to be possible. Needham's views on biological method soon began to change, however—indeed, the transformation is evident between the successive volumes of his *Chemical Embryology* of 1931. Needham began to see how he could use the organismic approach when he came to appreciate the potential of Hans Spemann's concept of the organizer, a chemical substance capable of affecting particular parts of the developing embryo in particular ways. This concept allowed chemistry to be incorporated into the search for structure because chemical gradients within the organism could be treated as fields. The structure of chemical molecules could be linked to organic structures in a way that at first seemed to offer fruitful hypotheses for research.

Various influences can be seen affecting Needham's thought at this time. He still remembered Haldane with sympathy, despite the latter's failure to realize that new levels of organization could generate new properties.[34] Whitehead's philosophy was now of real importance to science, along with the biological insights of J. H. Woodger and D'Arcy Wentworth Thompson, both of whom attempted philosophical critiques of the mechanist paradigm.[35] Most important of all was the move to Marxism, which had allowed Needham to see that ethics and politics had to be brought into science. It also allowed him to make sense of the hierarchies of organization that produced successive levels of new properties, thereby allowing science to have access to concepts that had been excluded by old-fashioned, nondialectical materialism.[36]

The Marxist view of history also gave Needham an interest in the theory of evolution because it presupposed that progress occurred through the successive stages by which new levels of organization were produced. In his Herbert Spencer Lecture of 1937, titled *Integrative Levels*, he noted that in his earlier life he had been attracted by the almost timeless universe of the mystics, as depicted by Dean Inge, although he now found Inge's middle-class spitefulness unworthy of a Christian.[37] The "Giant Vista of Evolution" led organisms to an increasing independence of the environment made possible

34. Needham, "Thoughts of a Younger Scientist on the Testament of an Elder One" (John Scott Haldane), in his *Time, the Refreshing River,* 121–40.

35. Needham, "A Biologist's View of Whitehead's Philosophy," in *Time, the Refreshing River,* 178–206. D'Arcy Wentworth Thompson's *On Growth and Form* was published in 1917, J. J. Woodger's *Biological Principles* in 1929.

36. Needham, "Metamorphosis of Scepticism," in *Time, the Refreshing River,* 1–27; see 21.

37. Needham, *Integrative Levels,* 8–9, 13. The lecture is reprinted in *Time, the Refreshing River,* 233–72.

by increased levels of organization, and this advance was inevitable in the long run.[38] His address to the Modern Churchmen's Union the same year, titled "History Is on Our Side," makes clear his desire to see evolutionary progress as a guarantee of social progress—the drive toward a more integrated society is also inevitable.[39] In the words of W. H. Auden's poem "Spain," the masses look to "History, the operator, the Organiser, Time the refreshing river" for salvation. Marxism was one obvious source of this new sense of historical development, but Needham also wrote an appreciation of Henry Drummond, perhaps the most influential late-nineteenth-century advocate of the view that evolution was a divinely inspired progress toward ethical values.[40] This may explain why he differed from fellow Marxist J. B. S. Haldane, who remained suspicious of evolutionary progress as a model for social history. But Needham and Haldane shared the view that—despite the reemergence of Darwinism—the eugenic policies advocated by Fisher, Inge, and others were not an appropriate way forward. For Needham, at least, this distaste rested as much on his Christian faith in the value of the individual as on his Marxist philosophy.

In the early 1930s Needham joined with Woodger, C. H. Waddington, and others to form the Theoretical Biology Club as a means of promoting the organicist approach in biology. He tried to get funding for an institute at Cambridge that would put the new science into practice, but without success. In general, the new approach failed to meet the challenge of increasingly successful mechanistic techniques, partly due to the inability of contemporary biochemistry to tackle the issues involved. The formation of the Theoretical Biology Club (or Biotheoretical Gathering) has been interpreted as the response of a group of biologists only too well aware of their professional marginalization.[41] Needham gradually turned from biology to the history of Chinese science, in which he saw parallels with the antimechanistic strands in Western thought. Waddington moved to genetics, although he retained an organicist worldview that left him constantly nipping at the heels of orthodox genetic determinism and Darwinism. Waddington, too, had been strongly influence by Whitehead, but he had no interest in the religious di-

38. Ibid., 40, 51.

39. Needham, "History Is on Our Side," in his *History Is on Our Side*, 22–34; see 23–28. See also "Liquidation of Form and Matter," in *History Is on Our Side*, 199–210, 208–9.

40. Needham, "The Naturalness of the Spiritual World: A Reappraisal of Henry Drummond," in *Time, the Refreshing River*, 28–49. On Drummond, see chap. 7 below.

41. Abir-Am, "The Biotheoretical Gathering."

mension of the philosophy of the organism that almost certainly attracted Needham.[42] He became deeply concerned for the ethical issues raised by science and in the 1950s supported Huxley's humanism.[43] The role of Needham's religious faith in shaping his science is thus problematic: his antimechanistic thought rested primarily on philosophical foundations that were compatible with belief, but could quite easily exist independently of it.

The wide range of backgrounds of the scientists supporting a vitalist or organicist critique of mechanistic biology did not prevent nonspecialist writers taking the whole movement as an indication that science was moving in a direction that would facilitate a reconciliation with religious belief. Rationalists, however, saw the threat posed by organicism's popular image and sought to discredit it in the same public forum. Their arguments merely continued a long-standing campaign of opposition to vitalism initiated when that philosophy was reintroduced into biology by Driesch. In 1903 E. Ray Lankester had responded to some comments by Kelvin, reprinted from an address to the Christian Association of University College in the Times. Kelvin had long been an advocate of divine guidance in evolution, and now he linked this position to a belief in a distinct vital principle, claiming that modern biologists were once more coming to accept the need for such a principle. Lankester queried whether such a vitalist theory had any relevance for true religion—the need for separate creations of matter and life was a relic of polytheism. But in any case, he insisted, he saw no evidence that the true leaders of biology were turning their backs on the mechanist program that had served them so well. The physicists may have found their own unobservable entities (atoms and the ether) very useful, but "we biologists, knowing the paralysing influence of such hypotheses in the past, are as unwilling to have anything to do with a 'vital principle,' even though Lord Kelvin erroneously thinks we are coming to it, as we are to accept other strange 'entities' pressed upon us by other physicists of a modern and singularly adventurous type."[44] The latter, he made clear, was a reference to Lodge's spiritualism, nicely linking both extremes of the antimechanist position in a way designed to highlight what for Lankester was the lunatic fringe of the movement. In his popular essays, Lankester also wrote against the vitalist philosophy, defending

42. Waddington, "Practical Consequences of Metaphysical Beliefs on a Biologist's Work"; on Waddington's antimechanistic approach, see Yoxen, "Form and Strategy in Biology," and Lewin, "Embryology and the Evolutionary Synthesis."

43. Waddington, *Scientific Attitude* and *Science and Ethics.*

44. Lankester, appendix to "Nature's Insurgent Son," in his *Kingdom of Man,* 62–65, see 65.

Huxley's claim that life was a property of the material functioning of proto-plasm.[45]

Lankester remained hostile to the organicism of the 1920s—we have al-ready noted his dismay at Arthur Keith's venture into the field (and this de-spite Keith's reputation as a fellow rationalist and supporter of a materialis-tic view of the human mind). In 1930 Lankester's friend, the zoologist Peter Chalmers Mitchell, gave a Herbert Spencer Lecture attacking vitalism in bi-ology. Mitchell acknowledged that his own preference for materialistic monism was now unfashionable:

> Certainly the opposite view has been proclaimed by some distinguished
> men of science and has been accepted from them with amusing eagerness
> by the successors of the religious and philosophical dogmatists of last
> century. These shorn lambs naturally rejoice to think that the bitter wind
> of science may have backed into a more temperate quarter. But let them
> not be too confident. It may only be an interglacial period.[46]

Mitchell went on to defend the reality of the material world on the grounds that it constantly surprised us, and to give a series of examples in which sci-ence had advanced by adopting a materialistic approach. It was only by rec-ognizing the importance of structure, from the molecular level upward, that biology was able to explain complex phenomena in terms of the laws of physics and chemistry.[47] Vitalism, holism, and emergence were all dismissed as manifestations of the attempt to get purpose back into nature. In fact, Mitchell claimed, the doctrine of emergence could be interpreted in strictly materialistic terms, because a machine can perform functions that cannot be reduced to the laws governing the operation of its parts.[48] He concluded by noting that the implications of materialism were by no means as pessimistic as its opponents claimed—if the human spirit had advanced this far from a "tangle of tropisms, reflexes, instincts and organic emotions," how much fur-ther might it rise in the future?[49]

Perhaps the most prominent defense of materialistic biology came from

---

45.   Lankester, "Protoplasm, Life, and Death" and "Chemistry and Protoplasm," in his *Science from an Easy Chair*, 180–86, 187–92.

46.   Mitchell, *Materialism and Vitalism in Biology*, 3.

47.   Ibid., 15–17. Mitchell noted that D'Arcy Wentworth Thompson, for all his opposition to Darwinism, saw living structures as determined by mechanical forces.

48.   Ibid., 28–29. Mitchell had advanced this view of emergence (not under that name) in his *Evolution and the War.*

49.   Ibid., 30.

a member of the younger generation, Lancelot Hogben. Hogben joined Needham and J. B. S. Haldane in the move toward Marxism, but adopted a far more hostile attitude toward the idealist resurgence in twentieth-century science. His *Nature of Living Matter* of 1930 developed from a projected symposium on the nature of life at the British Association involving Smuts, J. S. Haldane, Eddington, and Wildon Carr. All of the latter had books in print, so Hogben issued his own response, adopting a polemical tone that did not, he insisted, imply personal antagonism because Smuts and Haldane had asked him to provide "destructive criticism" of their position.[50] He also attacked J. W. N. Sullivan, a popular writer who expressed the dissatisfaction of those who wanted to defend art and religion from the incursions of scientism.[51] In the first two parts of the book, Hogben defended the mechanistic approach in psychology, physiology, genetics, and evolutionary theory. But his defense paralleled that used by T. H. Huxley and the younger Needham in that it saw mechanism as an essential tool of the scientific method, not as a philosophy telling us about the real nature of the world. Opponents such as Haldane and Smuts had misunderstood the nature of the argument: few mechanists thought that their approach offered a complete worldview, but in the arena of public discussion of these topics, especially the scientific arena, no satisfactory alternative had been found. Mechanism dealt with epistemological concerns (how we regulate our efforts to understand the world), not with ontology (the nature of reality).[52] In the third part of his book, Hogben generalized this point to accuse both the vitalists and idealists such as Eddington of trying to discredit science itself by erecting barriers beyond which only subjective viewpoints were thought to be relevant. Only intuition could reveal the true nature of life, mind, and reality, and the private world of the individual was more real than the public world of scientific discourse. But how can one be a philosopher—let alone a scientist—asked Hogben, if one could not communicate one's insights to others via the public domain? And in this domain, mechanism was the only means of providing publicly verifiable information about life.[53] Hogben admitted that science was no longer as dogmatically materialistic as it had been, but he thought this was a temporary retreat *pour mieux sauter*, and ar-

---

50. Hogben, *Nature of Living Matter*, foreword.

51. J. W. N. Sullivan, *Bases of Modern Science* and *Limitations of Science*. Sullivan's real interests lay with the new physics, however, not antimechanistic biology—although Mitchell too had singled him out as an example of the new and more limited view of science.

52. Hogben, *Nature of Living Matter*, 99–100.

53. Ibid., 261.

gued that in the wider world the secular and materialistic outlook had never been more widespread.[54]

MIND AND BODY

The debate over the need for nonphysical levels of explanation in biology was often associated with the equally contentious issue of whether mind or consciousness was a real entity capable of affecting the material world. Several writers, including Thomson, suggested that if there *was* a life force, it consisted of the role played by mind in controlling the behavior of animals and humans. The problem with this view was that it left one having to assume that mind existed in even the simplest forms of life (a position that Thomson was honest enough to endorse). Many thinkers found this assumption unacceptable, and the debate over the reality of mind as a nonphysical entity was conducted at least in part at a separate level from that over the nonmechanistic functions of life. Lloyd Morgan's emergent evolutionism postulated that the appearance of life and of mind were two distinct steps in evolution—it was only in the higher animals that one could talk about mind.

The existence of these two separate, but partly interpenetrating, debates was a function of the different scientific disciplines involved. The vitalist controversy was of most interest to biochemists, physiologists, and embryologists, while the debate over the nature of mind involved psychologists and the small number of ethologists (students of animal behavior in the wild). The latter debate also held more immediate interest for religious thinkers. In one sense, Lankester was right to argue that the mechanist-vitalist debate was irrelevant to religion—indeed, natural theologians had often treated animals as merely well-designed machines. But if the human spirit was to retain anything like the traditional status accorded to it by Christianity, mind had to be a nonmaterial entity capable of making real choices (for which, in human beings, the individual could be held responsible). Religious thinkers were quite used to arguing that the human mind was something more than the animal mind, and Morgan treated the emergence of spirit as yet another step in evolution. But in an age dominated by the model of evolutionary continuity, it was difficult to lay too much stress on such an absolute distinction, and the work of those scientists arguing for the reality of the animal mind figured prominently in calls for a new synthesis of science and religion. In fact, however, some of the opposition to mechanistic psychology came from scientists

54.   Ibid., 243 and 263.

with philosophical rather than religious concerns about the status of the mind.

To some extent, the rise of a psychology opposed to mechanism can be seen as a product of the discipline's bid for academic independence. If the mind were merely an epiphenomenon of the mechanical workings of the brain, psychology would always be seen as a branch of neurophysiology. Defining purely mental faculties that could nevertheless be studied by a form of the scientific method was thus crucial to the emancipation of psychology. But the great danger of this move was that it would always be difficult to sell the idea that studies of an autonomous, creative, and value-sensitive mental realm really were scientific. Those who persevered with this approach—and who were regarded as providing the most obvious support for a Christian view of human personality—ended up looking like old-fashioned intro-spective psychologists. They were perceived, and to some extent perceived themselves, as philosophers rather than scientists. The drive to create a truly scientific psychology very rapidly moved on in directions that were much less promising for those who wanted to see the science as supportive of tradi-tional religious beliefs. New movements came in from abroad, analytical psy-chology from Vienna and behaviorism from America, each with its own threats to the autonomy of the personality. Religious thinkers were actually less disturbed by Freud's ideas because, while they rejected his model of the mind as controlled by unconscious forces driven by biological urges, they found the unconscious itself a useful tool for exploring the darker side of the human mind traditionally associated with original sin. It was behaviorism that most thoroughly challenged the old psychology, because it once again denied the reality of the mind and reduced the organism (including the hu-man being) to a mechanistic system driven by predictable laws.

The limitations of the attempt to create a scientific psychology based on the autonomy of the mind can be seen most clearly in the career of James Ward, professor of mental philosophy and logic at Cambridge. In the late nineteenth century Ward emerged as a leading opponent of Huxley's scien-tific naturalism, defending the mind as a real entity that had to be studied through traditional categories such as feeling and volition. In the early years of the new century he was seen as the exemplar of a new wave in psychology, and his *Psychological Principles* of 1918 was hailed as a triumph, at least in the nonspecialist press. In the next decade, though, he was rapidly and almost completely eclipsed by the new trends from abroad. Significantly, Ward went out of his way to develop the broader implications of his science, giving the Gifford Lectures in 1896–1898 and again in 1907–1910. The first series, published as *Naturalism and Agnosticism*, was a direct assault on the philosophies of T. H. Huxley and Herbert Spencer. In fact, he argued, the two systems named in his title were not natural allies, naturalism being dogmatic while agnosticism was inherently skeptical.[55] Mechanics was not an adequate foundation for a view of nature because its laws were constructions of the mind. The mind could therefore legitimately be treated as an autonomous and active entity—Huxley's famous description of animals (and by implication humans) as merely conscious automata was dismissed on the grounds that the mind, if real, could not be just an epiphenomenon or passive spectator.[56] Ward argued not for dualism but for spiritual monism: he believed that there was a single substance that could be described from both mental and physical perspectives. This looks like exactly the position adopted by Ernst Haeckel—but where Haeckel subsumed the mental into a law-bound system, Ward used it to create a space for all the traditional moral and spiritual characters to function in nature. He was quite willing to see evolution being driven by such spiritual forces, perhaps via the Lamarckian mechanism.[57] The whole universe could thus be treated as a teleological system driven by mind. In his second series of Giffords he insisted that his "spiritualistic" position allowed the moral and ethical dimensions of life to move center stage, with the emergence of the freedom of the will being the Creator's chief goal in evolution.[58]

The pattern of Ward's career can also be seen in those of the two psychologists who became the leading exponents of an antimechanistic philosophy in the early twentieth century, C. Lloyd Morgan and William Mc-Dougall. Both made important contributions to scientific psychology around the turn of the century, but went on to become elder statesmen of science concerned more with philosophy than with actual research. Morgan had made important studies of learning and instinct in animals, and he had established the important principle ("Lloyd Morgan's canon") that animal behavior should be interpreted in terms of the lowest possible level of mentality, thereby excluding the casual attribution of higher powers so common in the anecdotal evidence of untrained observers. He had adopted a monistic philosophy that allowed the mental and the physical to be seen as two parallel and equally valid ways of describing the same basic reality. This left him

---

55.   J. Ward, *Naturalism and Agnosticism*, vol. 1, preface. On Ward's career and influence, see Turner, *Between Science and Religion*, chap. 8.

56.   *Naturalism and Agnosticism*, vol. 2, chap. 11; see also T. H. Huxley, "On the Hypothesis That Animals Are Automata and Its History," in Huxley, *Methods and Results*, 199–250.

57.   Ward, *Naturalism and Agnosticism*, vol. 1, chap. 10.

58.   Ward, *Realm of Ends*.

essentially dissatisfied with the efforts of those (including McDougall) who postulated an independent soul or mind animating an otherwise mechanical body.[59] At the same time, however, he was fully convinced that a purely mechanical explanation of life was impossible. We have seen (in chapter 4) how his philosophy of emergent evolution was presented as a divinely preordained program for the creation of life, mind, and finally spirit. But to Morgan, life was not a distinct force or entelechy,[60] just as mind was not the distinct spiritual force postulated by the dualists.

Morgan's monistic philosophy made him a organicist by requiring him to see emergent properties as more or less inexplicable higher-level activities manifested by structures that had evolved a certain level of complexity. *Emergent Evolution* had much to say on the nature of perception, memory, and volition, and often engaged with Bergson's views on these mental processes. *Life, Mind, and Spirit* dismissed J. B. Watson's behaviorism on the grounds that if its elimination of the mental level of explanation were accepted, no science of psychology could exist. For Morgan, it was always possible to tell the story of any event from the physical or the mental perspective, although for all higher-level activities the latter revealed far more of significance.[61] The emergence of animals' ability to learn from experience was a turning point in the history of life (Morgan had been a codiscoverer of organic selection, or the Baldwin effect, in which new behavior patterns shape the course of evolution). But the mental level of explanation (the mind-story) of every action could make sense only if it included values such as the enjoyment of the perceiving mind and, in conscious organisms such as ourselves, the ability to plan for the future on the basis of past experience. Any attempt to tell the whole story of behavior from the mechanist perspective, as in Huxley's efforts to portray animals as automata, would miss out on important aspects of what was going on, and in this sense the mind could be said to play an active role in governing the world.[62]

Unlike Ward and Morgan, William McDougall argued for an openly dualist philosophy in which the forces of life and mind actively interfered with the direction of the physical world. McDougall was Wilde Reader in mental philosophy at Oxford from 1903 until he moved across the Atlantic to Harvard in 1920. Like Morgan, he had played a major role in establishing psychology as an independent discipline in Britain. He had made important studies of sensory psychology and later of social instincts. His *Body and Mind*

59. Morgan, *Instinct and Experience*, 286–88.
60. Morgan, *Emergent Evolution*, 12.
61. Morgan, *Life, Mind, and Spirit*, 47–48.
62. Morgan, *Animal Mind*, 263–66.

of 1911 was subtitled *A History and Defence of Animism,* by which he meant the doctrine that living bodies were animated by a mental or spiritual force. William James had noted in an Oxford lecture three years earlier that "souls are out of fashion," but McDougall was determined to swim against the tide.[63] His concern to defend this position, however, was driven by philosophical rather than religious interests (see chapter 1).

Body and Mind opened with a long history of vitalism, animism, and the opposing mechanistic philosophy. McDougall's defense of animism included a chapter on physiology that drew substantially on Driesch, followed by a critique of Darwinism stressing the role of the mind as the guiding force of animal evolution, operating through either Lamarckism or organic selection.[64] To defend the mind as a distinct entity that could not be explained in physical terms, McDougall focused on the need for sensation to have meaning for the organism, the role of feeling, and the inadequacy of all efforts to explain memory in terms of brain mechanisms. He also included a chapter on psychic research, admitting that the attempted demonstrations by the spiritualists fell short of what would produce conviction, but insisting that telepathy was real and was itself enough to disprove the mechanist position because there was no sign that the impressions were transmitted via a physical medium.[65]

McDougall felt that *Body and Mind* had been largely ignored on its publication, but it was reprinted steadily over the next couple of decades, perhaps indicating a growing public interest in the question it addressed. He produced more surveys of the issue after his move to America, including *Modern Materialism and Emergent Evolution* (1929), which included a critique of Morgan's philosophy. He began by attacking his own profession: in a desperate attempt to show its scientific credentials, psychology had turned itself into a branch of biology rather than philosophy and had thrown the baby out with the bathwater.[66] The key to "the psychology we need" was the recognition that purpose plays a role in all mental activities, even those of animals. Even memory depends on purpose, because we remember those events that have a meaning for us, and the activity of remembering is a function of the brain as a whole—in this respect, Bergson had been right. Referring again to psychic research, McDougall insisted that the occurrence of telepathy had been

---

63.   McDougall, *Body and Mind,* xii; see also his biographical statement, "William McDougall," in Murchison, ed., *A History of Psychology in Autobiography,* 209.

64.   *Body and Mind,* chaps. 17 and 18.

65.   Ibid., chap, 25. In *Modern Materialism,* McDougall said that if he were to rewrite *Body and Mind,* he would put in more on psychic research; see vii.

66.   *Modern Materialism,* 19–20.

rigidly demonstrated, although the evidence for survival after death was still equivocal. Implying that the doctrine of emergence had made it popular once again to think of the mind as playing an active role in the world, he worried that the widespread reluctance of scientists to admit even the possibility that the mind might also survive the death of the body was an unthinking leftover of earlier materialistic prejudices.[67] His own examination of emergence claimed that Morgan and its other proponents (who disagreed among themselves over its major details) were "concerned chiefly to save God and the coherence of the evolutionary scheme."[68] They said nothing about the actual mechanism of evolution and ignored the case for Lamarckism as the best explanation of how mind guided the process.[69] More seriously, they provided no real explanation of how successive levels of emergence were produced— by stressing the novelty of each emergent event, they left the cause mysterious. There was no trend toward progress in the physical world, as shown by the second law of thermodynamics, so why should new qualities such as life appear at certain points? Whitehead's efforts to portray the physical universe as an organism were unconvincing precisely because life was different from the physical world it inhabited.[70]

McDougall's last book, published in the year of his death (1938), was called *The Riddle of Life*. It was another wide-ranging survey that took on all the latest trends in organismic biology, rejecting them because they still did not go far enough in conceding an active role for a nonphysical principle. McDougall was critical of Smuts, Haldane, Needham, and others: They admitted that the organism displayed properties not found in machines, attributing these properties to the complexity of organic structures. But unless their holistic principle was an active agent, it explained nothing, because there was no evidence of teleology in the physical world.[71] Whitehead's claim that mentality was latent in physical events was "a very dubious and ambiguous way of meeting the difficulty."[72] The book's concluding chapters offered another, even more positive appeal to psychic research as conclusive evidence for the role of nonphysical agents. In the last years of his career McDougall had moved to Duke University, where J. B. Rhine was conducting his pioneering attempts to produce verifiable demonstrations of telepathy and clairvoyance.

67. Ibid., 101.
68. Ibid., 111.
69. Ibid., 151–55.
70. Ibid., 130.
71. McDougall, *Riddle of Life*, 184, 186.
72. Ibid., 187.

McDougall felt that in ten or twenty years the evidence would become so strong that anything but a dualist position would become untenable.[73] Such a development would support belief in survival after death, in the influence of higher beings on our lives, in the validity of mystical experiences, and in the existence of higher forces directing evolution.[74]

Psychologists studied animal behavior in the laboratory, and the majority of them were certainly moving away from McDougall's position. At the same time, ethology was emerging as a scientific discipline within biology, focusing on the study of animal behavior in the wild. Julian Huxley made important contributions to this field, explaining behavior as far as possible in terms of instincts created by natural selection. This approach to ethology offered a perspective diametrically opposed to the behaviorists' assumption that all habits are learned by conditioning, although both techniques implied a reductionism that avoided the need to speak of the animal mind.

Both of these extremes were rejected by a small number of biologists who wanted to defend the view that animals were active, purposeful agents whose behavior could not be explained by either form of reductionism. The most widely cited biologist of this school was Edward Stuart Russell, who began his career studying invertebrate morphology with J. Arthur Thomson and became director of fisheries investigations for England and Wales in 1921. Russell is well known to historians of science for his pioneering history of animal morphology, *Form and Function*, published in 1916 and still useful today. This work traces the conflict between two schools of thought, one of which held that a species' structure is determined by inbuilt laws of form or growth, the other that it is determined by the functions to which the various parts of the body are adapted. Russell believed that the formalist school was normally associated with vitalism because it postulated purely biological developmental forces. Functionalism was all too often hijacked by the materialists, the best example of this being the Darwinian emphasis on adaptation by selection. Russell distrusted materialism and Darwinism, favoring a Lamarckian view because it more easily accommodated the creative power of animal behavior as a factor in evolution. The preface to *Form and Function* makes clear the direction that Russell would take in his later ethology: the key to liberating functionalism from its materialistic straightjacket was to stress that animals were active, purposeful agents, not passive genetic puppets at the mercy of natural selection.

73.   Ibid., 215, 235.
74.   Ibid., 273.

In 1924 Russell published his *Study of Living Things* with a subtitle that proclaimed its role: *Prolegomena to a Functional Biology.* He contrasted the mechanistic view of behavior adopted by most biologists (and, by now, many psychologists) with the intuitive sense that most people had of animals as individuals with their own interests and purposes. As Kipling wrote in his *Jungle Book:* "We be of one blood, ye and I"—yet scientists insisted on rejecting any attempt to judge animal behavior in terms of our own experiences.[75] Aristotle had already had to resist the claims of the materialists, and the same debate was still active today. Russell noted the critical aspects of Driesch's arguments with respect, but conceded that he had not made the case for a distinct vital force. J. S. Haldane and others had shown that there was a third way in physiology that rejected both mechanism and nonmaterial forces.[76] But Haldane did not go far enough in recognizing the psychological dimension of an animal's life—indeed, he had argued that the study of consciousness was not part of biology.[77] Russell wanted to do exactly what a whole generation of psychologists had rejected—he wanted to get inside the head of an animal and try to understand how it viewed the world and interacted with it. If we could do this with other human beings, why not with animals?[78] Animals were individuals whose lives were governed by their striving after goals—this was what Russell proposed to call "hormic" behavior. Hormic activity was goal-directed in a creative rather than a mechanical way, which Russell illustrated in several chapters detailing observations of animals in the wild. In his conclusion, he asked why biology remained enmeshed in materialism when even physics had gone beyond it to a new, nonmechanistic approach.[79] Russell defended his functional method in his presidential address to the zoology section of the British Association in 1934 and in a book on animal behavior published in 1938.[80]

Russell's efforts can hardly be said to have had much impact on ethology, which developed by stressing the observation of behavior without interpretation in psychological terms, relating animals' activities as far as possible to inbuilt instincts that still operated even when changes in the environment made them useless or harmful. Animals were not creative thinkers—Mor-

---

75.   E. S. Russell, *Study of Living Things,* ix–xiv. See Roll-Hansen, "E. S. Russell and J. H. Woodger."

76.   *Study of Living Things,* xix and part 1, chap. 3.

77.   Ibid., 46–48.

78.   Ibid., part 1, chap. 4, esp. 32–33.

79.   Ibid., 124.

80.   Russell, "Study of Behaviour" and *Behaviour of Animals.*

gan's point from years before. In this respect, ethology in biology paralleled behaviorism in psychology, although the two movements stressed opposing views of how behavior was mechanically determined. Once again, those non-specialists who picked up the latest strand of nonmechanistic thought in science were backing the wrong horse, although the links between Russell's position and the older organismic philosophy gave the impression of a coherent and still active movement. Russell himself never commented on the religious implications of his stand against mechanism, although his training with Thomson would have alerted him to the broader debates into which his work would all too easily be sucked. His biographer records that in his later life he turned toward the nihilistic philosophy of Schopenhauer, so he would not have actively joined in the efforts to present antimechanistic science as a support for religious faith.[81]

J. Arthur Thomson's efforts to identify the *élan vital* with the psychological creativity of the organism have already been noted, and there were other scientists willing to lend some support to the claim that mind should not be seen as a mere by-product of the brain. J. S. Haldane was sympathetic to this approach, his idealist philosophy being particularly suited to viewing psychological categories as essential to describing the functioning of the organism. Other neovitalists made the same connection. Driesch's *Mind and Body*, translated in 1927, offered a highly philosophical defense of this position and appealed openly to psychic research for evidence that the mind exists independently of the body.[82] Frederic Wood Jones supported Russell's views on purposeful behavior in his defense of vitalism and Lamarckism.[83] More surprisingly, Sir Charles Sherrington, after a career in neurophysiology in which he had steadfastly refused to acknowledge the concept of mind, took a less negative position in his Gifford Lectures for 1937–1938.[84]

The young Joseph Needham, too, seems to have been willing to concede the reality of the mind. While insisting that mechanistic physiology was not a threat to religion, he accepted that it could not be extended to cover mental activities. He opted for an openly dualistic relationship, not between body and mind, but between matter and spirit.[85] In an appendix on biochemistry and mental phenomena written for Charles Raven's *Creator Spirit* of 1927,

81.   R. S. Wimpenny, "Russell, Edward Stuart" (this *DNB* entry is written from personal knowledge).

82.   Driesch, *Mind and Body*, viii and 148.

83.   F. W. Jones, *Trends of Life*, 24–25.

84.   Sherrington, *Man on His Nature*, 200–210 and 232.

85.   Needham, "Mechanistic Biology and the Religious Consciousness" 250, 254–55.

Needham accepted that mental states were epiphenomena from the scientific point of view, but argued that since our whole image of the universe was created by the mind, both the mental and the physical descriptions of it had to be partial. At the same time, he suggested that some of the paranormal effects to which Raven had appealed might be explained by minute quantities of chemicals such as hormones being transmitted through the air from one individual to another.[86]

Needham's position at this early point in his career was not very clear—perhaps inevitably, given that he was trying to reconcile the success of mechanistic biology with the reality of religious experience. To most physiologists and biochemists, as Needham admitted, any concession of an active power to psychological agents was unthinkable. The most active presentation of the materialist position came in Arthur Keith's 1927 address to the British Association and a series of follow-up lectures, all of which were widely reported and were the subject of much hostility in the press. In his Ludwig Mond Lecture in 1928, Keith repeated his claim that the mind had a material basis, noting the importance of this claim for the effort to understand the role played by instinct in human behavior. At the close of the lecture, he was questioned by a reporter about his statement that "medical men can find no grounds for believing that the brain is a dual organ—a compound of substance and of spirit." The press transformed what had been a plea for the application of Darwinism to human nature into an attack on religion. "It was said that I had denied the existence of the human soul, whereas all I had done was merely to outline the conception held by all competent medical men concerning the constitution of the living human brain." For Keith, the soul was just the functioning of the brain.[87]

The attacks on this position were led not so much by biologists as by Oliver Lodge and the defenders of spiritualism. However marginal the antimechanist position may have been among working biologists, it seems clear that the general public were by no means as willing to accept the complete elimination of the soul. The existence of this fertile soil in which the seeds of a popular reaction against materialism could be sown has to be taken into account in assessing the significance of the neovitalist movement. While never able to gain much of a hearing in science, to the general public—led by the press and a wide range of nonspecialist publications—it could be presented as a significant change in the direction of scientific thinking since Victorian times.

86. Needham, "Biochemistry and Mental Phenomena," 291, 295–96.
87. Keith, "Darwinism and What It Implies," 21–22.

PSYCHOLOGY AND RELIGION

If biology threatened to take the mind out of living organisms, so did one branch of psychology itself. In their efforts to turn psychology into an experimental science, behaviorists such as John B. Watson dismissed introspection, and hence the whole realm of consciousness, as a chimera and turned both animals and human beings into learning machines whose habits could be determined by environmental stimuli. At the opposite extreme, Sigmund Freud and his followers reduced the status of the conscious mind by postulating an unconscious driven by primitive biological urges. Freud saw religion as a subject for psychological investigation, threatening to explain away the whole structure of the belief system as a by-product of unconscious urges generating emotions over which the rational mind had no control. In this respect, analytical psychology took over the role played by anthropology at the turn of the century. Popular works such as J. G. Frazer's *Golden Bough* had presented religion as primitive humanity's effort to rationalize psychological states such as dreams.[88] For Freud, religion was subsumed in a whole arena of unconscious emotions, along with a wide range of other psychological crutches evoked by the all-too-fragile conscious mind. This dismissive attitude was quite correctly perceived as a major threat by religious thinkers, although some were prepared to concede that the concept of the unconscious mind opened a window onto the complexity of human nature.

Fortunately for the British churches, the immediacy of the threat was blunted by the fact that most of the radical new ideas came from abroad. The first effort to treat religion as a subject for empirical research was published by the American educational psychologist Edwin Diller Starbuck in 1899. As William James—who had been skeptical of the project's value at first—admitted in his preface to the book, Starbuck's questionnaires elicited useful information about psychological processes such as conversion, which were of vital interest to all religious thinkers. Conversion was, in fact, a normal process in certain types of mental development associated with adolescence.[89] This kind of study was cited by Cyril Burt in the chapter on the psychology of religion in his *How the Mind Works* of 1933, originally a series of radio talks. But Burt would not drive home the most radical interpretation, which was, as he fully admitted, that religion was just a survival of primitive superstition re-

---

88. It was the third edition (1911–1915) of *Golden Bough* that brought Frazer's views to the general reading public, although by this time his evolutionary approach was already being eclipsed within academic anthropology; see Ackerman, *J. G. Frazer.*

89. Starbuck, *Psychology of Religion.* See also the work of another American psychologist, James H. Leuba, *The Psychological Origin and the Nature of Religion.*

tained in the modern world because people were still subjected to psychological pressures. Whatever the psychological reasons why people accepted it, religion might still be an appropriate response to the mysteries of the universe.[90]

Burt noted that the analytical psychologists had provided some insights into the unconscious forces that might drive a person to religious belief, thereby following in the footsteps of anthropologists such as Frazer. Freud's *Interpretation of Dreams* had appeared in 1900 and his *Three Essays on the Theory of Sexuality* five years later. In 1909 he made a lecture tour to America. By the 1920s no educated person in Britain could have been unaware of Freud's writings and of his contention that the conscious mind sat atop a hierarchy of unconscious levels driven by deep biological, especially sexual, urges. Christian apologists were appalled at the details of Freud's analysis, but by no means unreceptive to the claim that the unconscious might be a reservoir for the darker side of human nature long attributed to the legacy of original sin. They could not agree that the mind was a puppet in the hands of biological urges, but could accept those urges as the legacy of an evolutionary past that it was our duty to overcome.

Behaviorism had less of an impact, although the threat it posed to religion was in some respects even more blatant. Watson had launched his behaviorist manifesto in 1913, openly complaining about the use of terms such as "mind" and "soul" in traditional psychological texts.[91] But although his followers became very active in America, few British psychologists adopted the full behaviorist program as the basis for their research. Even so, no one could have been unaware of the behaviorists' claims that human beings could be conditioned in the same way as rats, their habits being imposed at the will of the experimenter. Free will once again became an illusion. This implication was especially clear after the publication of Aldous Huxley's *Brave New World* in 1932. But long before that, religious writers had rushed to defend the citadel of belief against these attacks from abroad. Efforts to nullify the impact of the new psychologies were pouring off the presses by the 1920s, some, at least, with inflammatory titles like *Is Christian Experience an Illusion?* and *The Menace of the New Psychology*.[92] Of all the sciences, psychology was perhaps the least easy to fit into the spirit of reconciliation preached in some quarters during the interwar years.

90. Burt, *How the Mind Works*, chap. 16.

91. See, for instance, Boakes, *From Darwin to Behaviourism*, chaps. 6 and 8.

92. Balmforth, *Is Christian Experience an Illusion?*; Conn, *Menace of the New Psychology*. For more details of these responses, see chap. 9 below.

Whatever the problems generated by the new psychology, the areas of biology dealing with the nature of life and mind had provided much comfort for those seeking a reconciliation between science and religion. The combination of progressionist evolutionism with the various efforts to defend either an explicitly vitalist, or a more subtle organicist, approach to biology had kept alive the hopes generated at the end of the nineteenth century that the life sciences had abandoned their commitment to materialism. The older generation of scientists who still took that vitalist and anti-Darwinian stance seriously continued to publish, even though they took no part in the latest research initiatives. Of the younger generation, most were aware that the direction of research increasingly favored a mechanistic approach in physiology and a Darwinian one in evolutionary theory. There were a few attempts to salvage something of the older viewpoint, including Needham's more sophisticated organicism and Fisher's very idiosyncratic view of selection theory. But as Needham himself admitted, most of his fellow biologists ignored his religious views, just as they tended to marginalize the research tradition he tried to found. A few welcomed biological materialism as a contribution to the old rationalist cause or to its successor, the rising star of Marxism.

By the 1930s, it was possible to defend the new natural theology only by ignoring the latest trends in the life sciences. In contrast to the physical sciences, where the new physics maintained the challenge to old-style materialism, biology thus threatened to undermine the attempted reconciliation between science and religion. Even in physics, there was little support among scientists for the explicit appeal to religious emotions by figures such as Eddington and Jeans. But most religious thinkers were ignorant of science and took their information from surveys of the kind that were dominated by the older generation in biology and by these skillful popularizers of the new physics. Even so, it must have become apparent in the course of the 1930s that there were fewer active endorsements of natural theology by scientists. The older generation was becoming less active, and the younger was either less inclined to support religion or, in the case of the Marxists, openly hostile to it.

PART TWO

The Churches and Science

CHAPTER 6

The Churches in the
New Century

For the various Christian churches, the early twentieth century marked the onset of a massive decline in influence. Over three months in 1904 the *Daily Telegraph* published an exchange of correspondence subsequently collected into a book under the title *Do We Believe?*[1] The very title attests to an awareness that doubts were eroding the status of the churches in society. Attendance at church services began to drop steadily, and everyone was aware of a growing public indifference to organized religion. By 1946 a survey titled *How Heathen Is Britain?* recorded that the majority of young men entering the army during World War II seemed to have had no exposure to the Christian faith at school.[2] Indifference was a more potent threat to religion than the hostility of the rationalists and materialists, although some religious thinkers assumed that if they could repackage their beliefs in a form more acceptable to educated people, there was at least a hope of stemming the ebbing tide of support.

The changing role of the Victorian churches has been widely studied by historians, although the early twentieth century has attracted less attention, perhaps because of this perception that religion was becoming a less significant fo-

1. Courtney, ed., *Do We Believe?*
2. Sandhurst, *How Heathen Is Britain?*

cus for national life. All the evidence from surveys of church attendance suggests that most churches saw a peak in the 1890s followed by a steady decline.[3] The decline in the Free Churches became noticeable after about 1910, a little later than that in the Church of England. Only the Roman Catholic Church kept up its membership; legal discrimination against it had been abolished by 1914, and it grew both by immigration and by a small number of conversions. There is much debate over the significance of such figures, since for many people the churches provided the ceremonies that marked "rites of passage"—christenings, marriages, and funerals—and contacts were thus kept up sporadically without indicating regular attendance at church or chapel. Sunday schools for children were popular, even among parents who did not normally attend church. Most accounts agree that the Anglicans kept a better hold on the middle classes than on the poor, despite their noble efforts in the slums. This observation is significant for the present study because much of the relevant literature was aimed at a middle-class readership (now much expanded by changing social circumstances). The Free Churches were stronger in some sections of the working classes, but their involvement with science was more limited. The Church of England certainly went out of its way to develop a social involvement, but the educated public would have been more aware, for instance, of the formation of COPEC (Conference on Christian Politics, Economics, and Citizenship) in 1924 than of the Anglo-Catholic priests' work in the slums.

The involvement of the churches in the debate over the implications of science has to be understood in light of the threat of declining membership and the disagreements within the religious community over how best to present their case to an increasingly indifferent public. The Modernists, who were anxious to forge a new theology purged of ancient dogmas, thought that the only way forward was to make Christianity compatible with science and other aspects of modern thought—even if this meant abandoning what most traditionalists saw as the essential foundations of their religion. The Gospel had to be preachable to a modern audience, they believed, and it was better to rework its message in fundamental ways than to leave an outdated church to wither on the vine. Traditionalists, whether Catholic or evangelical, felt that there was no point in preserving a church that was no longer truly Christian.

---

3. Discussions of the state of Christianity in twentieth-century Britain include Hastings, *History of English Christianity*; Spinks, Allen, and Parkes, *Religion in Britain since 1900*; Worrall, *Making of the Modern Church*; and Lloyd, *Church of England*; for a detailed local study based on industrial Yorkshire, see S. J. D. Greene, *Religion in an Age of Decline*; more generally, see Latourette, *Christianity in a Revolutionary Age*, vol. 4; on the early part of the century, see McLeod, *Religion and Society in England*, and Machin, *Politics and the Churches in Great Britain*. On secularization, see Gilbert, *Making of Post-Christian Britain*.

If faith in science and progress had obscured the awareness of sin and the need for redemption, then it was the church's duty to keep that ancient flag flying and to rally what few converts it could to the cause. In the end, the failure of modern science and thought to solve humanity's problems would become apparent, and the need for redemption might again become obvious to all. Both of these approaches were expounded with enthusiasm, but neither was ultimately successful. In the 1920s the Modernists thought they had a real chance of forging a popular new Christianity, but the next two decades saw their influence fall as the problems foreseen by their opponents materialized in the form of economic depression and the rise of totalitarianism in Europe. The traditionalists were then able to seize the initiative within the churches—and to some extent within the community of intellectuals—but their hopes of reviving Christianity's fortunes among the population as a whole were never realized. Despite a brief flowering of faith in the 1940s and 1950s, the churches were never able to recover their former influence.

There was no automatic link between Christianity's fortunes in the wider world and the strength of its attraction for intellectuals and literary figures. Nor should we imagine that the churches were declining from a position in which they had influenced the whole population. Even in their heyday in the late nineteenth century, the churches had never impinged on the lives of the vast majority of the poorest members of society. Despite the heroic efforts of some Anglican clergy in the slums, most of those helped by Christian charity remained indifferent to the underlying religious message. The Free Churches had more influence with the better-off working classes, but their popularity also began to decline. The churches' involvement with social issues may have helped them retain influence at first, but once other ideologies such as the Labour movement became available to provide the same support, people lost interest in religion as they shifted the focus of their social life.[4] Many working men had always preferred the pub to the chapel, and now the expansion of leisure opportunities on the weekend, ranging from football to seaside day trips by train, made the lure of the religious service even less attractive. The Anglican Church had flourished best among the upper and middle classes, and its decline in the early twentieth century reflects a similar growth in the popularity of secular entertainment among the educated and literate. Even where church membership was retained for social reasons, actual attendance declined, except for children's Sunday schools. The supply of well-educated men coming forward for ordination had already begun to decline in the late nineteenth century as the church became less attractive as a

4. This point is explored in Greene, *Religion in an Age of Decline.*

career, a trend that would have had worrying consequences even if popular membership had been sustained.[5] It was to the middle classes that most of those who wrote on science and religion addressed their work, and the decline of enthusiasm for organized religion among them created a fertile ground within which the seeds of many new ideas and attitudes could be planted. The Modernists hoped to recapture those with some scientific education for the church, but there were plenty of less conventional visions of spirituality available to them (see chapter 11). Perhaps some of the enthusiasm for religion had been transmuted rather than eroded away.[6]

The steady decline in church attendance certainly cannot be transferred directly onto a chart plotting changing attitudes toward religion among the intelligentsia. Even within the educated classes, opinions varied as to the true state of affairs at any one time. B. G. Sandhurst blamed the ignorance of Christianity among the young men he surveyed on the hostility of the teaching profession, which in turn reflected the cynicism and disillusionment of the 1920s.[7] This is the popular image of the twenties—Adrian Hastings, for instance, writes of the "confident agnosticism" of this era.[8] No doubt there were many intellectuals who adopted such an attitude, and the most famous of them—Bertrand Russell and the like—stand out in a way that allows them to be used as a symbol of the era's cultural identity. But we cannot assume that all well-read people shared the same cynical view of religion. Charles Raven recorded that skepticism was far more rampant among Cambridge undergraduates before World War I than it was when he returned to the university as a don in the postwar years.[9] It is possible that by concentrating on the elite intellectuals, writers such as Hastings have overestimated the degree of skepticism among the educated public. The rise of the "new natural theology" outlined above was made possible by the emergence of a conviction in at least some quarters that the influence of Victorian agnosticism was on the wane. Modernists in the churches believed that if Christianity could throw off its outdated dogmas, it could exploit this renewed willingness to admit a spiritual dimension in life and regain its lost status. This was one purpose of the church congress that Raven organized in 1926, and in 1930 the encyclical letter of the Anglican bishops to the Lambeth Conference recorded the growing sense that

5.   See Marrin, *Last Crusade,* 17–27.

6.   This point is stressed by Gilbert in *Making of Post-Christian Britain* and by J. Wolfe in *God and Greater Britain.*

7.   Sandhurst, *How Heathen Is Britain?* 10.

8.   Hastings, *History of English Christianity,* 221.

9.   Raven, *Wanderer's Way,* 43–44.

a reconciliation with science was now possible.[10] There was disillusionment in the 1920s, but there were still many who retained an interest in spiritual matters. The question was whether the churches could tap this interest to their advantage or would continue losing influence to less structured forms of belief.

This variation among intellectual attitudes continued in later decades. In the late 1920s some prominent intellectuals returned to the Christian churches—T. S. Eliot, who was baptized in the Church of England in 1927, is an obvious example. Popular writers such as C. S. Lewis were able to generate a Christian renaissance in the following decades.[11] Although the total number of people professing a Christian faith was declining throughout this period, there was still room for considerable debate both within the churches themselves and among a significant section of the educated public.

The Church of England was still regarded as an important national institution, even by those who did not attend its services. Because of its unique status as the established church, major changes in its liturgy had to be approved by Parliament and thus became matters of general public debate. The church's proposal to reform the Prayer Book was debated in Parliament twice, in 1927 and 1928, and was defeated both times, thanks to a massive campaign that whipped up popular anti-Catholic feeling with a claim that the new format gave too much away to Rome. The debate highlighted the tensions within the church itself (some Anglo-Catholics also opposed the revisions because they did not go far enough in their direction), but suggests that there was still a strong reservoir of public feeling to be tapped on such issues. The role of the archbishop of Canterbury, Cosmo Lang, in the abdication of Edward VIII in 1936 also attracted much public attention, again not all favorable to the church. Even so, the church still had considerable influence, and the disagreement between Modernists and traditionalists over how best to stem its decline was of wider interest to the country at large.

THE CHALLENGE OF THE NEW

What role did intellectual developments, including those associated with science, play in creating the sense that Christianity was on the wane? There were both cultural and social developments to which the churches had to respond if they hoped to remain relevant. The intellectual threats stemmed

10.   For an account of the congress, see Raven, *Eternal Spirit;* for the encyclical letter, see *Lambeth Conference,* 1930, 19.

11.   Hastings uses the examples of Eliot and Lewis to concede that the tide had begun to turn even among high intellectuals; see *History of English Christianity,* 235–39. Raven obviously noticed the tide turning among his undergraduates some years earlier.

from a complex of rationalist and critical perspectives that forced Christians to reevaluate their beliefs and the foundations on which they rested. If geology and evolutionary biology required the traditional doctrine of creation to be reinterpreted, a more critical approach to the text of the Bible left many wondering how much they could trust the accounts on which many other aspects of the faith were based. The two movements were related because science taught a distrust of the miraculous and thus encouraged a more skeptical attitude toward any biblical episode in which supernatural intervention was implied. The critical attitude toward Scripture was often described as "scientific" since it rested on the application of reason rather than faith.

Science also dovetailed to some extent with the social challenges that faced the churches. The expansion of the slums in the great modern cities was a consequence of the Industrial Revolution, and as such was attributable to a political philosophy that permitted or even encouraged material inequality. When science was recognized as an important factor promoting industrial development, it too came to be seen as an aspect of the social problems with which the churches had to deal if they were not to be marginalized.

The alleged materialism of Victorian science was certainly perceived as a major factor that had undermined popular Christian faith—although this image of science was itself a perception of hindsight that did not reflect the range of positions taken up by Victorian scientists. In any case, the popular emphasis on the nonmaterialistic developments within early-twentieth-century science fueled the hope that this perceived threat had now diminished. Science was now presented (again, somewhat unrealistically) as a vehicle for promoting a more spiritual view of the world—although the traditional churches were not the only movements hoping to cash in on this trend. This transformation in the popular image of science can be seen at work in the 1904 *Do We Believe?* controversy. W. L. Courtney's introduction to the collection points explicitly to science, especially evolutionary biology, as an agent that has undermined faith—yet a survey of the letters themselves reveals comparatively few references to Darwin, Huxley, and other scientists identified with the destruction of the traditional idea of creation.[12] The general materialism of the modern way of life, including commercial activity, was seen as far more dangerous to traditional beliefs than any specific doctrine of science. As the Congregationalist turned Anglican R. J. Campbell put it: "Experience compels me to affirm that it is not the difficulty of squaring Christianity with modern science that is in question, but rather the diffi-

---

12.   Courtney, ed., *Do We Believe?* 3–4. The index lists seven references to Darwin and eleven to Huxley, and relatively few letters refer directly to evolution.

culty of squaring its ethical precepts with the requirements of industrial and commercial practice."[13]

Even so, there were many (including Campbell in his early career) who saw coming to terms with science as an absolute prerequisite for Christianity to flourish in the new century. Physics might be losing its aura of materialism, but the destruction of the traditional idea of creation had to be taken into account. This was not just a question of accepting Genesis as an allegorical rather than a literal account of how creation unfolded, because if evolution were to be fully incorporated into Christian belief, the old idea of humanity's fall from a state of grace would have to be reinterpreted. There was a good deal of suspicion that clergymen were evading these questions, paying lip service to evolutionism without really thinking through what it would mean to abandon the story of Genesis in even an allegorical sense. Clergymen presented an air of confidence in their ability to accept the new ideas, yet many laypersons felt that they were refusing to talk about the issues that really threatened belief.[14] Modernist Anglicans justified their campaign on the grounds that young people with some knowledge of science were being alienated from the church because they realized that many clergy had not taken the new knowledge into account except at the most superficial level. Radical rethinking was necessary unless this whole generation was to be lost to Christianity altogether.

But even in Victorian times, science had been only one among several potentially disturbing intellectual developments for the churches. The controversy over Darwinism had coincided with the somewhat belated emergence onto the British scene of a far more divisive force, the Higher Criticism, which treated the Bible as a historical document to be evaluated like any other. The resulting reassessments of the texts' significance were all too easily taken up by hostile thinkers as a reason for challenging the Bible's authority as a source of supernatural revelation. Conservative Christians thus treated the new scholarship as a threat, and it was only at great risk to their careers that more liberal thinkers first began to argue that it could not be ignored. Bishop J. W. Colenso's interpretation of the Pentateuch and the contributions on similar themes to the 1860 volume *Essays and Reviews* were of far more concern to most Anglicans than *The Origin of Species.*[15] Biblical scholarship remained

13. Campbell, *Spiritual Pilgrimage*, 162.

14. This point is made in the case of Nonconformist clergy by John W. Grant in *Free Churchmanship in England*, 132.

15. On *Essays and Reviews*, see Altholz, *Anatomy of a Controversy*, and on later developments, see Hinchcliff, *God and History*, esp. chap. 5.

important to British clergymen, although it became increasingly remote from general culture, and most ordinary people remembered only the negative impact of the first wave of criticism upon traditional beliefs. A more liberal approach to the interpretation of the Bible began to emerge even in the Anglo-Catholic camp in 1889 with Charles Gore's edited volume *Lux Mundi*, although there were many still horrified at the new opinions. By the early twentieth century it was widely accepted even in evangelical circles that the sacred text often had to be taken as allegory or metaphor. Modernists used this freedom to argue that the old stories of miracles, now rendered implausible by science's devotion to the principle of the uniformity of nature, could be reinterpreted. But it was one thing to argue that Christ's miracles of healing could be reinterpreted in terms of psychological factors, and quite another to suggest that the virgin birth and the Resurrection had to be discounted or reinterpreted in nonliteral terms. The hostility of many traditional Christians to Modernism was based on the fear that by eliminating miracles altogether, the whole point of Christ's appearance on earth would be obscured. Science was certainly part of the rationalizing trend, but for most Modernist clergy it played a subsidiary role in comparison to the new interpretations of the Bible.

These intellectual challenges coincided with a growing feeling in some quarters that the churches were failing to respond to the social changes that were transforming the modern world, not always for the better. The Anglican Church in particular was seen as wedded to the old social hierarchy based on land ownership and hence unable to function adequately within an increasingly urbanized society. The new but increasingly powerful Labour Party was to a large extent hostile to the established church, although it was more sympathetic to Nonconformism. Some Anglicans responded to this challenge by promoting Christian socialism. Charles Gore was active in the moderate Christian Social Union, but a more radical Church Socialist League was founded in 1906 and included Conrad Noel, the "Red Vicar" of Thaxted who influenced Joseph Needham.[16] The Christian socialists believed that the church should endeavor to build the Kingdom of God in this world and should be willing to challenge its traditional social allies to do this. There was an obvious but superficial link between this program and theological Modernism, since both wanted to throw off the legacy of the past. But many Modernists refused to accept the need for social reform, in this respect sharing the more orthodox view that Christianity was more concerned with spir-

16. For a useful summary of the influence of Christian Socialism, see Marrin, *Last Crusade*, 37–41.

itual than with material welfare. Some Modernists, including Dean W. R. Inge, were actively hostile to socialism and supported what many regarded as the right-wing social philosophy of the eugenics movement. It was the Anglo-Catholic clergy, partly under the influence of Gore, that did most to alleviate suffering in the slum parishes.

Christian socialism was never more than a minority movement in the Church of England, leaving the bulk of the clergy to be associated in the public mind with reactionary forces. The overt hostility to religion expressed by the Soviet government that gained power in Russia made it difficult for a Christian minister to defend a socialist perspective (although a few certainly did). COPEC met under a blaze of enthusiasm, with William Temple presiding, in 1924, but its pronouncements on social affairs were vague, and it did little of note other than initiate a slow change of attitude among the clergy. Faced with the General Strike of 1926, the archbishop of Canterbury, Randall Davidson, issued a conciliatory statement that outraged many clergymen, whose natural inclination was to favor the establishment. Hensley Henson, whose appointment to the see of Durham had angered clergy opposed to his Modernist theology, wrote bitterly against the striking miners, although Durham was the center of a coalfield. Temple tried to act as a mediator but had no real understanding of the miners' problems, and COPEC (whose Continuation Committee had Charles Raven as its secretary) was similarly ineffective.[17] The strike thus did little to offset the image of the church as an ally of the wealthy. Many clergy were opposed not only to socialism but also to the whole network of technical and industrial development whose one-sided benefits were the source of the social unrest. It was only one year after the strike that the bishop of Ripon, Arthur Burroughs, preached his notorious sermon to the British Association calling for a moratorium on all scientific and technical developments for ten years—the "scientific holiday" (see chapter 9). Nor was there much evidence of social involvement by the Anglican Church during the depression of the 1930s, although a few radicals were now becoming more vocal, Needham and Raven both being involved with the publication of *Christianity and the Social Revolution* in 1935.[18]

What, then, of external threats to the social and cultural order? The churches were swept along with most other institutions into the maelstrom of World War I. Many clergymen joined in the popular enthusiasm for the war based on the perceived threat from German military might and the ruth-

17.   On the church's role in the General Strike, see Hastings, *History of English Christianity,* chap. 10.

18.   Lewis, Polanyi, and Kitchin, eds., *Christianity and the Social Revolution.*

lessness of German culture. A. F. Winnington-Ingram, the bishop of London, was active in promoting recruitment, and many church leaders denounced German atrocities as indications that civilization itself was under threat.[19] A few, including Lang and Temple, tried to stem the tide of anti-German hysteria, and later on Davidson argued against reprisals for German air raids, but all too often these moderates were vilified for their efforts. H. D. A. Major, a prominent Anglican Modernist, argued *for* reprisals, and many Modernists (including Charles Raven) tried to rationalize the war as part of the evolutionary process by which humankind was forced to raise itself in response to periodic challenges. To the soldiers in the trenches the church seemed remote, yet it was here that individual chaplains did much to rebuild some popular faith in undogmatic Christianity. Raven served with distinction in the war, and his willingness, along with that of hundreds of other chaplains, to share in and help relieve the men's suffering did much to gain the respect of ordinary people.[20] The relief work of Philip ("Tubby") Clayton in founding Toc H (Talbot House) at Poperinghe in the Ypres salient in 1915 also revealed a practical concern for the men's welfare. Toc H continued after the war to become an important force promoting Christian values among ex-servicemen and probably did far more than COPEC to show that the church had a human face.

The widespread belief that the war undermined traditional values and thus represented a disaster for the churches has to be set against these (admittedly limited) social transformations. Not everyone abandoned faith in progress, and some had their respect for human values reinforced. We should not be surprised by the temporary resurgence of interest in religious concerns after the war, nor by the enthusiasm of Raven and other Modernists, who argued that a Christianity that could rework its ancient creeds in the light of modern ideas could yet again have a wider appeal. It was a later combination of external factors that most clearly highlighted the failure of the Modernists and precipitated a new wave of enthusiasm for orthodox Christianity in some circles. Communism had not been taken seriously in the 1920s despite its success in Russia, but in the following decade its clash with the Fascist regimes emerging in Italy, Germany, and Spain defined an ideological conflict that religious thinkers could no longer ignore. The Left now became active and, ignoring the evidence for Stalin's atrocities, began to promote the atheist society now being forged in the Soviet Union as a model for all to follow. By

19. See Marrin, *Last Crusade*, and Wilkinson, *The Church of England and the First World War*.

20. Raven's experiences are recounted in "War and the Evolutionary Process," in his *Musings and Memories*, 165–74—although he, too, saw the war as a stimulus to moral progress.

contrast, the Roman Catholic Church openly sided with Mussolini and Franco and was very slow to recognize that Hitler represented a threat just as great as Stalin. A few Christians converted to Communism, but for the most part those who opted for the Left abandoned any religion they might once have had. Meanwhile, all too many Anglicans either joined the Roman Catholics in their enthusiasm for the dictators of the Right or were at least prepared to give them the benefit of the doubt and thus support the policy of appeasement. Curiously, it was Henson who spoke out most clearly against this capitulation to the new powers in Europe—he may have thought the miners were the bullies in the General Strike, but he recognized that Hitler and Mussolini were even bigger bullies who really did represent a threat to Western society. Some Christians became pacifists in the early 1930s, although that movement became much less popular in the years running up to 1939.[21]

What the 1930s did do for the Christian churches was to identify the failure of the liberals and Modernists to cope with the depth of irrationalism and hatred that was being revealed by the social developments of the time. The Great War had not totally shattered the expectations raised in the nineteenth century, leaving some more optimistic thinkers to hope for a restoration of rationality—and still convinced that such an analysis would reveal a universe moving toward a morally significant goal. But such hopes were now abandoned, leaving Modernists like Raven and E. W. Barnes marginalized in a church that was increasingly confident that if anything was to be salvaged, it was by reclaiming its ancient heritage, not by revising it beyond recognition. Some intellectuals returned to the churches, and it was often to the Catholic form of belief (in either its Anglican or Roman form) that they were attracted. The supernatural, once under attack as a barrier that prevented anyone taking Christianity seriously, was now back in favor, because it was only by looking outside the material universe that there seemed to be any hope of retaining meaning in life at all. Science and rational analysis became objects of distrust once again.

THE CHURCHES' RESPONSE

To some extent, the problems faced by the churches were a product of their own disunity—the fact that Christians could not agree among themselves over what their religion was supposed to mean was a potent argument

21.   On the developments outlined in this paragraph, see Hastings, *History of English Christianity*, chaps. 20 and 21.

for the rationalists.[22] The various denominations disagreed with each other, with the Church of England, as ever, trying to straddle the divide between evangelicals and Catholics. The Free Churches tended toward evangelicalism, although widely interpreted in a liberal fashion. They abhorred Catholic ritualism and thus found the Catholic wing of the Anglican Church almost as detestable as Rome. The Church of England was perpetually at risk of being torn in two by its evangelical and Catholic wings, and it took all the diplomatic skills of the archbishop of Canterbury, Randall Davidson, to hold it together in the controversies over issues such as the Prayer Book revision. Modernism flourished for a time in the Anglican Church, but it cannot be mapped directly onto the evangelical-Catholic division. Some Modernists were quite High Church in their style of worship, but could not stomach the veneration of the consecrated host that was the identifying mark of the Anglo-Catholic movement. Barnes called himself a liberal evangelical rather than a Modernist, and his enthusiasm for science was linked to a distrust of Catholic doctrine that endeared him to the Free Churches.

One answer to the divisions within the churches was the ecumenical movement, although it was only just beginning to gain influence in the early twentieth century and was as yet powerless to heal the bitter divisions between the churches. Not surprisingly, the most active calls for unity came from young people, often stemming from the Student Christian Movement. This movement had begun as an evangelical group in the 1890s, but widened its scope in the new century and played a major role in the Edinburgh Missionary Conference of 1910, which is often seen as the starting point of the ecumenical movement. The highly fragmented Free Churches now began to consolidate themselves, with the various Methodist denominations achieving effective unity in the late 1920s (formalized in 1932), despite much resistance from the Wesleyans. The reunification of the Free Church and the Church of Scotland in 1929 healed the most bitter rift among the Presbyterians. Some Free Church leaders hoped for a closer link with the Anglicans, but saw the influence of Anglo-Catholic clergy as a barrier to this. The Church of England had in fact issued an *Appeal to All Christian People* for greater cooperation at its Lambeth Conference of 1920, and the evangelical wing of the church hoped for unity with the Free Churches if only they would accept episcopacy (government of the church by bishops). But at exactly the same time, the Catholic wing of the church was engaged in the Malines conversations with leading Roman Catholics, hoping to overturn Rome's resolute refusal to rec-

22.   The title of Keith W. Clements's survey, *Lovers of Discord: Twentieth Century Theological Controversies in England*, is significant.

ognize the legitimacy of Anglican Holy Orders. Ultimately, the Roman Catholic Church moved to an even stronger position of isolationism, thwarting the hopes of the Anglo-Catholics—whose efforts only helped to whip up anti-Catholic feeling in the debates over Prayer Book revision. Serious moves to encourage cooperation between the churches did not begin until the Oxford Conference on Life and Work of 1937, which played a role in the founding of the World Council of Churches in the following year. William Temple's active role in these developments reflected his growing influence in the church (he became archbishop of Canterbury in 1942).[23]

The drive for unity thus came too late to affect the debates of the interwar years. The early twentieth century saw the Modernist trend introduced into a host of bickering churches, meeting resistance from both evangelicals and Catholics, who opposed it for very different reasons arising from their mutually incompatible ideas about how to defend the essence of traditional Christian teaching. The Roman Catholic Church offered the strongest claim for the church as a divinely ordained institution stretching back with unbroken continuity to Christ Himself.[24] Its rigid hierarchy and elaborate dogma based on the presumed interweaving of the supernatural with everyday life were seen by many British Protestants as an example of almost medieval superstition. The Catholic Church was thus never a strong presence in Britain, and indeed, was only just becoming free of the legal restrictions once placed on it. There had been a Modernist movement in the church in the late nineteenth century, led by the French priest A. F. Loisy, and influential in Britain through the work of George Tyrrell.[25] But the movement was rigidly suppressed by Pope Pius X in his encyclical *Pascendi gregis* of 1907. Clergy were forced to take an oath renouncing such tendencies, and Tyrrell died excommunicate in 1909. The message was clear: The church would maintain its ancient discipline and dogma, and would not (despite the Malines conversations) concede anything to the claims of rival forms of Christianity, which remained, formally, heresies. The attractions of such a church for the majority of the English population were minimal, but the hard-line attitude of Rome began to seem more plausible to some intellectuals concerned by the cultural fragmentation of the 1930s. If secularism had failed, and along with it all those forms of Christianity that had tried to compromise with it, perhaps it was only by returning to the oldest faith of all that Western civiliza-

---

23.   See Iremonger, *William Temple*, chaps. 24 and 26, and more generally, Hastings, *History of English Christianity*, chaps. 5 and 18.

24.   On Roman Catholicism, see, for instance, Hastings, *History of English Christianity*, chaps. 7 and 16.

25.   See Hinchcliff, *God and History*, chap. 7; Vidler, *20th Century Defenders of the Faith*, chap. 2.

tion could be saved. When promoted by able popularizers such as Hilaire Belloc and G. K. Chesterton, such a message had an impact far wider than the church's actual membership.

The Catholic movement within the Anglican Church drew its origins from the Tractarians of the early nineteenth century, who had begun to stress the church's historic roots and had thereby been led to adopt a form of worship using much of the ritual common in Roman Catholic churches. There was always the temptation to repudiate the effects of the Reformation altogether and return to Rome, as John Henry Newman had done, but for the most part the Anglo-Catholics preferred to remain Anglicans, although they hoped for some form of reconciliation.[26] They believed themselves to be part of a Catholic—that is, universal—Church founded by Christ and endowed with a historic mission to save humanity. Where evangelicals endorsed the Protestant emphasis on the Bible as the defining focus of Christianity, Catholics saw the church as an institution based on a real spiritual link back to Christ, the earthly representative of His saving grace. The Tractarians had been suspicious of science and of rationalism generally, and were horrified by the effects of the Higher Criticism because it seemed to threaten the legitimacy of the texts upon which the church was founded. But the next generation of Anglo-Catholics found it easier than the evangelicals to face up to the prospect of reassessing Scripture, and were also able to take on board the idea that God might have created the world through an evolutionary process rather than by miracle. Gore's Lux Mundi of 1889 did much to define this more forward-looking brand of Catholicism. Gore was still a powerful influence in the church in the early twentieth century (he was bishop of Birmingham and then of Oxford).

The Anglo-Catholics may have accepted evolution, but there were limits to how far they could go in accepting the scientific worldview because their theology rested on the assumption that spiritual forces really do interpenetrate this world in ways that affect our lives. They used elaborate rituals derived from Rome in their worship—although one could be "High Church" in this way without accepting the Anglo-Catholic interpretation of the sacraments' theological significance. For the Anglo-Catholics, these ceremonies highlighted the involvement of spiritual forces in the act of worship. This aspect of their theology became the center of bitter public controversy in the case of the "reservation" of the Eucharist or Holy Communion. For most

26. See, for instance, Hinchcliff, God and History, chap. 5; Mozley, Some Tendencies in British Theology; Ramsey, From Gore to Temple; and Reardon, From Coleridge to Gore.

Protestant thinkers, of whom Barnes was the most outspoken, the consecration of the bread and wine in Holy Communion had a purely symbolic role
in helping the worshipper come closer to Christ. But for the Anglo-Catholics,
the ceremony endowed the physical objects with a genuine spiritual power—
they literally contained the spirit of Christ and were worthy of reverence after the ceremony was finished. The church allowed the reservation of the
sacraments; that is, they could be kept by the priest for use by communicants
who were sick and unable to attend the service. Anglo-Catholics kept the consecrated host in a special chapel and gave it elaborate reverence—critics said
they worshipped it openly in a manner reminiscent of pagan ritual. Along
with the revised Prayer Book, the practice of reservation became a focus for
the tensions between Catholics and evangelicals that were threatening to tear
the church apart in the 1920s.

The Anglo-Catholic party was strong among the clergy, but not among
the lay Anglican communicants, many of whom preserved a long-standing
Protestant distrust of Romish ceremonies. Julian Huxley thought that many
ordinary people found it impossible to worship in churches where Catholic
ritual was used.[27] In one of her famous detective stories, Dorothy L. Sayers
has Lord Peter Wimsey solve a murder mystery in which the body is stolen
from a church under the cover of a Protestant "raid" to remove a consecrated
host being used for Anglo-Catholic ritual, and there were organized groups
that desecrated Anglo-Catholic churches or substituted unconsecrated wafers
for consecrated ones.[28] Paradoxically, it was precisely the Catholic form of
worship (Anglican or Roman) that began to appeal to those intellectuals of
the 1930s who despaired of rationalism.

The evangelical wing of the Anglican Church had strong grassroots support, but little leadership among the clergy. Some of the clergymen who embraced this more Protestant wing of the faith found it difficult to accept the
new, more critical outlook of the age. Although founded on an individualis-
tic reaction to the overinstitutionalized religion of Rome, the Protestant Reformation had thrown all the emphasis onto the individual's reading of the
Bible, and the text all too easily became the basis for a new dogmatism. If the
Word of God was the essence of Christianity, who could dare apply human
standards to the interpretation of the sacred text? As the liberal Congrega-

27.   Julian Huxley to E. W. Barnes, 17 October 1927, E. W. Barnes papers, Birmingham University Library, EWB 12/5/104.

28.   Sayers, "The Undignified Melodrama of the Bone of Contention," in *Lord Peter Views the
Body*, 87–145; see especially 103 and 113.

tionalist John Warschauer put it: "Having successfully got rid of the tyranny of an infallible church, Protestantism set up the more unprogressive and more unyielding tyranny of an infallible Book."[29] Evangelicals in both the Anglican and the Free Churches faced this dilemma when confronting the new science and the new biblical scholarship, and as in America, it was from the evangelicals that the antievolution movement was drawn. Yet there were many liberal evangelicals who now accepted some form of evolutionism while remaining hostile to Anglo-Catholic ritualism. Although on the defensive during the early decades of the century, evangelicalism in both the Anglican and the Free Churches revived in the 1930s as the optimism that sustained more liberal interpretations of Christianity faded.[30]

The Modernist movement within the Anglican Church had its origins in the Broad Church of the nineteenth century, that group of liberal-minded clergy who rejected the dogmatism of both the evangelical and Catholic wings and were willing to move as far as possible with the times.[31] Although not as numerous as the Anglo-Catholics, the Modernist clergy were well organized under the direction of H. D. A. Major. The Churchmen's Union for the Advancement of Liberal Religious Thought (later renamed the Modern Churchmen's Union) was founded in 1898, and the journal *Modern Churchman* first appeared in 1911. "Modernism" in the theological context is quite different from the "modernism" of the intellectual avant-garde, of course—indeed, the movements are polar opposites because theological Modernism was conceived to take on board exactly those aspects of late-nineteenth-century thought that more radical intellectuals were determined to undermine. Anglican links with Roman Catholic Modernism were tenuous, and Dean W. R. Inge wrote openly against the influence of Loisy and Tyrrell.[32] Ritual in itself was not a focus of Modernist attention. There was, in fact, nothing to stop a Modernist appreciating High Church ritual as a means of focusing the worshipper's attention on Christ as the way to salvation. But the Catholic interpretation of the sacraments was difficult to square with Modernism's desire to reconcile Christianity with the scientific worldview.

Science was not the only factor promoting interest in a remodeling of the traditional view of Christian theology; indeed, several leading Modernists

29. Warschauer, *New Evangel*, 15. The tensions between conservative and liberal Evangelicals are explored in Bebbington, *Evangelicalism in Modern Britain*.

30. See Bebbington, *Evangelicalism in Modern Britain*, chap. 6; on Evangelicalism in the Anglican Church, see Manwaring, *From Controversy to Co-existence*.

31. See Clements, *Lovers of Discord*, chaps. 3–5, and A. M. G. Stephenson, *Rise and Decline of English Modernism*. For a survey of Modernist views, see Major, *English Modernism*.

32. Inge, "Roman Catholic Modernism," in his *Outspoken Essays*, 137–71.

had little or no interest in science.[33] For those who tried to express the new approach in terms of traditional theology, Modernism worked by stressing the belief that God was immanent within nature, rather than a purely transcendental Being. The assumption that the spiritual domain could impose its will on the material universe took for granted the existence of a transcendent God who is in some respects separate from His creation. But there was a rival viewpoint that, while accepting that God exists on another plane, stressed that He had to some extent put something of Himself into His creation. If it is God's will that sustains the material world, then the forces of nature are to some extent the expressions of that will and can have a genuinely creative role. The claim that nature is merely a divinely constructed piece of clockwork that functions independently of His will is characteristic of deism, not of Christianity. At the same time, to stress that God is immanent within nature runs the risk of pantheism, of claiming that He is coextensive with the creative powers of this world and cannot exist apart from it. Opponents of Modernism frequently charged that it ended up as little more than pantheism, with Christ being reduced to merely a supremely good man who pointed the way for the rest of us to follow.

The temptation to stress the immanent nature of God was prompted not just by science, but also by the idealist philosophy that had dominated British academic life at the end of the nineteenth century.[34] If the world itself is a manifestation of spirit, its adherents argued, then we ourselves are part of the spiritual domain, and there is no reason to separate human life from the rest of the universe—nor to see Christ as someone wholly different from ourselves. Idealism thus had an effect very similar to that of scientific evolutionism, and indeed, many non-Darwinian evolutionists stressed the purposeful and moral nature of the forces bringing about the development of life. Mysticism, too, could be incorporated into the Modernist worldview—Inge's early reputation as a scholar was based on his sympathetic study of the early Christian mystics. Provided mysticism was seen as another form of experience in addition to the normal bodily senses, it provided yet another reason for arguing that human beings can access the spiritual dimension of reality directly and are thus not solely dependent on revelation or salvation provided by supernatural means.

The Modernist position generated controversy in two main areas: the ve-

33. Major's *English Modernism* makes only fleeting references to the new physics and to nonmechanistic biology; see 58–62 and 71–74. Hastings Rashdall and Hensley Henson were influenced almost exclusively by idealist philosophy rather than science: see Rashdall, "Ultimate Basis of Theism" and "Personality: Human and Divine"; Matheson, *Life of Hastings Rashdall;* Chadwick, *Hensley Henson;* and Henson's autobiography, *Retrospect of an Unimportant Life.*

34. See Hinchcliff, *God and History,* chap. 6; Sell, *Philosophical Idealism and Christian Belief.*

racity of the miracle stories in the Bible and the interpretation of original sin. Barnes was by no means the first to propound an interpretation of Christianity totally purged of the miraculous, although his last book did this quite explicitly and got him into a great deal of trouble with the church.[35] Already by 1903 the Rev. C. E. Beeby had attacked the Anglo-Catholics for insisting that the virgin birth was necessary to explain how Christ had escaped the taint of original sin.[36] In 1911 an Oxford don, the Rev. J. M. Thompson, published an account of the New Testament dismissing all the stories of miracles as later interpolations that should not be taken seriously. Thompson insisted that he still revered Christ as divine, but his work elicited agonized criticism from Christians who felt that the essence of their religion was its acceptance of the willingness of a supernatural God to intervene in human affairs. Thompson's license to preach was withdrawn.[37] Only a year later another Oxford-based volume, *Foundations*, raised a similar storm because the editor, B. F. Streeter—after surveying the latest biblical scholarship—conceded that the Resurrection might best be understood in terms of Christ still being able to communicate with His disciples at a spiritual level despite the death of His physical body.[38] In response, Frank Weston, the bishop of Zanzibar, began a campaign within the church that Randall Davidson had to work hard to prevent turning into an anti-Modernist witch hunt (fortunately, Gore, the leading Anglo-Catholic, was persuaded to soften his position in this instance). In 1917 an equally bitter controversy erupted over the appointment of another leading Modernist, Hensley Henson, to the see of Hereford.[39]

In the interwar years, the annual conferences of the Modern Churchmen's Union served as a highly publicized forum for the expression of Modernist opinions. The 1921 conference at Girton College, Cambridge, became notorious through press reports of the views expressed by Hastings Rashdall and J. F. Bethune-Baker, both of whom saw Jesus as a far more human figure than traditional dogma allowed. They were accused of denying the divinity of Christ, a charge against which Rashdall especially defended himself with vigor, protesting that the newspapers had both misquoted him and taken his statements out of context.[40] Yet the controversy was so fierce that Henson re-

---

35.   E. W. Barnes, *Rise of Christianity*; see J. Barnes, *Ahead of His Age*, chap. 10.

36.   Beeby, "Doctrinal Significance of a Miraculous Birth."

37.   J. M. Thompson, *Miracles in the New Testament*; for a typical protest, see Figgis, *Civilisation at the Cross Roads*, especially the appendix. See also Clements, *Lovers of Discord*, 50–51.

38.   See Clements, *Lovers of Discord*, chap. 3.

39.   Ibid., 76–85.

40.   Ibid., 93–98; Stephenson, *Rise and Decline of English Modernism*, chap. 5.

treated from the more extreme opinions he had expressed earlier and effectively abandoned the Modernist movement. As a result, his biographer argues, the Modernists came more strongly under the influence of Barnes, with disastrous results for the church.[41] This is perhaps unfair to Barnes, who was never a member of the Modern Churchmen's Union, although he attended the annual conferences and was widely regarded as the most prominent spokesman for the Modernist position.

If Barnes's rejection of the miraculous got him into hot water, he was already notorious for his "gorilla sermons," in which he urged the church to take the implications of evolutionism into account (see chapter 8). But here he was merely publicizing an interpretation already developed by other Modernist thinkers. Since the late nineteenth century, liberal Anglicans had accepted the general idea of evolution on the assumption that the progressive development of life could be interpreted as the unfolding of a divine plan. But while this position was compatible with a general theism, it was not widely appreciated that to accept the human race as improved animals was to undermine the foundations upon which the traditional notion of the Fall and the need for redemption were based. Putting it bluntly, even if evolution was conceived as the unfolding of God's plan, the element of progress made nonsense out of the idea of original sin (since there could be no Fall from an earlier state of grace), and if there was no original sin, one would have to ask what the point of the Atonement would be within the new theology. It would be easy enough to see Christ as a messenger from God pointing the way to future spiritual development, but what was the point of His death on the Cross if there was no need for redemption?

The necessity of rethinking the traditional view of the Fall was outlined by the Cambridge theologian Frederick Robert Tennant, whose early training had been in the natural sciences. In his Hulsean Lectures of 1901–1902, published as *The Origin and Propagation of Sin*, Tennant argued that, according to the new worldview based on a natural origin for humankind, the notion of an original sin for which we ourselves are somehow to be held accountable by our Creator was no longer tenable. He noted that the traditional idea of an inherited taint transmitted from generation to generation would—if the taint were acquired by our distant ancestors—involve the now increasingly suspect Lamarckian theory.[42] But his real aim was to challenge the notion that

---

41. Chadwick, *Hensley Henson*, 297–99. Henson expressed distaste for Barnes's *Rise of Christianity*, although he objected mainly to the book's inexpert handling of biblical criticism; see his letters to the Dean of Winchester and E. H. Blakeney in Braley, ed., *Letters of Herbert Hensley Henson*, 203–7.

42. Tennant, *Origin and Propagation of Sin*, 35–37.

sin could have suddenly entered the human condition at a certain point in time. Sin was real, a natural part of our constitution, but it was a normal result of the way in which our race had evolved from its animal ancestry. The Fall was not a turning away from our originally created nature, but a process by which we first became aware of the need to overcome the animal instincts buried in our character.

> But for us there has emerged an alternative view of man's original condition. What if he were flesh before spirit; lawless, impulse-governed organism, fulfilling as such the nature necessarily his and therefore the life God willed for him in his earliest age, until his moral consciousness was awakened to start him, heavily weighted with the inherited load, not, indeed, of abnormal and corrupted nature, but of a non-moral and necessarily animal instinct and self-assertive tendency, on that race-long struggle of flesh with spirit and spirit with flesh, which for us, alas! becomes but another name for the life of sin. On such a view, man's moral evil would be the consequence of no deflection from his endowment, natural or miraculous, at the start; it would bespeak rather in present non-attainment of his final goal.[43]

The same point was made equally forcefully in Tennant's *The Concept of Sin* of 1912.[44] Sin was real, but on Tennant's model, it was something we could and should be able to overcome in this world if we follow the teaching of Christ.

This new view of human sinfulness was hard to resist for those who were committed to the notion of development through God's immanent power in nature. It was endorsed by the anthropologist W. L. H. Duckworth and the chaplain of Trinity College in the volume widely known as the *Cambridge Theological Essays* of 1905.[45] Other Modernist theologians writing on the concept of sin also supported Tennant, including Peter Green and N. P. Williams.[46] Tennant's position was widely and quite reasonably cited as an attempt to reconcile Christian teaching with the scientific theory of evolution. Yet Tennant himself did not have a very positive view of science, his position being based much more solidly on his acceptance of an idealist philosophy. This is evident from his own contribution to

43.   Ibid., 11. Tennant noted that this idea had been sketched in briefly by Archdeacon Wilson at a church congress in 1896, see 82.

44.   Tennant, *Concept of Sin*, e.g., 155.

45.   Duckworth, "Man's Origin, and His Place in Nature"; Askwith, "Sin, and the Need of Atonement." *Cambridge Theological Essays* is the half-title of the book whose full title is *Essays on Some Theological Questions of the Day by Members of the University of Cambridge.*

46.   Green, *Problem of Evil;* N. P. Williams, *Ideas of the Fall and of Original Sin.*

the *Cambridge Theological Essays*, which undertook a sustained critique of the natura-listic metaphysics that the previous era had uncritically derived from science.[47] Our knowledge of nature is really derived from our sensations, Tennant argued, and he noted science's own tendency at the time to move toward a more idealist position. He would repeat this critique in articles published in the early 1920s.[48] In the end, though, Tennant succumbed to the hope that a new natural theology would be possible if science could become more firmly based on antimechanistic and teleo-logical principles. He noted, if only briefly, the vitalist movement in physiology and the efforts being made to establish a teleological view of evolution by sup-posing that variations are not random, as Darwin had held, but directed along pre-determined paths.[49] Here we see an indication of the way in which Modernism would influence the debate on science and religion. With a few notable exceptions, most of its supporters who explored the link to science opted for the new natural theology being promoted by those biologists who endorsed the vitalist and non-Darwinian viewpoints and by the physicists who stressed the nonmechanistic in-terpretations of their theories (see chapter 8).

This view of the Fall was hard for many traditionalists to accept. Coupled with the Modernists' tendency to demystify Christ's own activities on earth, it confirmed in many eyes their rejection of any true notion of sal-vation. Criticism was launched equally from the evangelical and the Catholic wings of the church (see chapter 9). Support for Modernism diminished rap-idly in the 1930s as the mood of the country darkened and intellectual atti-tudes became more polarized. Those who took up a rationalist position wanted no compromise with Christianity, and those who returned to the faith did so precisely because they wanted to regain the traditional concept of the supernatural. By the time Barnes published his *Rise of Christianity* in 1947, he and the few who still thought like him were largely marginalized within the church.

Similar tensions can be seen within the Free Churches, although here there was a two-sided conflict between liberalism and evangelicalism. In those denominations that prized enthusiasm and respect for the Word of God, there was not much room for liberal opinion. But those that aimed at a more middle-class worshipper had to take account of current intellectual trends, and here the equivalent of Modernism reared its head. The so-called Down Grade controversy within the Baptist Union had been sparked in 1887–1888

47. Tennant, "The Being of God in the Light of Physical Science," 70.

48. Tennant, "Present Relations of Science and Theology."

49. Tennant, "The Being of God in the Light of Physical Science," 89, 93. Note, however, that Tennant concedes that if natural selection were accepted as the sole mechanism of evolution, it would not necessarily undermine theism.

by John Clifford's expression of liberal views, and although his chief critic, C. H. Spurgeon, left the Union in disgust, a resolution was passed reemphasizing the authority of Scripture as the Word of God. In the interwar years T. R. Glover, a noted authority on ancient history at Cambridge, inflamed the same passions among the Baptists by his expression of a more human view of Christ.[50] The Congregationalists were the most liberal of the Free Churches in England, their City Temple in London serving as almost the equivalent of an Anglican cathedral. Here the first major outburst of what would later be called Modernism came when R. J. Campbell began to promote his "New Theology" in 1907. The controversy that erupted between Campbell's supporters and opponents brought the tensions between liberals and conservatives out into the open—and led Campbell himself to repudiate his book and turn to Anglo-Catholicism. Subsequent expressions of a more liberal position by Congregationalists were considerably muted by comparison, and in the less optimistic atmosphere of the 1930s, the more conservative position adopted by many evangelical thinkers began to seem justified.

In Scotland the Free Church had been thrown into controversy in 1881 when William Robertson Smith, the professor of Hebrew and Old Testament criticism in its college at Aberdeen, expressed support for the new biblical scholarship. Robert Rainy, the influential principal of New College, Edinburgh, tried to steer a moderate course, but Robertson was eventually deprived of his chair.[51] Unlike the Baptists south of the border, however, the Scottish Presbyterians were stung into action by this debate, and support for more liberal views soon became widespread. In the 1920s James Young Simpson, professor of natural science at New College, published widely on the new natural theology, in effect accepting the Modernist position on the relationship between science and religion. As was the case in England, though, the tide of liberalism began to ebb in the 1930s and 1940s. The early decades of the twentieth century thus marked the high point of Modernist influence within both the Anglican and the Free Churches, and it was the link between this movement and the new natural theology being promoted by some religious scientists that formed the axis upon which the debates studied in the following chapters turned.

---

50.   Clements, *Lovers of Discord*, 15–16 and 109–29. On Evangelicalism in the Free Churches more generally, see Bebbington, *Evangelicalism in Modern Britain*.

51.   Vidler, *The Church in an Age of Revolution*, chap. 15.

CHAPTER 7

The New Theology in the Free Churches

In 1907 the Congregational Church was rocked by a controversy over the "New Theology" being preached by R. J. Campbell at the City Temple in London. This controversy was the first major outburst in the new century over what was termed Modernism within the Anglican Church (borrowing the term already used to describe a Roman Catholic modernizing movement, suppressed by that church in the same year). In fact, Campbell, along with several other ministers, had been exploring the same themes in sermons for several years. His book *The New Theology* may have precipitated the public controversy, but it was written in response to tensions already rising within the Congregationalist ministry. For Campbell and many others, this new direction in theology had purely religious roots—it emphasized God's immanence within the world and within humankind. The human race could be seen as an agent of the divine will and Christ as the forerunner of what we or our descendants may become in the future. But this spiritual message was entirely congruent with the hopes of those who wanted to reconcile religion and science. If the activity of the human spirit could be seen as an integral part of a divinely ordained universe, the product of organic evolution and the spearhead of a spiritual evolution to come, then the project to reconcile religion with the nonmechanistic trends in science could be realized. The New Theology was the Free Churches' most active attempt to present a face compatible with the new image of a purposeful universe being promoted by many scientists.

It was no accident that Charles Gore bracketed Campbell with Sir Oliver Lodge when he came to attack the New Theology for its rejection of traditional Christian beliefs. Campbell, however, began a correspondence with Lodge only after he had begun preaching his new message, suggesting that his ideas were not driven directly by the influence of developments in science. Several commentators have noted that the New Theology—like most Modernist Anglicanism—was a product of the idealist philosophy that dominated British academic philosophy in the late nineteenth century.[1] The neo-Hegelianism of Edward Caird and F. H. Bradley generated a worldview that was to a surprising extent congruent with the non-Darwinian evolutionism promoted in biology. Campbell was also attracted to George Bernard Shaw's idea of creative evolution. Idealism and progressionist evolutionism both stressed continuity and the purposefulness of the forces that drive nature: according to both, the human mind is both an integral part of and the culmination of the spiritual activity of the world, which produced it and will ensure its future perfection. Campbell and Lodge were coming at the same position from different directions. Theologians were more likely to be influenced by philosophy than by science, but they made common cause with the scientists exploring the nonmechanistic worldview in their own areas of specialization. For the Christian, however, the temptations of idealism were as dangerous as those of science: by blending humankind into an ever-improving, spiritualized nature, idealism obscured the distinction between Creator and creature, minimized the extent of sin and the need for salvation, and threatened to reduce Christ to a model of the perfection to which we can all aspire.[2] If Lodge had de-Christianized Christianity, so (at least potentially) did the New Theology.

Evangelical theology was no longer wedded to biblical literalism or to Calvinism, and many liberals stressed Christ's identification with suffering humanity rather than His atonement for our original sin. With such a theology, scientific ideas on the evolution of life were not perceived as an immediate threat, although the complete submersion of Christ in the evolutionary process was unacceptable. Liberal evangelicals were thus reasonably comfortable with the new natural theology. Writing in 1901, W. F. Adeney saw the liberalizing movement within evangelicalism as the dominant theme of nine-

---

1. See, for instance, Clements, *Lovers of Discord*, 28.

2. On the dangers of idealism, see Sell, *Philosophical Idealism and Christian Belief*; Hinchcliff, *God and History*, chap. 6. See also Den Otter, *British Idealism and Social Explanation*.

teenth-century theology and included a discussion of science, stressing the acceptability of evolution.[3] He supported the work of Henry Drummond, whose *Ascent of Man* of 1894 had become a best-seller with its efforts to portray evolution as a divinely planned ascent toward moral values. In this sense, Campbell merely pushed an existing trend of thought to a conclusion that went beyond what all but the most liberal could tolerate.

More conservative thinkers opposed the whole idea of humanity achieving God's purpose in this world and wanted to retain the traditional meaning of the Fall. Campbell was assailed by a wide range of opponents, all convinced that he had gone too far in his desire to shift the emphasis of the Christian message. Eventually he abandoned his immanentist theology and returned to what he regarded as his true spiritual home, Anglo-Catholicism (he was received back into the church by Gore himself). Some Free Church ministers, including E. Griffith-Jones, made continued efforts to develop a liberal theology based on evolution, but their ideas stopped short of the complete de-supernaturalizing of religion advocated by the more active Modernists. One Free Church writer who did advocate a complete reconciliation with science was James Young Simpson, whose work formed an important component of the new natural theology promoted in the interwar years.

PRECURSORS OF THE NEW THEOLOGY

Drummond was by no means the first voice in Scottish theology to advocate a reconciliation with new developments in science. As early as 1876, Robert Flint, professor of divinity at the University of Edinburgh, had advocated a theistic evolutionism in which the appearance of higher organisms was preordained by the Creator's will. Flint was even prepared to accept that suffering and death were aspects of the creative process.[4] Drummond went much further in his bid to forge an alliance with new ways of thought. His *Natural Law in the Spiritual World* used quotations from Herbert Spencer to argue that spiritual matters were not outside the realm of law.[5] In the end,

3.   Adeney, *A Century's Progress in Religious Life and Thought,* chap. 4. See Bebbington, *Evangelicalism in Modern Britain,* chap. 6, and the same author's "Science and Evangelical Theology in Britain from Wesley to Orr."

4.   Flint, *Theism;* see esp. 188 and 202–4. On theistic evolutionism, see Bowler, *Eclipse of Darwinism,* chap. 3.

5.   On Drummond, see, for instance, J. R. Moore, "Evangelicals and Evolution" and *Post-Darwinian Controversies,* esp. 224; also Kent, *From Darwin to Blatchford,* 21–28. The latter source stresses the failure of late-nineteenth-century efforts such as Drummond's to preserve the central teachings of Christianity.

though, his vision of evolution completely overturned Spencer's emphasis on struggle as the driving force of progress. Drummond's *Ascent of Man* followed Flint in implying that the energies driving evolution are really spiritual and moral, rather than material. The element of intelligent design seen in progressive evolution is thus still directly attributable to the Creator. But instead of welcoming struggle as a progressive force, Drummond argued that it was steadily replaced in the course of evolution by a moral power: the struggle for existence was gradually superseded by the struggle for the life of others. This was inevitable because—as the case of parental care shows—cooperation and altruism have positive survival value. Drummond argued that evolution and Christianity were one: each strove to produce more perfect human beings through love.

> Evolution and Christianity have the same Author, the same end, the same spirit. There is no rivalry between these processes. Christianity struck into the evolutionary process with no noise or shock; it upset nothing of all that had been done; it took all the natural foundations precisely as it found them; it adopted Man's body, mind and soul at the exact level where Organic Evolution was at work on them; it carried on the building by slow and gradual modifications; and, through processes governed by rational laws, it put the finishing touches to the Ascent of Man.[6]

Drummond's vision of evolution driven by altruism chimed with the thesis advocated in Peter Kropotkin's *Mutual Aid*, their popularity reflecting a widespread turn against the ruthlessness associated with the Darwinian and Spencerian models.[7] Drummond's book offered a tempting model for those Christians who sought a reconciliation with science, although its teleological model of evolution would prove increasingly difficult to square with developments in early-twentieth-century biology. It also brought its author into conflict with conservative forces in his own church, many of whose members recognized that its evolutionary theism had abandoned traditional Christian teaching on questions such as original sin and salvation through Christ.

If some in the Free Church of Scotland recognized dangers in Drummond's approach, there were many who found his views appealing, at least up to a point. Such active figures at the end of the nineteenth century included W. H. Dallinger, W. F. Adeney, and James Orr. Orr's *Christian View of God* had argued that evolution extended the argument from design, while his *God's Im-*

---

6. Drummond, *Ascent of Man*, 342.
7. See Bowler, *Eclipse of Darwinism*, 55–56, 81, 87.

age in Man of 1905 saw God as active within evolution.[8] These writers reveal how the basic idea of evolution as God's method of creation at the physical level had been adopted by liberal evangelicals in the late nineteenth century, thus paving the way for further developments after 1900. Orr, however, was concerned that the application of evolutionism to the origin of the human mind advocated by Drummond might undermine our moral and spiritual status. He wanted a supernatural origin for the human spirit and defended the reality of the Fall. He thereby identified the key stumbling block that would limit all but the most liberal in their efforts to work out a reconciliation with the new scientific way of thought.

There were ministers in other Nonconformist churches willing to explore the liberal position. The most successful was Ebenezer Griffith-Jones, whose *Ascent through Christ* was published in 1899 and went through a number of editions over the following decade. Griffith-Jones was a Congregationalist minister in Balham, London, and was subsequently appointed principal and professor of dogmatics and apologetics at the Yorkshire United Independent College, Bradford.[9] He wrote *Ascent through Christ* quite explicitly to go beyond Drummond's position, arguing that the latter's account of the ascent of man had left the outcome problematic for the Christian because it sketched in what we might become, but did not address the question of why we had failed to reach that goal. What was needed was for someone to address the question in the spirit of the Jehovistic writers, dealing with humankind not as we were meant to be, but as fallen creatures.[10] Yet the account had to be "in the light of that great principle of Development which has taken possession of the mind of to-day, and which seems destined, in its broader aspects, permanently to affect human thought in all its departments."[11] Griffith-Jones had no scientific training, but he read widely in the works of evolutionists, especially Spencer, Darwin, and the American Joseph LeConte, and believed that it was possible for the nonspecialist to address the issues competently on this basis.[12] He was able to produce a complex, and in some respects quite subtle,

8. James Orr, *God's Image in Man*, e.g., 96–99. On these turn-of-the-century attitudes, see Bebbington, "Science and Evangelical Theology in Britain."

9. For a brief discussion of Griffith-Jones, see R. T. Jones, *Congregationalism in England*, 348–49, although Jones does not really address the complexity and limitations of his subject's confrontation with evolutionism.

10. Griffith-Jones, *Ascent through Christ*, 159.

11. Ibid., vii.

12. Ibid., 46, and for a list of Griffith-Jones's major sources, xi–xii. LeConte was an important member of the American school of neo-Lamarckian evolutionists. His most important work in the area of science and religion was his *Evolution: Its Nature, Its Evidences, and Its Relation to Religious Thought* (1899); see Stephens, *Joseph LeConte.*

application of evolutionary ideas to theological questions, going beyond the simple progressionism of Drummond to explain the Fall by analogy with biological degeneration. But in the end there were major limitations in his ability to take the whole evolutionary perspective into account because he preserved the traditional view that certain great events in history required supernatural intervention.

Griffith-Jones's concern with sin and the Fall enabled him to see the limitations of existing efforts to incorporate scientific knowledge into Christianity—they always stopped just short of the crucial issues.[13] He accepted Drummond's view that the laws of nature show that we live in a purposeful cosmos, and that its purpose is to develop spiritual awareness. LeConte and many others had insisted that when viewed in this light, evolution did not destroy teleology, because the universe did not have to be viewed as a soulless mechanism.[14] Natural selection was just a negative process for weeding out the less favored variations, and science was only just beginning to understand how variations were formed and to take into account the possibility that their production was not aimless.[15] Some evolutionists, most notably Alfred Russel Wallace, had insisted that there were episodes in the upward march of life that transcended natural forces and seemed to imply an intervention from outside. But where Wallace limited the intervention to the origin of the human mind, Griffith-Jones took the view that the origin of life, the origin of consciousness, and the origin of self-consciousness all required such a helping hand.[16] Yet he was unwilling to promote these breaks in continuity as interruptions in the normal course of events—in effect, there was a deeper unity in nature that allowed higher forces to accelerate the normal processes from time to time. In the case of self-consciousness, biological evolution had prepared the way by leading up to the ape, which had then been "touched by God": "Is the coming of the spiritual principle in Man less Divine and impressive because, in the higher animals, organism was already becoming sensitive to something nobler than physical stimulus, and was, as it were, trembling forward towards finer issues, till ready to be kindled at the sacred touch of God's breath, and become the tabernacle of an immortal soul?"[17] Here humankind became both the goal of biological evolution and its final limit—all further development would be at the spiritual level.

13.   Griffith-Jones, *Ascent through Christ*, viii.

14.   Ibid., 9 and 18.

15.   Ibid., 42.

16.   Ibid., 34.

17.   Ibid., 69.

Where Drummond had left the situation at this, Griffith-Jones saw that for the Christian, there was still a major aspect of the human situation missing. Something had gone wrong with our spiritual evolution, introducing sin into the world and subverting the Creator's plan. There was no longer any difficulty in accepting that Genesis was an allegorical rather than a literal account of creation, but that account still told us about its purpose, and what had gone wrong with it.[18] Even so, the Fall could be understood in evolutionary terms because biologists knew that not all branches of the tree of life were progressive. Griffith-Jones cited E. Ray Lankester's well-known account of evolutionary degeneration, along with August Weismann on the same topic and Darwin on atavism.[19] Without ever describing why our distant ancestors had turned aside from the upward path, he postulated a sinful tendency having entered human nature, where it was preserved by heredity in a way that made it impossible for the individual to overcome its influence.[20] LeConte was cited to argue that once a branch had diverged from the upward path, it could never recover and was doomed to degeneration. This meant that divine intervention was needed to restore human nature to its intended form—nature by itself would allow no reversal of the downward trend.[21]

Griffith-Jones conceded that it was possible for us to progress in some areas while degenerating in others, but was suspicious of the assumption that modern civilization was technically more advanced than the ancients—did not the pyramids suggest that the Egyptians had access to higher powers now lost to us?[22] Many human races had never risen beyond a level little removed from the animal—they were equivalent to what T. H. Huxley had called persistent types of life.[23] Only those exposed to the influence of Christianity had showed any sign of spiritual development, a clear indication that Christ was the divinely created inspiration sent to turn us from the path of sin and degeneration and return us to our rightful place in God's plan:

> The whole creation would become a vast anomaly and enigma if Man were to be hopelessly and finally lost: for the Universe only becomes intelligible in Man; he is its coping-stone, its goal; without him, the whole history of the past would lose its cohesion and rational sequence. In or-

18. Ibid., 77 and 115–16.

19. Ibid., 153, 186 and 205–6; see Lankester, *Degeneration;* Weismann, *Essays on Heredity,* vol. 2, 3.

20. Griffith-Jones, *Ascent through Christ,* 135 and 195.

21. Ibid., 221.

22. Ibid., 150 and 162–63.

23. Ibid., 204; see Huxley's essay "Geological Contemporaneity and Persistent Types of Life" in his *Discourses Biological and Geological,* 272–304.

der, therefore, that the general system of things might not prove to be a wreck, instead of a drama which has its developing plot and its gradual culmination in a glorious and convincing spiritual climax . . . it was necessary that an outpouring of redemptive power should take place through the same agency as that by which the Universe came into being at the first.[24]

If Christ was a supernatural being, the virgin birth was not implausible. His intervention might seem arbitrary to the biologist, but it was essential to restore the course of evolution to its intended goal. Griffith-Jones conceded that Christ's sacrifice on the Cross was difficult to fit into the evolutionary perspective, but offered as a hint Drummond's image of self-sacrifice (including parental devotion to offspring) as the driving force of progress.[25] Christ was the perfect man, and thus served as a model for us to follow in our lives, but He was also far more than that because His intervention had established a church that would supervise the redemption of humankind and open up the possibility of future spiritual evolution. His Resurrection in a glorified body gave us hope of our own immorality, a possibility now being conceded by science thanks to the work of psychic researchers such as Henry Myers.[26]

Ascent through Christ was popular among Nonconformists because it tried to preserve elements of the traditional doctrine of sin and redemption instead of sweeping them under the carpet in the name of progress. Progress toward man was still taken as a sign that evolution had a divinely ordained goal, but the need for salvation was retained by appealing to other biological concepts such as degeneration and heredity. Such a complex synthesis of scientific ideas would have had an appeal to those Christians who wanted a reconciliation with modern ideas but were worried that Drummond's approach threatened to rob the faith of its original purpose. The package would have been less appealing to those approaching a reconciliation from the direction of science, however, because it implied the need for divine intervention at key points in history. Griffith-Jones's vision of God still portrayed Him as transcendent, capable of intervening from outside in a universe that He had endowed with only limited powers to progress in its own right. To those whose Modernist tendencies led them to reject any appeal to the supernatural, such a compromise was unacceptable. God's immanence within the world had to

24. Griffith-Jones, *Ascent through Christ*, 276–77.

25. Ibid., 292.

26. Ibid., 353; on Myers's efforts to prove survival after death, see Oppenheim, *The Other World*, 152–55.

be complete—but this would re-create the problem already obvious from Drummond's attempt to deny the discontinuities between human beings and their Creator.

Griffith-Jones would return to these issues in the 1920s, but others were already pushing harder toward the more immanentist view of God's relation to the world. Campbell himself was preaching openly on this topic for years before he wrote his book, and the same was true of another Congregationalist minister, Thomas Rhondda Williams. Like Griffith-Jones, Williams came from Wales (the two met at college and remained friends). He was minister in Bradford from 1888 to 1909 and then at Brighton until his retirement in 1931. Williams came to prominence following his response to the popular socialist writer Robert Blatchford's account of Ernst Haeckel's *Riddle of the Universe*. Blatchford had used the deterministic implications of Haeckel's monistic philosophy as the basis for an attack on religion. Williams replied in three immensely popular lectures, subsequently published as a pamphlet under the title *Does Science Destroy Religion?* Williams records that he got a letter from Wallace praising his rebuttal of Haeckel's position.[27] Haeckel's deterministic monism was a favorite target for exponents of the new natural theology because it inverted their own position. Their belief that God was immanent in nature, and hence in the human spirit, also generated a form of monism— but it was a monism in which the spiritual aspect of the universal substance was more important than its restriction by the laws of nature.

Williams was denounced by his critics as a Unitarian, and he delivered a series of sermons under the title "The New Theology and Unitarianism" defending himself against this charge. Unitarians denied the Trinity in order to proclaim the unity of God, but the New Theology unified God and His creation by stressing the extent to which He remains immanent within it. Williams argued that such a position was hardly new because similar beliefs had been held by the church during the first three centuries of the Christian era.[28] He was quite happy to accept that God had revealed Himself to other cultures besides that of the Jews, since a God who pervades all nature would hardly limit Himself to a single group of humans.[29] But although God reveals Himself, both through nature and directly to humans, He also conceals parts of His purpose. Thus science can be reconciled with religion, but can

27.   T. R. Williams, *How I Found My Faith*, 67–68. I have not seen a copy of *Does Science Destroy Religion?*—neither the British Library nor Cambridge University Library has a copy, although one is listed for the Library of Congress. On Blatchford, see below, chap. 10.

28.   Williams, *Evangel of the New Theology*, 13, and *How I Found My Faith*, 69.

29.   *Evangel of the New Theology*, chap. 4: "If the Bible is indebted to Babylon, what then?"

never expect to tell us everything about the human situation. Evolution is a divinely instituted process, but "if anyone supposes that our *knowledge* of evolution accounts for Jesus, I would remind him that it does not account for a single one of us."[30] Williams thus wanted a reconciliation with a science that knew its own limitations. Responding to criticism that had been aimed at a young preacher for appealing to scientists when there were none within a mile of his church, he wrote:

> It is unfortunately true that the men of science are not within a mile of our churches. It is further true that they never will be until the pulpit squares its approach with all the sure results of science. So long as the pulpit insists on the proclamation of obsolete opinions as fundamentals, the cultured men will pass by on the other side.[31]

The New Theology aimed to remove that obstacle as far as possible and allow a reconciliation with those scientists who were willing to take into account the nondogmatic aspects of theism. Williams made it clear that he saw no supernatural involvement in the Eucharist, which was a purely symbolic act intended to help the worshipper feel closer to God.[32] His collection of sermons, *The Evangel of the New Theology*, appeared in 1905, confirming Campbell's claim that his own book merely precipitated public debate over a topic that was already under discussion within the churches.

CAMPBELL AND THE NEW THEOLOGY

Campbell himself had been teaching the substance of what became known as the New Theology for some years before a confrontation within the Congregational Church forced him to publish in haste the book that sparked a major public controversy. His thought reflected all the liberal tendencies of the late nineteenth century and chimed with what had become a major strand of contemporary culture. The main influence on Campbell and his fellow theologians was the idealist philosophy of British academia, but the New Theology movement also coincided with the latest developments in biblical criticism and science. For those who embraced idealism and evolutionism, the notion that humanity was separate from the universe or its Cre-

30. Ibid., 98.
31. Ibid., 110.
32. Ibid., chap. 16.

ator was unthinkable. Continuity reigned as the supreme unifying principle, and hence God must be the ongoing foundation of the universe, not a Creator who remained external to His work. Humanity must be a product of, and hence an integral part of, that universe—which meant that we must also be pervaded by the divine spirit. God was not mysterious because He was contained within the world and within our own spirits. Campbell insisted that this was not pantheism—he preferred the term "spiritualistic monism"—but the claim that we are all part of the divine substance reinstated all the problems for Christianity that Griffith-Jones had tried to avoid. There were obvious discontinuities built into the traditional view that humanity is distinct from its Creator and its Savior. Sin became a problem once those discontinuities were denied because it was difficult to imagine how one element of the divine whole could rebel fundamentally against its overall purpose. And if the limited evil emerging in the world was not a rejection of its Maker, then what would be the purpose of an Atonement for a Fall that had never taken place? Christ became just the supreme example of humanity, an illustration of a potential inherent in all of us, sent to show us the way forward. The church was not an institution founded to project His saving grace into a hostile world—it must be merely the community of all those who acknowledged His lead in humanity's spiritual evolution.

Reginald John Campbell was born in Northern Ireland and raised as a Presbyterian. When he went to study at Oxford he was exposed to the Anglo-Catholicism of Charles Gore, which he afterward remembered as the most fundamental influence on his spiritual development.[33] He was also attracted by the neo-Hegelian philosophy of Edward Caird and the writings of James Ward and Henry Jones; the latter was especially important because he presented idealism as a practical creed. Although he subsequently claimed that there was no serious Nonconformist influence on his thought, Campbell came eventually to feel that Gore's Anglo-Catholic insistence that the Free Churches were excluded from the true Christian tradition was unacceptable, and turned to Congregationalism. In effect, he recognized in Gore's teaching the call of Rome, and since he could not go to Rome, he turned his back on the Anglican Church as well, although he remained on friendly terms with many Anglicans.[34] He read German biblical scholarship avidly for fifteen

33. See Campbell's autobiographical *Spiritual Pilgrimage*, chap. 3. On the New Theology, see Clements, *Lovers of Discord*, chap. 2; J. W. Grant, *Free Churchmanship in England*, 131–50; Hinchcliff, *God and History*, chap. 9; R. T. Jones, *Congregationalism in England*, 349–52; K. Robbins, "The Spiritual Pilgrimage of the Rev. R. J. Campbell"; and Vidler, *20th Century Defenders of the Faith*, 24–31.

34. Campbell, *Spiritual Pilgrimage*, 60.

years, leading to the gibe that one could always tell what would be in his next sermon from the most recently published translation of a German text.[35] Like Williams, he was concerned about the influence being exerted on young readers by Haeckel's *Riddle of the Universe* and invoked the Christian mysticism being highlighted by W. R. Inge as a counterweight to its materialism.[36]

Campbell became an extremely successful preacher, first at Union Street Congregational Church at Brighton and then from 1903 on at the City Temple in London, the most influential Nonconformist pulpit in the land. He also developed a close link with Christian socialism and often spoke on socialist platforms. He hoped to achieve Christian goals in this world as well as in the next, and saw the exploitation of science and technology by the commercial elite as a force that was generating human misery instead of relieving suffering: "Science has turned 'procuress to the lords of hell,' and is filling the world with grief and despair."[37] But this complaint was directed against the practical applications of science. Campbell adopted a very different position on the question of science's effect on the intellectual climate of the time. The New Theology was "a restatement of the essential truth of the Christian religion in terms of the modern mind" based on "a re-emphasis of the Christian belief in the Divine immanence in the universe and in mankind." As a result, many aspects of what had come to be regarded as orthodox doctrine would have to be rethought, especially the Fall and the concept of blood-atonement by Christ. Such doctrines were misleading and unethical and "go straight in the teeth of the scientific method, which, even where the Christian facts are concerned, is the only method which carries any weight with the modern mind."[38] Whatever the sources of his own desire to focus on immanence, Campbell saw that his beliefs needed to be congruent with the popular ideas promoted by science—the principle of continuity and the rejection of the supernatural in the sense of an external God interfering with the world.

> Again, the New Theology is the religion of science. It is the denial that there is, or ever has been, or ever can be, any dissonance between science and religion; it is the recognition that upon the foundation laid by modern science a vaster and nobler fabric of faith is rising than the world has

35. Ibid., 116; for the critical comment, see Grant, *Free Churchmanship in England*, 137.
36. Campbell, *Spiritual Pilgrimage*, 106.
37. Ibid., 158.
38. Campbell, *New Theology*, 3, 4, 9.

ever before known. Science is supplying the facts which the New Theology is weaving into the texture of religious experience.[39]

One consequence of the new focus on immanence was a full acceptance of evolution as the method by which the divine spirit works its purpose in the world. Cosmic evolutionism allows us to see a relationship between "the human soul and the great whole of things of which it is the outcome and expression." The tables could thus be turned on Haeckel, for all his insistence that science destroys belief in God, freedom, and immortality. By proclaiming the purposeful nature of evolution, argued Campbell, Haeckel in fact declares a belief in God on every page of his *Riddle of the Universe*, because he gives us an insight into the great Power that expresses itself in the universe.[40] Campbell's was an idealist, or spiritualistic, monism in which the universe, including the human race, was a means of the self-realization of the infinite. "God is ceaselessly uttering Himself through higher and ever higher forms of existence; or rather, which is the same thing, He is doing it in us: 'The Father abiding in me doeth His works.'"[41] From this perspective, the emergence of human personality is a key step in the realization of the divine plan, and our intuitions about the freedom of the will and survival of bodily death are valid. Such a view was far older than modern science, but science had helped to confirm its validity.[42]

Campbell had to explore the implications of this position for the doctrines of the Fall and the Atonement. He did not deny the reality of sin, but held that evil was merely a failure on the part of the individual to recognize his or her role in the universal scheme. The old idea of a fall from a state of grace that was somehow our fault was both unethical and contrary to the teachings of science: "It is almost superfluous to point out that modern science knows nothing of it [the Fall], and can find no trace of such a cataclysm in human history. On the contrary, it asserts that there has been a gradual and unmistakable rise; the law of evolution governs human affairs just as it does every other part of the cosmic process."[43] Campbell devoted several chapters of *The New Theology* to showing that the popular view of the Atonement was misleading, although not actually false. We did not need to be forgiven for an

39. Ibid., 15.
40. Ibid., 17.
41. Ibid., 24.
42. Ibid., 65.
43. Ibid., 60.

ancestral sin, and atonement was at work in every good deed performed in the world. He was particularly hostile to the writings of St. Paul on this topic, whose language, he claimed, had been a source of mischievous misinterpretations.[44]

What, then, was Christ's role in human salvation? Campbell made it clear that he did still see Christ as the divine man, unique and above all others. But since we are all expressions of God, His perfection is not beyond our reach—it is a goal toward which we must all strive in order to complete the divine plan.[45] The concept of Jesus as a deity was a late development in Christian thought. To proclaim the unity of Christ and humanity was not Unitarianism, however, because the Unitarians had always followed the deists in teaching a separation between God and humanity, which the New Theology denied. There was still room for the concept of the Trinity, however—even physical science implied this because it left room for a distinction between God, the universe, and God's activity in the universe (Campbell singled out E. Ray Lankester as his target in making this claim).[46] The doctrine of the virgin birth of Christ was not essential for Christianity, nor was it necessary to believe in a physical resurrection of Christ's body. Yet on the basis of a monistic idealism, the body itself was just a thought, and the Resurrection was a plausible way of explaining how Christ communicated with His disciples after the Crucifixion.[47] Campbell was convinced that the human soul survived the death of the physical body, referring to the work of scientists such as Crookes and Lodge to substantiate the claim that it would soon be impossible for anyone to deny the evidence from spiritualism. Psychic research promoted the credibility of miracles of healing, showing the power of mind over matter.[48] Campbell remained an advocate of spiritualism long after he abandoned the New Theology, and wrote the preface to William Barrett's *Personality Survives Death* (1937).

Under the circumstances, it is easy to see how the New Theology could be taken as equivalent to the kind of denatured Christianity being promoted by Lodge. In an article in the *Daily Mail*, Campbell himself said that "From the side of Science, the New Theology is typified in the work of men like Sir Oliver Lodge," and one letter in the subsequent controversy called it the

44. Ibid., 176.
45. Ibid., 70–78.
46. Ibid., 86.
47. Ibid., 221–23.
48. Ibid., 230 and 259.

"New Theoliver Lodgery."[49] Campbell first wrote to Lodge in 1905, telling him that the congregation at the City Temple was already used to hearing his ideas quoted. The following year he asked Lodge to speak at the Temple and tried to get him to take up the vice-presidency of his Progressive League.[50] More seriously, Campbell had also expressed support for George Bernard Shaw's philosophy of "creative evolution" driven by a blindly progressive life force. In 1906 he had introduced Shaw when he lectured at the City Temple and commented afterward that they shared the view that God was a continuously persistent will striving for self-expression in the universe.[51] This was all very progressive, but few Christians could have failed to be aware that Shaw was calling for the complete destruction of the old religion and its replacement by his own form of pantheism.

The outcry from conservative forces was predictable. Campbell had been preaching the New Theology for years, and although he had received some support from other Congregationalist ministers, there had also been growing opposition. This opposition came to a head in September 1906 when he gave an address to the London Board of Congregationalist Ministers, which was extensively covered by the newspapers. He pleaded for a reconciliation within the churches at the National Free Church Council in March 1907, but also felt obliged to write his book in order to defend his position. Campbell himself later admitted that *The New Theology*, because it was written in response to the controversy, was crude and polemical. Most Free Church ministers thought the new version of Christianity had effectively abandoned the old vision of humankind's relationship to God through Christ. Lancelot Hogben recalled how his father, a Fundamentalist Methodist, attacked the New Theology and how his views chimed with the wider campaign from many Nonconformist pulpits.[52]

Of the many attacks on the New Theology, the most broadly based was a collection of sermons edited by Charles H. Vine of Ilford under the title *The Old Faith and the New Theology*. This collection was designed to reassure Con-

49. See Clements, *Lovers of Discord*, 30–31.

50. Campbell to Lodge, 13 June 1905, 26 March 1908, and 24 November 1910, Lodge papers, Birmingham University Library, OLJ 1/74/1, 5 and 6; on his contacts with Lodge, see *Spiritual Pilgrimage*, 156.

51. Comments by Campbell recorded following Shaw's "Religion of the British Empire" in Shaw, *Religious Speeches*, 1–8, see 8. Campbell also introduced Shaw when he delivered his lecture "The Ideal of Citizenship" and included it as an appendix to the 1909 edition of *The New Theology*; see Shaw, *Religious Speeches*, 20–28. On Shaw, see chap. 11 below.

52. See *Lancelot Hogben: Scientific Humanist*, 9 and 11.

gregationalists that the challenge to the traditional view of Christ as Savior was being resisted by the majority of ministers, but it was a pedestrian compilation—Alec Vidler recorded that when he looked at the copy in Cambridge University Library many years later, the pages were still uncut.[53] Much more effective was Charles Gore's *The New Theology and the Old Religion,* based on lectures originally given in Birmingham Cathedral (see chapter 9). Campbell was not surprised by these attacks, and conceded that he had gone too far in his enthusiasm—although the critics were in error when they claimed that he had effectively undermined Christianity. He insisted that he did still revere his Savior and was being misinterpreted when he was accused of denying Christ's unique role. He noted that his acceptance of the Resurrection disgusted many liberals, but conceded that he had pushed monism so far that he had confused the human and the divine and had weakened the sense of sin and hence the need for salvation.[54] He soon began to realize that even when his hasty extravagances were eliminated, he had gone too far. By 1910–1911 he was arguing openly against those who proclaimed that Christ was a purely mythological figure, and soon found himself returning to his spiritual roots in the Catholic view of Christ. He turned his back on Nonconformism and returned to the Anglican Church in 1915, being ordained by Gore in the following year. Campbell's book *The War and the Soul* (1916) contained chapters rejecting "The Illusion of Progress" and attacking Haeckel's monism as the cause of the spiritual malaise underlying German ruthlessness.[55]

Thanks to the involvement of the *Daily Mail,* the controversy over the New Theology was intense, if short-lived. The popular newspapers evidently thought that religious questions were still of interest to the public and were willing to publicize theological debates—especially when they could so easily be presented in terms of a war between orthodoxy and modern life. The press soon lost interest as the New Theology collapsed as a concerted movement within the Free Churches. One of Campbell's early supporters, W. E. Orchard, became gradually disillusioned with the attempt to reconcile the faith with science and eventually converted to Roman Catholicism.[56] A few liberals, including Rhondda Williams, maintained and extended their position as Campbell and most others abandoned it.

53.   Vidler, *20th Century Defenders of the Faith,* 30. For these attacks on the New Theology, see chap. 9 below.

54.   *Spiritual Pilgrimage,* 191. For Campbell's immediate response to the controversy, see his "Aim of the New Theology."

55.   Campbell, *The War and the Soul,* chaps. 11 and 24. Campbell contrasts Haeckel's vision of the world with Alfred Russel Wallace's selflessness.

56.   See Orchard, *From Faith to Faith.*

The most effective attempt to sustain the teaching of the New Theology came from John Warschauer of Anerley Congregational Church in south London, whose book *The New Evangel* of 1907 went a good deal further than Campbell along the Modernist path. Like Campbell, Warschauer was concerned that traditional Christian doctrines were "in conflict with what thoughtful modern men and women well know to be true."[57] Protestantism had tied itself to the word of the Bible, with disastrous consequences when scholarship and the new sciences of geology and evolutionism had shown that the Word could not be taken at its face value.[58] The rationalists had then been able to claim that they had smashed both the Bible and Christianity. Spencer's philosophy had left only the Unknowable, but no one could worship that.[59] In fact, though, science had shown that we *can* know something about the power behind the universe, if only because the fact that we can make sense of the world confirms that it was made by an intelligence—here Warschauer quoted Lodge's *Substance of Faith*.[60] Intelligence was inconceivable without consciousness, and hence purpose, will, and directivity, all of which must be reflected in the Creator's involvement in the world. Beyond this we must turn to Christ for information about God's purposes. Christ was the embodiment of God's love—but He was not a superhuman figure, and in particular, He should not be seen as a worker of miracles. Science carried a strong presumption against the plausibility of miracles, and this meant that the lack of evidence for the virgin birth had to be taken seriously.[61] Warschauer went further than Campbell, however, in denying the physical Resurrection. Clearly the disciples had been given some reassurance of Christ's continued presence, but even eyewitness accounts of His appearances could not be trusted—to confirm this, one had only to look at the contradictory reports given by eyewitnesses of the surrender of Napoleon III to Bismarck.[62]

Like Campbell, Warschauer had to reinterpret the Atonement while still retaining the validity of the feeling essential to many evangelicals of having been personally saved by Christ. His solution was to present Christ as the ideal of humanity—an ideal of which we all fall short, yet which liberates us by showing what may ultimately be possible.[63] Unlike Griffith-Jones, he saw

57.   Warschauer, *New Evangel*, 13.

58.   Ibid., 21–26.

59.   Ibid., 50–54.

60.   Ibid., 61, quoting Lodge, *Substance of Faith*, 8.

61.   Ibid., 84–85.

62.   Ibid., 170–71; see also 188–89.

63.   Ibid., 149, 151, 156.

heredity as having only a limited power to determine human behavior—there are no "born criminals," and inherited good is far more likely to be transmitted than inherited evil.[64] He also defended immortality, sharing Campbell's hope that the work of Crookes, Lodge, and others was now bringing it within the realm of scientific fact.[65] But there was another argument from science to support the hope that the spirit survived the death of the body. If the human soul was the intended goal of evolution, it was unthinkable that its duration should be limited by the physical body.

> Having regard to the fact that the whole orderly progress of evolution from lower to higher indicates the work of a planning intelligence and a directive purpose, we can hardly escape the conclusion that the entire drama of creation had this end in view, viz., the emergence of self-conscious, free agents, knowing themselves related to God and capable of returning His love. Is it conceivable that this age-long process has been gone through, this last and highest product of evolution been slowly matured through millions of years, simply in order to be annihilated again? We can only say that such an assumption strikes us as wildly improbable; for in that case the very purpose of creation would have been brought to fruition only to be instantly defeated, and the pyramid of evolution, after all the aeons of building, would be found to be crumbling at the apex.[66]

Here the progressionist evolutionism so characteristic of the new natural theology was linked to the hope of immortality now receiving apparent support from the study of spiritualism. The New Theology once again showed its links not with traditional Christianity, but with Lodge's optimistic theology of progress.

MODERNISM IN THE FREE CHURCHES

The position that Campbell staked out and then abandoned would continue to be advocated by Modernist Anglicans into the 1920s. The popular press would still be headlining Modernist attacks on the idea of original sin twenty years after it had stirred up the confrontation over the New Theology. But the Free Churches would play comparatively little role in these later de-

64. Ibid., 129–30.
65. Ibid., 195 and 199.
66. Ibid., 206.

bates. Evangelicalism closed its ranks against the interpretations advocated by the enthusiasts for modern ideas, leaving much less room for their articulation by ministers. There was no mass assault on evolutionism because few evangelicals adopted a literal reading of Genesis, but there was little incentive to explore the more controversial applications of scientific theories in any detail. There is little evidence of positive engagement with scientific ideas within the Free Churches in the interwar years, at least in England. Griffith-Jones made a new effort to explore the position he had suggested in his *Ascent through Christ*, and the widely respected Presbyterian theologian John Oman supported a non-Darwinian evolutionism. But Oman was a Scot, and the most important new voices all came from north of the border, where James Young Simpson and others began to promote a reconciliation between science and religion very much along the lines being explored by the Anglican Modernists.

In the preface to his *Providence—Divine and Human* of 1925, Griffith-Jones noted that *Ascent through Christ* was currently in its tenth edition. His intention was still to explore the thesis that evolution had been intended to produce a perfect human race, but had been disturbed in a way that threw humanity off course until redeemed by a supernatural influence. But he now sensed that the old faith in progress had been eclipsed, leaving a fear of degeneration.[67] In part this was because of the materialism rampant in the late nineteenth century, and Griffith-Jones once again returned to the theme of evolution as an inherently purposeful process. He was now prepared to admit that the process depended on the interaction between life and its environment. The first humans had evolved in response to environmental stress in the plains of central Asia, and the great races of the world were those that had had to struggle against adversity.[68] From this point on, Griffith-Jones fought a losing battle to retain a role for his Christian principles within a worldview increasingly dominated by his enthusiasm for the idea that struggle promotes progress. He minimized the extent of animal suffering, but argued that the higher human faculties are all the products of a response to external challenges. While still accepting the need for Christ to redirect the course of human spiritual evolution, he now believed that it would be possible for us to complete the process through the future development of the human race.

This latter theme was explored in a second volume entitled *The Dominion of Man*. Significantly, Griffith-Jones saw Bergson's *élan vital*, now mediated by

67. Griffith-Jones, *Providence—Divine and Human*, vol. 1, 19.

68. Ibid., 204–5, 211. In the 1920s it was still widely believed that the first humans had evolved in Asia, not Africa: see Bowler, *Theories of Human Evolution*, 173–85.

our moral sense, as the basis for human creativity.[69] He then launched into a classic exposition of the race biology then current among many physical anthropologists. There were three great races of humankind, Caucasian, Mongolian, and Negroid, which had experienced different degrees of progress from their Stone Age beginnings. Only in the Mediterranean region had a happy combination of the Mediterranean, Alpine, and Teutonic stocks produced a human constitution able to move ahead toward modern civilization. The Hebrew spiritual capacity had awoken European culture and allowed Christ's message to be appreciated. The chapter titles give some indication of Griffith-Jones's thinking: "Racial Struggles in the Dark" and "The Breaking Day."[70] The final section of the book looked to "The Future—Man's Unfinished Tasks," and here Griffith-Jones charted the opportunities and problems facing the white race as it struggled to complete God's plan by creating the perfect form of humanity idealized for us by Christ. While praising the efforts made toward enhancing international cooperation, he nevertheless made it clear that only the white race could complete the program. The biggest obstacle to the whites' dominion over the world was birth control, which was now beginning to limit the population expansion that had fueled the dispersal of Europeans to America and other continents.[71] This development not only cut back on the total numbers of people, but was increasingly being employed by those who represented the race's best biological potential, while the worst continued to breed unchecked. Griffith-Jones was thus led to Francis Galton's program of eugenics (selective breeding within the human race) as the only way of ensuring that the divine plan could be completed.

> We must breed eugenically in future in the interests of humanity at large, or we will most surely breed dysgenically, and the species *Homo* will revert to a lower type, on the way possibly to extinction. If man is in the world to subdue its natural forces and agencies in the interests of the race as a whole, he must not abdicate this, surely his highest function, as the sub-agent in the Providential Order. That function is to help and not defeat God to produce the highest type of man this planet is capable of realizing.[72]

69.   Griffith-Jones, *Providence—Divine and Human*, vol. 2, 43–44 and chap. 4.

70.   Ibid., chaps. 2 and 3. On the popularity of such ideas about race origins, see, for instance, Bowler, *Theories of Human Evolution*, esp. chap. 6.

71.   Griffith-Jones, *Providence—Divine and Human*, vol. 2, 277–78.

72.   Ibid., 285. There is a vast literature on eugenics; for a general survey, see Kevles, *In the Name of Eugenics.* Works on British eugenics that include some mention of involvement by religious figures

This was a social application of the Modernist program that would become popular among many of its Anglican followers. If Christ came only to show us a model on which to create a perfect humanity in this world, it was our duty to take control of evolution and direct it toward the goal that we believe the Creator wants to achieve. The science of heredity was invoked as the key to an artificial completion of the natural process of evolution instituted by God and redirected by Christ. The involvement of religious thinkers such as Griffith-Jones in eugenics reveals how that movement was able to present itself as the fulfillment of a moral duty. There were many Christians, of course, who objected, seeing the program to restrict the breeding of the "unfit" as a violation of their teaching's central focus on the value of each human individual. But the desire to bring the faith into line with modern attitudes and modern biological technologies could easily lead in the direction sketched out by Griffith-Jones.

The other Nonconformist churches in England showed little enthusiasm for the proposed reconciliation with science. The Baptists, with their largely working-class congregations, favored a conservative evangelicalism, although two contributions to the *Baptist Quarterly* in the late 1930s did argue for a more positive attitude. In 1935 R. L. Child proclaimed that science was "a friend of religion" thanks to its expansion of our vision of the universe.[73] In 1938 L. H. Marshall argued that science was now closer to religion thanks to the work of the physicists, cosmologists, and antimaterialist biologists. He believed that science could purify religion of magic, but could not provide any meaning for the phenomena it described.[74] In the following year there was an appeal for evangelicals to make a greater effort to come to grips with modern science by Arnold Aldis, himself an applied scientist teaching at University College, London. Significantly, Aldis identified Karl Barth's theology as a factor that was now making it more difficult for Christians to follow this course—the pendulum had once again swung back from an age of reason to an age of faith. In part, he believed this had been encouraged by the scientists' own willingness to admit how tentative their theoretical conclusions were.[75]

The small Presbyterian Church in England exerted an influence beyond its size, especially through the writings of John Oman, principal of its Westminster College at Cambridge. Oman had attacked Campbell's New Theology as

include G. Jones, *Social Hygiene in Twentieth-Century Britain*, Searle, *Eugenics and Politics in Britain, 1900–1914*, and Searle, "Eugenics and Politics in Britain in the 1930s."

73. Child, "Science, a Friend of Religion."
74. Marshall, "Religion and Science."
75. Aldis, "The Present Position of Evangelicals in Relation to Theology and Science."

pantheism,[76] but in the interwar years made some effort to bring his faith into a form that would allow a reconciliation with nonmaterialistic science. He contributed a chapter on the religious viewpoint to Joseph Needham's *Science, Religion, and Reality.*[77] This said very little about science, but sought to defend the claim that the supernatural realm of values interpenetrated the world of the senses. A few years later Oman included a chapter on evolution in his *The Natural and the Supernatural.* This chapter complained about the mechanistic interpretation of Darwinism and the consequent elimination of the mental activity of the living organism from any role in evolution. If the physicists had now abandoned the mechanical model of the atom, recognizing that we know nature only through our mental images of it, should not mind be extended back down the scale of animal organization? "And if life be thus developing toward mind and purpose, and if nothing can be known concerning it except from this high achievement, is it not more rational and convincing to carry mind and purpose as far down as we can than to carry up mechanical explanations to the utmost limits of plausibility?"[78] Mind had to be counted as an active principle in evolution, even without the inheritance of acquired characters. This did not necessarily require a return to vitalism, but it did imply a *tertium quid* that gave form and unity to the living body.[79] Oman thus threw his considerable theological weight behind the organismic, non-Darwinian biology promoted by an earlier generation of scientists such as J. Arthur Thomson.

The most effective voices promoting the new natural theology came from the Scottish Presbyterians, of whom James Young Simpson was the most articulate. Simpson was a trained biologist—he had studied with the embryologist Adam Sedgwick[80] at Cambridge and taught biology at New College, Edinburgh, and at the university there. But his position at the Free Church's college led him increasingly to stress the broader implications of biology at the expense of practical work. He also built a second career as a diplomat, being especially concerned with Russia and the Baltic region.[81] As a young man he had met Henry Drummond and subsequently wrote a biography of him—it was Drummond's influence that confirmed his support for

---

76.   See Clements, *Lovers of Discord,* 33–34.

77.   Oman, "The Sphere of Religion."

78.   Oman, *The Natural and the Supernatural,* 264.

79.   Ibid., 276.

80.   Not to be confused with the geologist of the same name who had opposed both Lyell and Darwin.

81.   There is a memoir on Simpson's life prefaced to his posthumously published *Garment of the Living God,* 15–78.

a non-Darwinian version of evolution that would be easier to reconcile with the Christian faith.

Simpson's first major work was his *Spiritual Interpretation of Nature* (1912). Here he began by challenging Lankester's assertion that there was no relationship between science and religion—if we believe we are living in a divinely created world, he argued, we must take that into account.[82] The critical spirit of science could purge religion of nonessential beliefs, but nothing in the vast development of scientific knowledge had disproved the essential truth of religion. Much of the book was taken up with a detailed survey of biology in general and evolutionary theory in particular. Simpson was at pains to stress that evolutionism should not be equated with Darwinism, so that the development of life could be seen as the divine method of creation.[83] Evolution was a law-like process "directed by an overruling yet indwelling purpose" representing divine immanence.[84] Although there was much non-progressive evolution, the whole process did reveal an overall advance toward higher types of life. Like many other writers, Simpson tried to minimize the significance of the struggle for existence, while at the same time insisting that the lower animals did not, in any case, suffer pain in the same way as humans.[85] The true directing agent of evolution was the production of new characters by variation, something that was still not understood despite new advances such as the mutation theory. Theories of orthogenesis suggested that there were factors directing variation along predetermined channels, and this confirmed that there was a purpose at work in nature.[86] The Lamarckian effect did not seem to work, but this did not rule out some deeper effect of the environment in directing variation.

Simpson was sure that the emergence of mind was a central feature of evolution. He saw mind as leading inevitably to self-consciousness and the moral sense—Darwin's efforts to reduce morality to the social instincts had missed the point.[87] The mind clearly transcended material nature, and this left open the possibility of miracles by which extraordinary minds could control matter and energy.[88] Immortality was also plausible, since to rule it out would rob the whole process of evolution of its meaning. The true goal of

82. Simpson, *Spiritual Interpretation of Nature*, 2–4.

83. Ibid., 106–7.

84. Ibid., 110 and 112. See also chap. 12.

85. Ibid., chap. 6.

86. Ibid., chaps. 7 and 9.

87. Ibid., chaps. 13 and 14.

88. Ibid., chap. 16.

progress was the production of souls perfectly adapted to the spiritual world.[89] Simpson did not mention Lodge, but the parallel with the latter's model of spiritual evolution is obvious. Like Griffith-Jones, however, Simpson was concerned about the social implications of the new knowledge of heredity. In the absence of any selective effect on the human species, the worst individuals were breeding freely, and this was lowering the standard of the race. The answer was eugenics, although Simpson was cautious enough to note that we could far more easily identify the harmful characters that needed to be eliminated from the population than the good ones that should be encouraged.[90] Reproduction ought to be a matter of individual responsibility, but the unfit should be isolated in farming colonies where they would be discouraged from breeding.

In 1925 Simpson published his *Landmarks in the Struggle between Science and Religion.* The title of this book is misleading, since the preface makes clear that its purpose is to challenge the negative attitude toward religion popularized by J. W. Draper, A. D. White, and other exponents of the claim that science must inevitably come into conflict with it. Simpson minimized the debate over evolutionism, insisting that the mechanistic worldview was now defunct, leaving the way open for a spiritual interpretation of the world process.[91] The same theme was developed in his *Nature: Cosmic, Human, and Divine* (1929), based on a series of lectures given at Yale University. Here he stressed that all across the spectrum of scientific activity, "Victorian cocksureness and arrogance have been superseded by Georgian hesitancy, or, shall we say, open-mindedness and humility."[92] We should not expect to prove the existence of God from nature, but if we could not show at least some plausibility for natural theology, the situation would be serious indeed.[93]

Simpson began with a brief overview of the new cosmology, expressing suspicion of Jeans's efforts to show that the solar system is unique in the universe and a preference for Millikan's claim that matter was being continuously created in the depths of space from the energy of cosmic rays. But he soon turned to evolution, giving a detailed survey of the latest thinking on human origins, with an emphasis on Grafton Elliot Smith's claim that the earliest hunter-gatherers had lived without conflict, war being a product of the first

89.   Ibid., 317.

90.   Ibid., chap. 9, 195.

91.   Simpson, *Landmarks in the Struggle between Science and Religion*, 213.

92.   Simpson, *Nature: Cosmic, Human, and Divine*, 6.

93.   Ibid., 12–13.

great agricultural civilizations.[94] This vision of human nature originating without any predisposition toward violence had obvious implications for the story of the Fall—far from suggesting that human beings were inherently sinful, it implied that we are essentially pacific and are driven to evil only by environmental circumstances. Simpson went out of his way to challenge St. Paul's and St. Augustine's views of original sin on the grounds that they had known nothing of prehistory and were not representative of the early church.[95] He drew out the optimistic message that human nature could be improved by removing the social factors that elicited the unnatural selfish and violent behavior still prevalent today. His concluding section explored the theme that the cosmic process of evolution is a divinely ordained ascent toward higher things, with the human race being an essential step in that progression: "We are part of a process that is a rational, orderly Whole, with these tremendous possibilities in it, emerging at every stage through struggle and suffering and service."[96] Religion would survive the traumas of the present superficial age if it harmonized itself with science, thereby exploiting popular discontent with the emptiness of materialism and the dogmas of official theology. Along these lines, he believed, scientists such as J. Arthur Thomson and Oliver Lodge had done more than most professional theologians.[97]

Simpson's last book was his *Garment of the Living God* of 1934, also based on lectures given in America (this time at Richmond Theological Seminary) and published posthumously. Here he argued that the universe is "a self-limiting manifestation—the Garment of the living God."[98] The book was a survey of the many efforts being made to show that life is something more than mere mechanism and that evolution is a purposeful process aimed at the production of higher spiritual faculties. Simpson referred to the work of Anglican Modernists, including George Adami and Charles Raven, to Robert Broom's ideas on the irreversibility of evolution, and to Smuts's philosophy of holism.[99] He again stressed the ethical character of evolution and the claim that the mind is something more than the material activity of the brain. Cosmic evolution could thus be given an idealist interpretation: "Indeed the

94. Ibid., 93. On Elliot Smith's views, see Bowler, *Theories of Human Evolution*, 214–18.

95. Simpson, *Nature: Cosmic, Human, and Divine*, 95–96.

96. Ibid., 119.

97. Ibid., 130–31.

98. Simpson, *Garment of the Living God*, 215.

99. Ibid., 123–24, 153, 162.

question begins to take shape as to whether in the world-process as a whole, there may not be proceeding the gradual creative elaboration of a great idea."[100] Christ was the turning point of human history, showing us the path toward perfection.[101] Significantly, however, Simpson was still worried about the possibility of hereditary weakness holding back the progress of the race. He twice referred to the Nazis' views on race and the compulsory sterilization of the unfit, expressing concern at the extreme nature of their policies but insisting that it was important to recognize a duty to the future of the race.[102] Here again the belief that science and religion should cooperate in the drive toward spiritual progress showed its darker side.

The year 1927 saw the publication of works by two Scottish authors, both echoing Simpson's ideas. Hector MacPherson was a member of the Free Church and a friend of Simpson's who wrote popular works on astronomy and on the history of the Church in Scotland. His *The Church and Science* offered another historical survey of the relationship between religion and science, minimizing the impact of the conflict and supporting the kind of non-Darwinian view of evolution endorsed by Thomson and Simpson. A chapter on miracles argued that the new worldview offered a way out of the old dilemma: there was no need to postulate interruptions of the laws of nature now that those laws could be seen as manifestations of divine power.[103] The concluding chapter argued that the old dogmas were breaking down on both sides, heralding a new era of reconciliation.

In his *Nature and God*, William Fulton, professor of systematic theology at Aberdeen, supported both Simpson and MacPherson in their efforts to show how the division between science and religion was being healed.[104] Fulton exploited the whole range of antimechanistic interpretations of science, citing Lloyd Morgan, Bergson, J. S. Haldane, Thomson, and MacDougall in the process. In his view, the history of the universe was a great adventure in which mind was being generated at ever higher levels, and to which every human being could contribute—but only when their actions fitted in with the great plan of creation.[105]

The works of all of these writers show that the Modernist impulse was gradually moving further away from the traditional Christian interpretation

100.   Ibid., 150.
101.   Ibid., chap. 7.
102.   Ibid., 121–22 and 209.
103.   MacPherson, *The Church and Science,* chap. 11.
104.   Fulton, *Nature and God,* 79.
105.   Ibid., 283–85.

of humanity's relationship to God. For all that Campbell had tried to retain a role for Christ as the Savior, his critics could see that the New Theology tended to eliminate the Fall, and hence the need for salvation. Christ became only a blueprint for what humanity could hope to achieve when the process of spiritual evolution was eventually—and with our help—brought to its intended goal. Simpson and his supporters in the Free Churches, like the Anglican Modernists with whom they made common cause, ended up pushing the impulse of the New Theology to its obvious conclusion: the almost total submergence of Christianity in the progressionist optimism of the new natural theology.

One writer, at least, saw the dangers of this trend and made a serious effort to indicate a new direction that might be taken by the effort to reconcile science and Christian thought. J. H. Morrison's *Christian Faith and the Science of Today* was based on a series of lectures given at New College, Edinburgh, in 1936. In tune with the more pessimistic outlook of the thirties, Morrison sought to exploit the antimechanistic trend in science without falling completely into the progressionist worldview of non-Darwinian evolutionism. Like Simpson, he thought that theology could not afford to ignore what science said about God's creation, but he saw that many now doubted science's claims to give certain knowledge.[106] A survey of the new physics exploited the views of Jeans, Eddington, and Whitehead, claiming that matter had now been shown to be based on mind or spirit. But Morrison saw a new possibility here for the Christian view of the Resurrection: the Scriptures tell us that "we shall be changed," and this might imply a transition to a new state of matter.[107] The new cosmology had extended our sense of being isolated in a vast universe—Morrison noted Eddington's claim that we are probably the only intelligent race at this point in time—but this only reinforced the need to see that human values were uniquely worthwhile. A chapter on "Nature and Supernature" combined the new physics and antimechanistic biology in a defense of the freedom of the will. The debates among the physicists over determinism were noted: Morrison admitted that the demise of strict determinism had been "acclaimed in theological circles as a charter of ethical liberty," but warned against such "unrestricted licence" because the theist needed to believe in a world of law.[108] Nevertheless, the new physics endorsed a role for unobservables and saw cause and effect as buried in a mystical underworld. This blurring of the distinction between the natural and

106. Morrison, *Christian Faith and the Science of Today*, 2–3.
107. Ibid., 33–35.
108. Ibid., 118–19.

the supernatural allowed us once again to see ourselves as active agents in the world, a point that was also being made by antimechanistic biologists such as Haldane. At the same time, the sense of a supernatural realm underlying the laws of nature made it possible to accept the plausibility of miracles, permitting a new attitude toward the historical evidence of such events in the Gospels.[109]

Morrison's chapter on evolution seemed at first to endorse the progressionist vision of the non-Darwinians, with Smuts, Thomson, Bergson, MacDougall, and Simpson quoted (among others) to endorse the claim that the "evolutionary process has intelligent purpose at the heart of it."[110] But in fact Morrison was doubtful about the confident image of humankind as the "hero of the cosmic saga, a valiant and aspiring creature who has fought his way upward from the lowest depth and has now won the heights."[111] Christianity had gone overboard in its enthusiasm for evolutionism and had failed to notice the growing sense of pessimism that was visible in the writings of many philosophers and highlighted in Dean Inge's emphasis on the universal degeneration predicted by thermodynamics and cosmology. When applied to human origins, this tension between progress and degeneration allowed us to ask whether we are heroic, aspiring creatures or fallen ones. If people are derived from apes, should we cheer them on or preach the Gospel to them?[112] The complex process of human evolution depicted by paleoanthropologists such as Arthur Keith had overthrown the idea of a simple ascent from the apes—the human stock was very ancient, and it could even be argued that the apes are degenerate humans.[113] It was thus possible that the first humans had been created by some great mutation, their mental and moral faculties appearing suddenly rather than by a gradual extension from the animal level. And since evolution showed many examples of degeneration, the traditional idea of the Fall could now be reconciled with science.

> There is a great body of evidence to show that at every stage all along the line of ascent living creatures seem to have missed the upward road, and turned aside into blind alleys where they remained stagnant or even began to degenerate. Does this not suggest that in some mysterious way, deep embedded in nature, there is a principle hostile to progress, some in-

109.   Ibid., 136–39.
110.   Ibid., 158.
111.   Ibid., 158.
112.   Ibid., 176–77.
113.   Ibid., 179–81.

scrutable power which continually obstructs and diverts and frustrates
the upward movement?[114]

The Fall, so long dismissed by evolutionists as unthinkable, was thus merely
the latest example of this negative power's influence. The human race had not
quite hit the mark, and its history had thus been polluted near its fountain-
head—although Morrison would not condone the notion of total depravity and criticized those who attributed it to Calvin. Even Lodge admitted that
we had been tempted by false pride, in this respect coming closer to the Christian tradition than the supposed evangelical Drummond, who had converted
so completely to optimism.[115] There was still a need for Christ to save us
from the effects of this false start and convince us that God is something
more than a mathematician or an architect.

Morrison returned, in effect, to the position sketched out in Griffith-
Jones's *Ascent through Christ,* although this position was now supported by reference to a new generation of scientific work on the complexity of human
origins. The slide into a purely optimistic vision of humanity's destiny, so
obvious even in Griffith-Jones's later work, let alone in that of Simpson and
the Modernists, was checked—and not by a total repudiation of science and
its influence. To many conservative Christians, however, the Modernist impulse had gone so far that the link with even a nonmaterialistic science had
to be repudiated. Morrison had referred at one point to Karl Barth's negative
image of science, and while not accepting it, had clearly seen the need to retain some sense of humankind as a lost species desperately in need of salvation from a source outside this world. It was the Modernists' loss of any sense
of sin and the need for redemption that came increasingly to disturb those
Christians who saw in the disasters of the contemporary world only too clear
evidence that the prophets of progress had been led astray (see chapter 9).

114. Ibid., 185.
115. Ibid., 193–95.

CHAPTER 8

Anglican Modernism

While Campbell's New Theology was rapidly eclipsed in the Nonconformist churches, Modernism was gaining strength within the Church of England. Anglicanism had long tolerated a wide range of opinions within its ample remit as the established national church, although it walked a constant tightrope as it strove to balance the potentially divisive forces unleashed from time to time by enthusiasts for one position or another. In the mid-nineteenth century the Oxford movement, forerunner of Anglo-Catholicism, had created major tensions as it strove to assert itself against the more evangelical tradition favored by those who stressed the church's origins in the Protestant Reformation. The later nineteenth century saw the emergence of a liberal viewpoint usually referred to as the Broad Church, and it was this liberalizing trend that gave rise to twentieth-century Modernism. The Modernists wanted to make Christianity compatible with the latest trends in philosophy and science, and they feared that unless some accommodation could be reached, the ongoing decline in church membership and influence would become precipitate and end with the de-Christianization of the country. They believed it was vital that the church not reject the overtures being made by the new nonmaterialistic trend in science, and that the only way to ensure that these scientists were not rebuffed was to purge Christianity of ancient superstitions and dogmas that did not square with the modern worldview.

Modernists were drawn to teleological evolutionism and holistic biology, and many welcomed the new idealism of Eddington and Jeans. But Modernism was not a unified movement, and some of its supporters had little interest in science. Some of the most widely publicized controversies surrounding Modernism arose from its challenge to the orthodox view of Christ's divine status, and this challenge was only peripherally related to the influence of science. The true source of Modernism for many theologians was the idealist philosophy popular in British academic circles in the late nineteenth century. This movement was out of date in the intellectual world of the 1920s and 1930s, but many theologians remained unaware of the latest trends in philosophy. Most Modernists were also unaware that their view of science, especially biology, was no longer up to date. Their perception of the latest developments was distorted by the activities of scientists-turned-popularizers such as J. Arthur Thomson and C. Lloyd Morgan, who convinced many nonspecialists that the antimechanistic trend of the late nineteenth century was still active. Only the few Modernist theologians who kept in touch with the latest developments in research were aware that the tide was turning.

Modernism flourished in the first three decades of the new century, achieving a significant level of institutionalization with its society, the Modern Churchmen's Union, and its periodical, *The Modern Churchman.* Its supporters entertained the real hope that they might be able to shift the balance of opinion within the church sufficiently in their own direction that the hoped-for revival would become a reality. In fact, however, the Modernists succeeded only in generating the most serious tensions the church had experienced in a century or more. The gulf between evangelicals and Anglo-Catholics was already wide enough to generate massive controversies such as that surrounding the revision of the Prayer Book. Modernism simply added a new source of fragmentation, its views anathema for entirely different reasons to both evangelicals and to Anglo-Catholics. Liberal evangelicals were some of the most active Modernists, but those with more conservative views watched with horror as the new enthusiasm for evolution swept aside the tradition of the Fall and denatured the whole concept of the Atonement. Anglo-Catholics, who had made some effort to accommodate the liberalizing trends of the late nineteenth century, saw their reverence for the sacraments challenged by a new wave of hostility to ritualism, based now on supposedly rationalist grounds.

There were many both inside and outside the church who saw Modernism as a betrayal of Christian principles. The Anglican priest who has "Doubts" in Evelyn Waugh's *Decline and Fall* of 1928 eventually stays in the church because "he has been reading a series of articles by a popular bishop and has discovered that there is a species of person called a 'Modern Churchman' who draws the full salary of a beneficed clergyman and need not commit himself to any religious belief."[1] Waugh was hardly a neutral observer, since he was received into the Roman Catholic Church two years later, but his conversion heralded the more general transformation of values in the 1930s, which saw Modernism enter a period of catastrophic decline. Far from encouraging young people into the church, Modernism created a denatured and abstract form of Christianity that had little attraction for anyone who wanted a religion of personal salvation. If science gave any credence to religion, it was to a form of pantheism that offered no real focus for the kind of belief that might transform a person's life. Those who wanted to climb more firmly on board the liberal bandwagon lost interest in religion altogether, while those who wanted something to believe in rejected both rationalism and science. Meanwhile. the temptation of Marxism for intellectuals disillusioned with the social upheavals of the time grew ever more powerful. Even without the resurgence of Darwinism and materialism in biology, Modernism was doomed.

MODERNISM AND THE NEW NATURAL THEOLOGY

The Modernists were anxious to ensure that science be portrayed in a manner that would not upset their own efforts to make Christianity compatible with modern culture. If religion was to hold its hand out to science, it was in the expectation that science would reciprocate by framing its vision of nature in a way that was compatible with, and preferably supportive of, at least the basic theistic position. For humanity to be the product of nature, it was necessary that the divine purpose be seen at work in nature's activities. The Modernists naturally endorsed the "new" natural theology, and many collaborated actively with scientists who endorsed this position. The Lamarckian botanist George Henslow, himself an ordained minister, was active in the Modern Churchmen's Union. In 1925 the Union's conference was de-

---

1.   Waugh, *Decline and Fall*, 141.

voted to "The Scientific Approach to Religion" and was addressed by E. W. MacBride, J. S. Haldane, and Lloyd Morgan. In 1931 the theme was "Man," and the speakers included J. Arthur Thomson and Oliver Lodge.[2] Thomson also translated Rudolph Otto's *Naturalism and Religion,* a work that impressed many Modernist theologians with its emphasis on nature's purposeful activity.[3]

Yet the new natural theology's emphasis on evolution required a massive transformation of Christian belief. As theologians such as F. R. Tennant pointed out, even a teleological view of evolution undermined the traditional concept of the Fall and original sin (see chapter 6). Tennant's new concept of sin, coupled with the Modernists' denial of the supernatural, allowed humanity to portray itself in a new light—no longer a sinful race facing judgment, but the agent of God's power on earth, with the ability to push spiritual progress forward by its own efforts. Christ became the perfect human, a model for what we may all become in the future if we follow his teachings, not a supernatural agent with miraculous powers. Bishop E. W. Barnes's efforts to popularize this new version of the Christian message were still attracting newspaper headlines in the 1920s, along with much criticism from clergy with more conservative views. The hope of human progress encouraged a number of Modernists, including Barnes, to take a very hard line on a number of social issues. For them, it became a Christian duty to further the spiritual progress of the race, and if this meant following a eugenic policy to weed out the genetically unfit, so be it.

Acceptance of teleological evolutionism was already commonplace at the turn of the century, but for many Modernist clergymen this implied no more than an extension of the idealist philosophy in which they had been trained. A chapter in a 1902 volume titled *Personal Idealism* suggested that Darwin had merely followed in the footsteps of Aristotle.[4] Examples of uncritical acceptance of evolutionism from this period include the Rev. W. Profeit's *Creation of Matter* (1903), James Hope Moulton's *Is Christianity True?* (1904), and Vernon F. Storr's *Development and Divine Purpose* (1906). Profeit's book offered a kind of idealistic gloss on science as a whole, in which the properties of matter indicated the Mind of its Creator, while evolution was progressive, orderly, and beautiful.[5] Moulton's lecture was one of a series given to defend

---

2. Details of the conferences are given in Stephenson, *Rise and Decline of English Modernism,* appendix A.

3. Otto, *Naturalism and Religion,* chaps. 4–9. Otto was professor of theology at Göttingen.

4. Underhill, "Limits of Evolution," 220.

5. Profeit, *Creation of Matter,* 152–53.

the faith against an assault by the socialist writer Robert Blatchford, and has been taken by one later writer as a classic illustration of how Christians failed to realize the extent to which the liberalizing trend had undermined elements of the traditional faith.[6] Moulton was both an ordained minister and a professor of Hellenistic Greek at Manchester; he used his lecture to attack anthropologists such as James Frazer and argue that the development of religion is a progressive revelation of God's truth. The whole evolution of life and human civilization could thus be seen as a progression culminating in the appearance of Christ and the spiritual progress of humanity.[7] Storr's book explored the theme of an evolutionary natural theology, stressing its foundations in a philosophy of divine immanence.[8]

These themes were developed in a little book called *Science and Religion* published in 1906 by William Gascoyne Cecil, bishop of Exeter. Cecil admitted that he had no qualification to speak on science but argued that, in an age increasingly distrustful of authority, even "one of the crowd" had a right to express an opinion. Science itself now admitted that its discoveries offered no final truth, and this was most obvious in evolutionary theory. Darwinism was perhaps part of the story, but the evolutionists themselves did not dare put its teachings into practice, or they would be advocating the death penalty for anyone deemed unfit.[9] There was no real conflict with Genesis, which no one supposed to be a historical account of creation. For Cecil, evolution was a purposeful process with the human race as its intended goal; the concept of natural selection could be equated with the Christian belief that only the elect would be preserved.[10] The process was still going on within the human species, with the weaker races constantly disappearing—a lesson that should be learned by the British Empire itself, which would not survive unless those who governed it maintained a spirit of self-sacrifice. They must also maintain their birth rate, or there would not be enough leaders for the next generation.[11] Cecil insisted that scientific evolutionism could only explain the pro-

---

6. Kent, *From Darwin to Blatchford*, 32–33; this is a perceptive account of the failure of late-nineteenth-century theology to confront the underlying problems of the acceptance of evolutionism.

7. Moulton, *Is Christianity True?* 16, 19. A series of lectures under the same title as that used by Moulton was given in Manchester in 1903–1904.

8. Storr, *Development and Divine Purpose*; see esp. chaps. 6 and 7. On Storr's career, see Harris, *Vernon Faithful Storr*.

9. Cecil, *Science and Religion*, 9–12. Cecil contributed little else on science beyond a response to Arthur Keith in 1929; see Cecil, "Darwinism and What It Implies."

10. Cecil, *Science and Religion*, 37.

11. Ibid., 48.

gress of intelligence up to the human level—but intelligence by itself found the universe meaningless, and it was the appearance of the moral and aesthetic faculties that revealed that "this world is one in which God's creation is being gradually perfected."[12] The book ended with an argument for the immortality of the soul based on the fact that matter, energy, and (at the cellular level) life itself were all eternal, so it was improbable that individuality was the exception.

Enthusiasm for the progressionist vision of evolution can also be seen in the contributions of Anglican theologians to collected works on this theme. The 1917 collection *Evolution in Modern Thought* contained a chapter on religion by Philip Napier Waggett, a member of the Society of St. John the Evangelist (the "Cowley Fathers" of Oxford). Waggett had already tried to minimize the gulf between science and religion in a book published in 1905, in which he had noted the extent of anti-Darwinian ideas in biology.[13] In his 1917 chapter he again sought to limit the damage done by earlier controversies over evolution by stressing the existence of alternatives to the Darwinian explanation.[14]

A more actively Modernist position was taken by James Maurice Wilson, canon of Worcester, in the 1925 survey *Evolution in the Light of Modern Knowledge.* As a young man, Wilson had temporarily lost his faith until he realized that theology could be reformulated to take the new ideas into account. Evolution was the method of creation, and we are its highest products, "the incipient vehicle of self-expression, the incarnation in some degree, of that Unseen Creator Spirit Himself."[15] Wilson conceded that the most important consequence of this transformation was the elimination of the traditional concept of the Fall, sin now being seen as a survival of the lower elements of mind from our animal ancestry. He hoped that the rejection of the old idea of the Fall would remove the chief obstacle to the wider acceptance of Christianity by the modern generation.[16] He acknowledged, however, that in transforming our interpretation of sin, evolutionism challenged the traditional view of Christ's role as Savior. Evolution was approaching "the citadel of our Christian Faith"—but was it an enemy to be repelled or a welcome reinforcement for the defenders? Wilson argued for the latter, but accepted that there would have to be a transfor-

12. Ibid., 72–73.
13. Waggett, *Scientific Temper in Religion*, e.g., 96, 136.
14. Waggett, "Influence of Darwin upon Religious Thought."
15. J. M. Wilson, "Religious Effect of the Idea of Evolution," 493.
16. Ibid., 498.

mation in our ideas about Christ, who now had to be seen as the supreme manifestation of a divine spirit that was present in the whole of nature.

Wilson's views were quoted with approval by one of the leading theologians of the Modernist movement, J. F. Bethune-Baker, Lady Margaret Professor of Divinity at Cambridge.[17] Bethune-Baker, who became notorious as
one of the clergy who openly rejected Christ's miraculous powers, linked this
position to the worldview based on evolution. He welcomed the insights of
J. Arthur Thomson and other scientists who saw evolution as a process designed to generate higher levels of morality. Progress was not continuous, but
there was a main stem of evolution leading toward humankind, a "continuous process of creative synthesis productive of higher orders of being from
the amoeba to Man."[18] Christ was the highest product of the whole scheme,
sent to teach us how to overcome our baser instincts, but humankind must
now see itself as working and struggling to achieve the divine purpose—future progress depended as much on humanity as on God.[19]

Similar views were expressed by another leading Modernist, Burnett
Hillman Streeter, who was canon of Hereford and a fellow of Queen's College, Oxford. In 1919 Streeter edited a volume called *The Spirit,* containing a
very Bergsonian article by A. Clutton-Brock on the life force and progress.[20]
He also edited *God and the Struggle for Existence,* a product of the soul-searching
going on among many evolutionists in the aftermath of the Great War. His
own contribution accepted that humanity was not doing what it could to
eliminate suffering, but argued that since God is in the world and suffers with
it, we can be sure that the pain is necessary for further development to take
place.[21] A chapter by Lily Dougal addressed the question of whether evolution tended toward anything that could be called "good." Accepting that humanity was "Nature's masterpiece," she asked where future evolution could
be taking us: if there was an intelligence behind nature, what was its future
purpose with us?[22] If there was to be a higher form of humanity, this would
require the elimination of inferior races and individuals: "The dream of the
eugenist, or indeed of any other scientist, can never be fully realized until the
stupid, weak or unwholesome human beings harboured by our present civi-

17.   Bethune-Baker, *The Way of Modernism,* 51, 81, 86; see also "Evolution and the New Theology."

18.   *The Way of Modernism,* 71.

19.   Ibid., 134.

20.   Clutton-Brock, "Spirit and Matter."

21.   Streeter, "The Defeat of Pain." The same theme was taken up by C. F. D'Arcy's contribution to the same volume, "Love and Omnipotence," discussed below.

22.   Dougal, "Survival of the Fittest," 67–68.

lization have left the earth."[23] The answer at one time would have been to allow the struggle for existence to carry them off, but this was no longer an option, and so eugenic policies of selective breeding would have to be instituted.
Dougal accepted that moral reform was also necessary, but hard decisions
would have to be made about the unfit—to evade this responsibility was to
turn our backs on the Creator's plans for humanity and, indeed, on Christ's
teachings, which had always been concerned with our future welfare. Here the
Modernists' vision of the human race as the supreme expression of God's
will on earth translates directly into a plan of action in the name of progress.
We are the agents of His will and must take what steps are necessary to improve the quality of the race. Eugenics becomes a moral duty fully compatible with the Modernist vision of humanity's place within the divine scheme.

Beginning in 1924, Streeter met at Dougal's house with a group of fellow
thinkers (including, on occasion, Julian Huxley) for weekend conferences.
The result of their discussions was a volume titled *Adventure: The Faith of Science
and the Science of Faith*. Here scientists, philosophers, and theologians united to
stress that they were all searching for knowledge in the same dynamic and adventurous way. The attraction of this claim for the Modernists was obvious:
not only would theology become compatible with scientific knowledge, but
its willingness to transform its outdated dogmas would be legitimized by
analogy with the scientific method. The link between science and religion was
developed further in Streeter's *Reality*, published in the same year (1927). Here
he celebrated the demise of the old mechanistic viewpoint and the new philosophy of Eddington and others in which science offered only abstractions,
leaving room for other areas of knowledge with their own very different
modes of representation.[24]

Streeter noted the neovitalist viewpoint of Hans Driesch, without com-
mitting himself to it, but went on to discuss Henri Bergson's theory of the
élan vital at length. Life was creative and dynamic and represented the only organizing principle we know in the world. But it could not be the blind impulse postulated by Bergson: there was an organizing power behind it, and
hence behind the universe as a whole: "Ultimate Reality is certainly no less
(and, if that, probably far more) alive and fully conscious than the highest of
its products of which we have any knowledge—the mind and heart of
man."[25] Peter Kropotkin and others had shown that from the creative strife

23.   Ibid., 73. On Modernism and eugenics, see G. Jones, *Social Hygiene in Twentieth-Century Britain*,
47–50.

24.   Streeter, *Reality*, 28–30.

25.   Ibid., 125.

of evolution came the principle of love, reinforced in the higher animals as they gradually rose above the struggle for existence.[26] Life was a dangerous game, but it was a game of character building, a "school of manhood."[27] In human beings, at least, mind took on a power of its own, confirmed by the experiments suggesting that telepathy was real: we are each a center of "psychic radio-activity" capable of influencing others through prayer.[28] In these circumstances, it was unthinkable that God would permit the minds thus created to perish with the body.

The link between the paranormal and the new natural theology had been made plain by scientists such as Lodge, and Streeter was by no means the only theologian to take up the theme. John Charlton Hardwick, who had been a chaplain at Ripon Hall, Oxford, wrote his *Religion and Science from Galileo to Bergson* (1920) to provide an overview of the antimechanistic trends in science, noting in his final chapter the work being done in psychic research.[29] Hardwick's pamphlet *Religion and Science* (one of a series called Papers in Modern Churchmanship) focused on the problem of miracles, arguing that such supernatural events were a problem from both the scientific and the religious viewpoint. Like many Modernists, Hardwick felt uncomfortable with the image of Christ as a being exerting supernatural powers to challenge the laws of nature. In a much later pamphlet, *The Bible and Science* (1938), he put these two positions together in a way that preserved the Bible's veracity without requiring supernatural agencies. The marvels attributed to Christ, especially His powers of healing, could be explained as the products of psychic energy emanating from within the personality—an area largely ignored by science until the modern period. According to Hardwick, the efforts of the Society for Psychical Research had now put the study of these phenomena on a scientific footing.[30] The future would see whole new areas of science opening up that would transform our understanding of the world while allowing us to appreciate the role played by the paranormal powers in human spiritual development.

An influential work devoted solely to the problem of God and evolution was published in 1929 by W. R. Matthews, Dean of King's College, London,

---

26. Ibid., 159.

27. Ibid., 223.

28. Ibid., 294–95.

29. Hardwick, *Religion and Science from Galileo to Bergson*, chap. 12; in his preface, Hardwick acknowledges the support of the leading Modernist H. D. A. Major.

30. Hardwick, *Bible and Science*, e.g., 12. The claim that psychic research could explain the apparently miraculous events recorded in the Bible had been made by some of the early supporters of spiritualism; see, for instance, Wallace, *On Miracles and Modern Spiritualism*, esp. 207–9.

and chaplain to the King. The book was a celebration of the concept of emergence centered around an evaluation of Lloyd Morgan's theory and the philosophy of Samuel Alexander. Matthews acknowledged that evolution had deeply affected the modern mind. He admitted that he had little knowledge of science outside psychology, but insisted on the right of all thinking persons to assert an opinion.[31] The question of Genesis was explicitly set aside; as far as Matthews was concerned, matters had moved on so far that it was now only a question of asking whether evolution could be understood in a way that would make it compatible with religion. The mechanistic neo-Darwinian model of evolution was obviously a problem, however. If natural selection were the sole driving force of evolution, it would, in William Mc-Dougall's words, "exclude any influence of intelligent purpose upon the course of development."[32] But the mechanist program had already been challenged by the vitalism of Driesch and J. S. Haldane, while the Lamarckian mechanism of evolution had still not been ruled out. More important was the recognition that in the course of evolution new values had emerged. This was not a matter of mere complexity because a more complex machine was not preferable to a simpler one that was equally efficient, but everyone accepted that human life was more valuable than animal life.[33] If evolution generated higher values, then it could be seen as "the progressive realization of the Divine purpose." Bergson had been on the right track, but his blind, striving force did not, in the end, guarantee freedom because there was no transcendent purpose involved. It was necessary to look to the writings of Morgan and Alexander to gain a better understanding. Their insistence that there was no external interference at the points where new values emerged was crucial to the formulation of a nonmechanistic evolutionism. At the same time, Matthews was critical of writers such as H. G. Wells and George Bernard Shaw who tried to make evolution the basis of a new religion, and the same criticism had to be extended to Alexander, despite the latter's more positive contributions. Alexander was "the Aquinas of the religion of evolution," but his notion of a nisus toward perfection that actually created God was unacceptable. After all, Huxley stressed the progressive role of evolution, yet was opposed to theism.[34] The Christian could accept the nisus, but saw it as the expression of the purpose of a preexisting Deity: "Though in words it may be denied, clearly 'nisus' and 'direction' in evolution are teleology thinly dis-

---

31. Matthews, *God and Evolution*, 3.
32. Ibid., 9.
33. Ibid., 16–17.
34. Ibid., 31–34.

guised."[35] Emergent evolutionism was a new view of creation as a continuous process that encouraged "a nobler and more hopeful view of God" and allowed us to see the world as "a great adventure of creative will in which we are privileged to cooperate."[36]

Matthews's book shows how Modernist theologians were able to react positively to what were being presented to them as the latest developments in nonmaterialistic science. There is no shortage of other examples to confirm the extent to which Anglican theologians exploited this movement to uphold the expectation of a reconciliation between science and religion. Stewart A. McDowall of Trinity College, Cambridge, wrote his *Evolution and Spiritual Life* to endorse the philosophy of Bergson.[37] E. H. Archer-Shepherd, vicar of Avebury, praised the writings of Thomson, although his enthusiasm was qualified by a suggestion that we have indeed fallen from an early state in which our ancestors lived in harmony with nature.[38] Vernon F. Storr, canon of Westminster and an influential liberal evangelical, wrote a contribution to the popular Lambeth series in 1931, picking up the theme of his much earlier *Development and Divine Purpose* and stressing how evolution had developed along predictable lines toward higher levels of spiritual awareness.[39] R. O. P. Taylor, vicar of Ringwood in Hampshire, opted for a theistic version of Bergson's philosophy, in which God expressed himself indirectly in the evolutionary process, requiring struggle and effort by individual creatures for progress to be possible.[40] Taylor was also enthusiastic about the move toward an idealist interpretation of physics promoted by Jeans and Eddington, which he saw as an endorsement of the new emphasis on immanence by Modernist theologians.[41] An article by W. Robinson used teleological evolutionism to challenge the pessimistic philosophy of Bertrand Russell.[42] Russell's nihilism was also a target of A. C. Bouquet's *Doctrine of God* (1934), which attacked materialism by appealing to the "spontaneity" of the atom as revealed by modern physics.[43] Bouquet favored emergent evolutionism, although he opposed

35.   Ibid., 47.

36.   Ibid., 52–53.

37.   McDowall, *Evolution and Spiritual Life*, xi. See also McDowall's *Evolution, Knowledge, and Revelation*, esp. 15–20.

38.   Archer-Shepherd, *Orthodox Religion in the Light of Today*, 21–22 on Thomson and 52 on the Fall.

39.   Storr, *God in the Modern Mind*, 16–21.

40.   R. O. P. Taylor, *Meeting of the Roads*, chap. 9. See also his *Does Science Leave Room for God?*

41.   *Meeting of the Roads*, chaps. 3 and 4.

42.   W. Robinson, "Christianity and Evolution."

43.   Bouquet, *The Doctrine of God*, 10–11; on Russell, see 17, 26, and 88.

those who saw God emerging from the universal progress on the grounds that Mind must preexist to explain the purposefulness of nature's development.[44] Thomson was invoked as an authority for a vision of evolution in which there was a minimum of suffering beyond what was necessary to allow nature the freedom to experiment in the drive toward higher forms.[45]

These examples show how readily the Modernists were to make common cause with the new natural theology being promoted by the more conservative members of the scientific community. Their desire to stress God's immanence within nature was a natural foundation on which to build support for the neovitalism of J. S. Haldane and McDougall, the progressionism of Bergson, Thomson, Morgan, and Lodge, and the idealist view of the philosophy of science running from Haldane to Jeans and Eddington. This small group of scientists, often out of touch with the majority of their community, completely dominated the Anglican clergy's image of what contemporary science had to offer. Their influence reveals the power of the popular press and the publishing industry to create a significantly distorted impression of science, which, in this case at least, misled a whole generation of the clergy into thinking that the latest developments were running in favor of their own hopes for a reconciliation. In fact, of course, the Modernists' image of modern science was already out of date, except perhaps in the case of the new physics. Tensions would arise as soon as those clergy more directly in touch which what was actually happening in science realized that materialism and Darwinism were once more on the march in biology.

In 1930 F. Leslie Cross published a volume titled *Religion and the Reign of Science*, a volume in the "Anglican Library of Faith and Thought." He accepted that the church was losing ground because it was out of touch with science and realized that the scientific way of thinking undermined the old reliance on dogma. But Cross saw science as a metaphysical wilderness that could offer little guidance. The new physics of Eddington and Jeans opened up the prospect of a reconciliation through idealism, yet their views were highly individual and could hardly be taken as the basis for a dialogue with science as a whole.[46] The collapse of neovitalism had left biology in a state of confusion, leaving the new psychologies as the more innovative trends in the study of life.[47] Here was one theologian, at least, prepared to acknowledge the point made by Joseph McCabe and the Rationalists—that the new natural theol-

---

44. Ibid., 65.
45. Ibid., 91–92.
46. Cross, *Religion and the Reign of Science*, 36.
47. Ibid., 49 and chap. 4.

ogy was based on an interpretation of science that was at best idiosyncratic and at worst out of date.

There were surprisingly few Modernists, however, who acknowledged the new directions being taken in science. The few who did were increasingly forced to confront the new developments in biology, either by attempting to take them into account or by openly repudiating them. To illustrate this tension, we shall explore the work of four prominent Modernists who took a detailed interest in science throughout their careers and played prominent roles in the attempt to reconcile science and religion. They illustrate the divisions that existed within the Modernist movement itself over the role of idealist philosophy in Christian thought and over the social policies to be derived from the progressionist ideology. Their careers also reveal the changing fortunes of the Modernist movement within the church as the more conservative mood of the 1930s undermined both confidence in progress and the desire to reforge a natural theology.

Of the four, Archbishop Charles D'Arcy of the Church of Ireland remained closest to the mainstream version of Modernism outlined above, including support for nonmechanistic biology and the eugenics movement. E. W. Barnes achieved preferment by a stroke of luck, being offered the see of Birmingham by the new Labour government in 1924 and remaining a thorn in the side of the church throughout the rest of his career. He attracted wide publicity for his emphasis on the need to take evolution seriously and for his attacks on Anglo-Catholic ritualism. He was more in touch with current developments in biology than most Modernists, and in grappling with the new Darwinism of R. A. Fisher, he, too, became an enthusiast for eugenics. W. R. Inge, the Dean of St. Paul's, was known to thousands through his writings in the daily newspapers, but his views reflected a degree of eccentricity that only the Church of England could tolerate. His Modernism made him a target for many in the church, while his increasingly right-wing political views made him a controversial figure on the national stage. Inge used the cosmologists' image of a universe running down toward an eventual "heat death" to oppose progressionism—but his gloomy prognosis for the future of humanity made him all the more committed to the eugenics movement's campaign to check the breeding of the unfit. The last example is Charles Raven, who achieved fame through his popular writing on natural history but was for a long time marginalized within the church. Raven, too, knew what was happening in biology, but he simply rejected it, along with the social implications of hereditarianism, remaining a lifelong proponent of non-Darwinian ideas of evolution. He explicitly criticized the eugenicists within the church and supported those who sought to realign Christianity with left-wing politics.

CHARLES F. D'ARCY

Charles F. D'Arcy was born in Dublin and built his career in the Church of Ireland, an offshoot of the Anglican community that identified itself with the evangelical wing of the mother church. In Ireland, hostility to Roman Catholicism was an integral part of the ideology of the Protestant ruling class. D'Arcy was educated at Trinity College, Dublin, and served as vicar of Belfast from 1900 to 1903. He was appointed bishop of Down and Connor in 1903, archbishop of Dublin in 1919, and then almost immediately archbishop of Armagh, that is, the Anglican Primate of Ireland. As a young man he had studied natural history, especially botany, and had no difficulty understanding the evidence for evolution. But his education led him toward the idealist philosophers as a bulwark against the materialistic interpretations of Huxley and Spencer. His politics were staunchly loyalist; he was a signatory of the Ulster Covenant of 1912, which proclaimed the Unionists' opposition to home rule for Ireland. He also served as the first president of the Belfast Eugenics Society.[48]

D'Arcy first achieved some attention when he spoke on Christianity and modern thought at the 1908 Lambeth Conference. He contributed a substantial chapter, called "Love and Omnipotence," to Streeter's *God and the Struggle for Existence* (1919). The Great War had caused many to ask why an omnipotent God does not end suffering in the world. For D'Arcy, Darwinism was but one of a number of factors that had destroyed the eighteenth century's vision of a benevolent God: we now had to see God as operating within the world, perhaps suffering along with it.[49] Science reveals the unity of nature and hence of the God who sustains it, but the nineteenth century placed too much reliance on Spencer's claim that the underlying Power was inscrutable or unknowable. D'Arcy believed that we can know something of this Power; most obviously, the existence of a fundamental order means that the universe is trustworthy: "Science is indeed man finding himself at home in the universe, and finding that, within limits, he is safe."[50] At first sight it

---

48. See G. Jones, "Eugenics in Ireland." On D'Arcy's education and career, see his autobiography, *Adventures of a Bishop.*

49. D'Arcy, "Love and Omnipotence," 16–18.

50. Ibid., 36.

might seem that the Power is concerned more with beauty than with goodness. But God can produce beauty without the help of His creatures, whereas goodness involves the participation of these finite personalities, which must first evolve and then strive to transcend their selfish inclinations. Pain and suffering are the driving forces of progress, but in a world in which the individual has free will, there is no guarantee of progress other than the faith of the Christian that God is good and that good will triumph in the end.[51] D'Arcy concluded by hailing Bergson and McDougall as the key influences in destroying the mechanistic view of the universe and restoring our sense that we have the freedom to choose and the power to influence the direction of events.[52]

D'Arcy subsequently wrote two major studies proclaiming the new natural theology as the basis for a synthesis between science and religion. In the preface to his *Science and Creation* (1925), he noted that Jesus "taught that the work of God in creation never ceases."[53] Darwin's name was still the symbol for the idea of evolution, an idea that had taken over the popular mind even though his detailed theory was no longer accepted by scientists. D'Arcy traced the "epic of creation" revealed by cosmology, geology, paleontology, and prehistoric archaeology, stressing the overall progress toward higher values. In biology, natural selection was no longer seen as a creative force: "Natural selection is a sifting process and a fixing process. It is nothing more. All that is really creative must be pre-supposed before the process begins."[54] Noting that biologists had now rejected the Lamarckian process of the inheritance of functionally acquired modifications, D'Arcy turned instead to Arthur Keith's claim that processes within the organism shaped the consequences of germinal variations along purposeful lines. But where Keith had intended only to suggest that these processes blunted the force of the old arguments directed against "random" variation, D'Arcy thought that they implied divine preordination of what would appear. The features of the human body "were worked out in detail and prepared countless ages before [we] had need of them," so that the characters once explained by recapitulation of ancestral stages were really more of a prophecy of what was to come.[55] This was true even for the specific qualities of the highest human race, and here D'Arcy nailed his ideological colors firmly to the mast:

51. Ibid., 42–46.
52. Ibid., 54–55.
53. D'Arcy, *Science and Creation*, v–vi.
54. Ibid., 20.
55. Ibid., 28.

It appears, in fact, that the Nordic race, the fair-haired Achaean, the great adventurous race, which in time subdued the fairest region of the earth and created the noblest civilizations, was shaped and coloured not amid Scandinavian snows, but far back in the dim womb of time, and preserved, through countless generations until the epoch for his birth had come.[56]

It is this sense of racial superiority that explains D'Arcy's involvement with eugenics. Charles Raven, who was aware of D'Arcy's sympathies with that movement, complained that his vision of predetermined change leaves us "with a concept of the deity appropriate perhaps to the Calvinism of Ulster, but hard to reconcile with any other form of Christian faith."[57]

D'Arcy's real intention was to promote a nonmechanistic view of life that would sustain the idea that evolution is a genuinely creative process, but then to subvert the Bergsonian notion of a "blind" *élan vital* by arguing that all the achievements of the life force had been foreseen by its Creator. He cited the American paleontologist Henry Fairfield Osborn's evidence for orthogenesis, in which evolutionary lineages were driven as though toward some predetermined goal. For D'Arcy, these trends confirmed the existence of a deeper controlling force that operated on a broader scale than the activity of the individual and hence revealed that it was the product of a "Supreme Universal Intelligence."[58] The true vitalists who saw life as something striving against brute matter were "modern Manicheans," whereas science had now shown that the laws of nature themselves were able somehow to embody a creative design.

Providence and the World-Order (1932) engaged much more closely with the ideas of Smuts and Morgan. D'Arcy also noted Jeans's idealist interpretation of physics as an indication of the new directions being taken in science, although he made clear his own early commitment to idealism through exposure to the writings of T. H. Green and the brothers Caird.[59] The book's survey of biology and psychology made frequent references not only to holism and emergent evolutionism, but also to the writings of Bergson, Keith, Haldane, and McDougall on the creativity of living organisms. The problem with all these scientific efforts to explain the emergence of purposeful activ-

56. Ibid., 31.

57. Raven, *Creator Spirit*, 47.

58. D'Arcy, *Science and Creation*, 62–63. On Osborn's ideas, see Bowler, *Eclipse of Darwinism*, 131–33 and 174–80.

59. D'Arcy, *Providence and the World-Order*, 15 and 34. The conclusion, chap. 8, also highlights Eddington's views on the new physics and the destruction of determinism.

ity was that they were all little more than descriptions of what had actually been achieved by evolution: "Creative evolution, emergence, holism: these are all attractive, picturesque, and more or less true as presentations of the order of events, and of the way in which the stages of the creative process succeed one another, but not one of them tells us anything of the true nature of the activity which has been at work in the universe."[60] In the end, we have to accept that the creative Power behind nature intended to produce minds with the freedom to act upon the world. Moving in conclusion into the arena of human prehistory, D'Arcy again hailed the European racial stock as the high point of creation: "It is surely not too much to claim for Western civilisation that it forms the central stream of advancing human progress."[61] Here again we see how easily the teleological form of evolutionism could generate support for hereditarian social policies.

E. W. BARNES

Ernest William Barnes was perhaps the most controversial exponent of the Modernist position. Like D'Arcy, Barnes was a eugenicist, but in his case, support for the policy of selective breeding for humanity was bolstered by a recognition that the Darwinian theory, far from fading away, was now gaining ground in biology. He became notorious through what the press called his "gorilla sermons," in which he stressed the need for the church to go beyond paying lip service to the idea of evolution. Barnes was also attacked for his stand against Anglo-Catholic ritualism, which he saw as a survival of primitive magical superstitions. His position represented the extreme Protestant view of the Eucharist, in which the sacrament has a purely symbolic role, but it was also an integral part of his attempt to bring Christianity into line with modern science. For Barnes, the claim that a spiritual change took place in the consecrated wafers was an affront to the scientific worldview. His opposition to ritualism and his "gorilla sermons" thus stand side by side in the campaign to forge a rationalized Christianity that would appeal to the modern way of thought.

Barnes's interest in science and his evangelical viewpoint can both be traced back to his early career.[62] His father was a schoolteacher with strong Baptist beliefs, although Barnes himself began his move toward Anglicanism

60. Ibid., 127.

61. Ibid., 157.

62. Barnes's biography was written by his son, John Barnes, under the optimistic title *Ahead of His Age*; see also Bowler, "Evolution and the Eucharist."

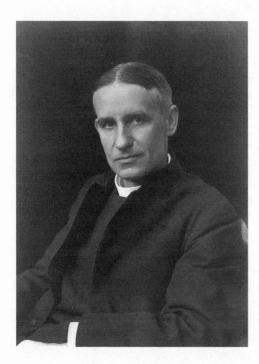

E. W. Barnes. Photograph by Walter
Stoneman, 1918. By courtesy of the
National Portrait Gallery, London.

before going up to Trinity College, Cambridge, on a mathematics scholarship
in 1892. He became a don at the same college, where he taught Eddington,
Fisher, and Hogben, among many others. He worked in mathematical physics
and was elected a Fellow of the Royal Society in 1909. Although Trinity was
not a very religious college, Barnes deepened his involvement with Anglicanism and was ordained in 1902. He had radical political opinions and was a
pacifist during World War I—he was one of the few college fellows who supported Bertrand Russell during the campaign to oust him, also on the grounds
of pacifism. In 1915, however, he left academic life when he was appointed Master of the Temple (incumbent of the Temple Church of the Inns of Court).
In 1920 he was appointed canon of Westminster, and it was during the following decade that he became notorious for his sermons and addresses, collected in book form in 1927 under the title *Should Such a Faith Offend?* In 1924 the
Labour Prime Minister Ramsay MacDonald offered him the see of Birmingham, and Barnes remained bishop there until his retirement in 1953.

Barnes first spoke at the Modern Churchmen's Union in 1914, and did so
frequently thereafter, although he did not formally join the Union and always
preferred to think of himself as a liberal evangelical. He knew that he was regarded by many as the de facto leader of the Modernist movement, but felt

rather uncomfortable with this distinction. In particular, he was opposed to the idealist philosophy that formed the basis of many Modernists' theology. His background as a scientist inclined him to a realist viewpoint, best illustrated in his refusal to support the new idealism promoted by Eddington and Jeans.

On the question of ritualism, his scientific background coincided with his evangelical sympathies. He attacked Anglo-Catholic practices openly and forbade them in his diocese after he became bishop of Birmingham. In the most controversial case, at St. Aidan's, Small Heath, he refused to visit the church while the curate, Alec Vidler, defied him, and would not allow Vidler back to preach in the diocese after he left for another post.[63] Following a series of widely reported sermons on the topic in 1927, Barnes received a vast correspondence, mostly supportive, including many letters from Nonconformists and even from some Orange Lodges. The paleontologist Robert Broom wrote from South Africa to say that if all Anglican clergy were like Barnes, reunification of the Protestant churches would be easy.[64]

For Barnes, Catholic sacramentalism was a survival within the church of primitive magical beliefs derived from the ancient mystery religions. Like reliance on miracles, it had no place within a church that had to function within the modern intellectual environment. The consecration of the bread had a purely symbolic role—after the service, it remained just bread and should not be the object of reverence. If some kind of spiritual change took place in the bread, it ought to be detectable by any human being with spiritual perception, and Barnes asserted "that there is no man living who, if a piece of bread were presented to him, could say whether or not it had been consecrated."[65] He was criticized for proposing the simpleminded view that some kind of chemical change ought to be detectable, but in fact Barnes insisted the test was a psychological one, depending on an alleged human ability to detect spiritual qualities associated with the bread.[66] Not surprisingly, some correspondents wrote to say that they *could* detect the difference, but Barnes remained convinced that any test held under rigorously controlled conditions

63.   Vidler, *Scenes from Clerical Life*, chap. 5.

64.   Broom to Barnes, 16 January 1925, E. W. Barnes papers, Birmingham University Library, EWB 12/5/104; see also the note from Julian Huxley to Barnes, 17 October 1927, EWB 10/2/128. Files 10/2–4 and 11/1 of the Barnes papers contain vast numbers of the letters he received on this topic.

65.   Barnes, "Sacramental Truth and Falsehood," reprinted in his *Should Such a Faith Offend?* 318–23; see 321.

66.   Barnes, *Should Such a Faith Offend?* preface, xxi.

would falsify the Catholic position.[67] Looking into the origins of the popular assumption that something did happen to the consecrated bread, Barnes developed the view that it had been imported into the faith by the early Christians from mystery religions such as Mithraism.[68]

The Protestant view of the Eucharist was part of the process by which the church had purged itself of primitive beliefs in order to move into the modern world. For Barnes, the rejection of what he regarded as a relic of ancient superstition was vital if young people were to be attracted to Christianity. They were no longer willing to accept beliefs that smacked of magic, just as they were no longer willing to accept doctrines that were plainly incompatible with the scientific worldview proposed by geology and evolutionary biology.[69] The public response in the form of the letters written to Barnes suggests that his position enjoyed considerable support, but within the church it was highly controversial, and the archbishop of Canterbury wrote publicly to Barnes, warning that the tone of his sermons was giving offense to many.[70]

It is in the light of Barnes's determination to forge a Christianity acceptable to the modern world that we should interpret the "gorilla sermons." He felt that many churchmen had accepted evolution only in the most superficial sense. When not discussing the subject directly, they continued to speak as though Genesis were literally true, and in this sense they had not really thought out the implications of accepting the new scientific worldview. More seriously, they had not confronted the need to rethink the question of original sin along the lines already marked out by theologians such as Tennant. When asked by an undergraduate why he did not say publicly that Genesis was no longer relevant, he replied that he had indeed spoken out, and received the response "Cannot you say it a little more loudly?"[71] Say it loudly he did, and the result was a flurry of headlines on several different occasions during the 1920s.

Barnes first drew attention to the issue in a sermon preached at the Cardiff meeting of the British Association on 29 August 1920. Here he attacked the traditional notion of the Fall quite briefly, then went on to develop a pro-

67. For two letters making such a claim, see E. W. Barnes papers, EWB 10/2/82 and 89.

68. Barnes, "The Eucharist," reprinted in *Should Such a Faith Offend?* 209–29, and Barnes, *Rise of Christianity*, chap. 16.

69. *Should Such a Faith Offend?* xviii–xix. Numerous letters in the Barnes papers indicate his desire to involve young people in the church.

70. Davidson's open letter to Barnes is reprinted in Bell, *Randall Davidson*, vol. 2, 1322–24.

71. Barnes, "Evolution and the Fall," reprinted in *Should Such a Faith Offend?* 10–17; see 13.

gressive evolutionism in which the emergence of the human spirit was planned by God. Spirit emerged from mind as mind emerged from matter, and since the process could not be all for nothing, we may look forward to survival in a spiritual world that transcends the material.[72] The *Times* report of this sermon initiated a public exchange of letters between Barnes and General Bramwell Booth of the Salvation Army, with the latter defending the traditional view of the Fall.[73] In his reply, Barnes wrote that for those Christians who accept the scientific worldview, "the Christ-Spirit is the supreme and final power in the evolution of man."[74] He was also attacked in the pages of the evangelical journal *The Record*.[75] In a sermon preached in Westminster Abbey on 5 September 1920, Barnes responded to criticism by lamenting the unwillingness of many clergy to admit that Genesis must be taken as an allegory. The old concept of the Fall was an attempt to explain sin that was ultimately unsuccessful, although sin itself was real and had to be accounted for.[76]

It was in 1927, again following the British Association meeting, that the most intense controversy arose. This was the meeting at which Arthur Keith gave his widely reported presidential address on the topic of human origins. On 25 September Barnes preached a sermon in Westminster Abbey to the boys of Westminster School, in which he referred to Keith's address and asked the boys to welcome new discoveries such as those embodied in the evolutionary explanation of human origins. The story of Adam and Eve should be reduced to the status of folklore, and the "horrible theory of the propagation of sin, reared on the basis of the Fall by Augustine" could be rejected.[77] Some still tried to argue that the mind, if not the body, was a special creation, but this position was untenable. Evolution may have advanced by small discontinuities, and new functions thereby developed, but the legacy of earlier stages could not be obliterated. Biology showed "that much that is

72. Barnes, "Christian Revelation and Scientific Progress," reprinted in *Should Such a Faith Offend?* 1–9.

73. "Doctrine of the Fall: Canon Barnes on Science and Faith," *Times*, 30 August 1920, 7; letters from Booth, *Times*, 31 August 1920, 6, and 3 September 1920, 6. In his autobiography, Booth seems to endorse a kind of theistic evolutionism, but does not mention Barnes; see Booth, *These Fifty Years*, 208–9. The evangelical response to Barnes is discussed briefly in Bebbington, *Evangelicalism in Modern Britain*, 207–9.

74. Barnes, "The Fall of Man," *Times*, 1 September 1920, 6.

75. Tisdall, "Dr. Barnes and the Fall" and "Canon Barnes Again."

76. "Evolution and the Fall," 16; see *Times*, 6 September 1920, 7, and a critical letter signed "A. S." on the same day, 6.

77. Barnes, "Religion and Science: The Present Phase," reprinted in *Should Such a Faith Offend?* 309–17; see 311.

evil in man's passions and appetites is due to natural instincts inherited from his animal ancestry. In fact, man is not a being who has fallen from an ideal state of innocence: he is an animal slowly gaining spiritual understanding and with the gain rising far above his distant ancestors."[78] Christianity must accept this fact—but it left Christ's teaching unaffected. God's creative activity was shown through the process of emergent evolution, even though it was difficult to account for the amount of suffering involved. Unlike Keith, Barnes argued that mind was not a mere by-product of material activity; it was a new factor introduced into the world because God's creative power was at work in all nature. This sermon was widely reported, the *Manchester Guardian* using perhaps the most provocative headlines: "Outspoken Sermon by Dr. Barnes—Evolution a Fact—Darwin's Destruction of Theological Scheme."[79] There was even a rumor that Barnes had brought a monkey into the pulpit to emphasize his belief in evolution.[80]

This controversy coincided with, and was to some extent overshadowed by, that sparked by Barnes's attacks on Anglo-Catholicism. Barnes certainly saw himself as fighting against ancient superstitions on all fronts. Significantly, the archbishop of Canterbury's public rebuke to Barnes focused mainly on the question of ritualism. Randall Davidson tried to defuse the issue of the "gorilla sermons" by arguing that hardly anyone still held the old-fashioned ideas that Barnes satirized.[81] Others were not so sure that Barnes's warnings were unnecessary. One commentator later noted that many who liked Davidson remembered his rebuke with shame, because he should have known how little the implications of evolution were accepted by the clergy.[82] Alec Vidler, who had come into conflict with Barnes as a young Anglo-Catholic clergyman, later conceded that the latter had been right to claim that many Anglicans were not being candid about the implications of evolutionism.[83] The failure to assimilate the new science may indeed have been a factor in the growing disillusionment of many young people with the church.

78. Ibid., 312–13.

79. *Manchester Guardian*, 26 September 1927, 7; see also *Times*, 26 September 1927, 9, and *Daily Telegraph*, 26 September 1927, 12.

80. This was reported to Barnes by G. Walker, who attributed it to a "Plymouthite preacher"; see his letter of 7 March 1932, Barnes papers, EWB 10/4/458, and Barnes's reply denying the rumor, EWB 10/4/459.

81. Bell, *Randall Davidson*, vol. 2, 1322. William Temple later wrote to Barnes saying he thought there was much still of value in the old myths that was being obscured by the Modernists' attacks; see his letter of January 1930, reprinted in Iremonger, *William Temple*, 490–92.

82. Richardson, "Bishop Barnes on Science and Superstition," esp. 367.

83. Vidler, "Bishop Barnes: A Centenary Retrospect," 90.

Barnes's sermons and addresses of the 1920s say little about the mechanism of evolution, although they adopt the vaguely progressionist tone popular among Modernist commentators. But Barnes was becoming more involved with eugenics, and this seems to have alerted him to the role of selection in evolution. In 1926 he was invited to give the annual Galton Lecture to the Eugenics Education Society. Here he explicitly accepted that the divinely instituted process of evolution involved random changes in which some variants were more successful than others: "God's judgement on this random process of change is expressed by the subsequent action of the environment in which it occurs. By what is termed 'the ruthlessness of Nature' He weeds out the less valuable products of His plan."[84] Barnes accepted the hereditarian view that many characters, once produced, could not be modified. He was aware that nature normally exerts a selective effect, and believed that artificial selection might be necessary in the human population. It was the Christian's duty to the human race to support measures that would prevent future generations being burdened by the transmission of harmful characters. Barnes welcomed efforts to limit the reproduction of the feebleminded and other undesirables, although he held that Christians must oppose compulsory sterilization. He also stressed the need for social reform to improve slum conditions.

In 1930 R. A. Fisher wrote to Barnes saying that he would send a copy of his *Genetical Theory of Natural Selection* for comment. Barnes replied that he had already bought a copy of the book and had found some of the mathematics in it difficult. Fisher then offered to discuss any mathematical problems of interest, and raised the question of a scheme to encourage clergy to have more children by offering them family allowances (a typical eugenic ploy, on the understanding that professional people were of superior genetic stock).[85] In 1931 Fisher and Barnes, along with Julian Huxley and the geneticist Reginald Ruggles Gates, attended a conference on eugenics intended to prepare a presentation on the topic to the Convocation of Canterbury, with Fisher's family allowance proposal as a major item on the agenda.[86] Fisher and Barnes remained in touch, although much of their surviving correspondence dates from the 1950s. Barnes also became the godfather of Fisher's youngest daughter.[87]

84.   Barnes, "Some Reflections on Eugenics and Religion," 10; the address is reprinted in *Should Such a Faith Offend?* 273–88. His correspondence on eugenics is preserved in the Barnes papers, EWB 9/16.

85.   Fisher to Barnes, 25 September 1930, E. W. Barnes papers, EWB 11/2/186; Barnes to Fisher (carbon copy), 29 September 1930, EWB 11/2/185; Fisher to Barnes, 4 October 1930, EWB 9/16/19. Part of the latter is transcribed in Bennett, *Natural Selection, Heredity, and Eugenics,* 182.

86.   Minutes in Barnes papers, EWB 9/16/29.

87.   Barnes papers, EWB 10/7/233–25; Bennett, *Natural Selection, Heredity, and Eugenics,* 181–82, the latter transcribing parts of eighteen letters held by the University of Adelaide, Australia. On Barnes as godfather to Fisher's daughter, see J. F. Box, *R. A. Fisher,* 279.

In his letters to Fisher, Barnes mentioned that he had dealt with some topics of mutual interest in his Gifford Lectures of 1927–1929, which he was currently revising for publication. The resulting book, *Scientific Theory and Religion* (1933), must rank as one of the most extensive treatments of the theme published during the interwar years. The opening lectures focused on the new physics, of which Barnes was able to offer a detailed mathematical treatment far beyond the scope of most lay readers. Yet these lectures challenged what many theologians saw as the most interesting trend in twentieth-century physics, the new idealism of Eddington and Jeans. Barnes adopted an explicitly realist philosophy that allowed him to repudiate the claims of those who thought that the distinction between matter and mind had now been abolished. He conceded that Heisenberg's uncertainty principle offered, at least in theory, a means by which determinism could be circumvented within the physical activity of the human brain, but did not elaborate the idea.[88] He received a great deal of correspondence from ordinary people about these issues, and always actively opposed the suggestion that the new physics made matter "more spiritual."[89]

In the area of cosmology, Barnes noted the development of theories in which the universe expands from an initial singularity, but found little comfort in the claim that divine intervention could be called upon to explain the original creation. He asked, "Are we to bring in God to create the first current in Laplace's nebula or to let off the cosmic fire-work of Lemaître's imagination?" For Barnes, the best evidence of the Creator was in the orderly structure of the universe as it can now be seen in operation.[90] He expressed doubts about Jeans's theory of the creation of planetary systems by the near collision of stars. While accepting that this was currently the best-established theory, he noted Jeans's conclusion that, if true, the theory implied that planetary systems must be very rare. His own inclination was to believe that the universe had been created to make life possible, and hence that many planetary systems must exist. In a 1931 symposium on the topic, he even suggested that extraterrestrial civilizations might be contacted by radio.[91]

Turning to the biological sciences, Barnes used a chapter on evolution to repeat his arguments against those who resisted the theory on religious grounds. He also managed to slip in his claim that the belief in spiritual pres-

88. Barnes, *Scientific Theory and Religion*, 308–10.

89. See, for instance, his response to a letter by J. H. Peel, 2 May 1930, Barnes papers, EWB 11/2/392–93.

90. *Scientific Theory and Religion*, 409–10.

91. Ibid., 398–404. On the radio communication idea, see Barnes, "Contributions to a British Association Discussion on the Evolution of the Universe," and Dick, *The Biological Universe*, 413–14. See also "Life on Distant Worlds: Bishop of Birmingham's Theories," *Times*, 21 October 1933, 12.

ences within inanimate objects is a survival of primitive superstition.[92] From an account of Mendelism, he went on to insist that since mutations occurred in the human species, characters such as feeblemindedness were due to defective genes.[93] A chapter on the machinery of evolution shows clear evidence of Fisher's influence, including several references to *The Genetical Theory of Natural Selection*. Barnes repudiated Lamarckism, noting the geneticists' challenge to the experiments of Paul Kammerer and others.[94] Mutations were the raw material of evolution: there was no inner urge pushing life in a purposeful direction. Barnes argued against the view championed by most Modernists that nature was creative because God was immanent within it. God was essentially transcendent, apart from His creation, and it was for this reason that some aspects of the evolutionary process could seem amoral. Yet evolution is a revelation of God's creative activity, and the evil within it must occur with His permission: "For some unknown reason He permitted death, disease, struggle, the instincts which have led to selfishness and lust in man, because He willed that higher moral, intellectual and emotional development which in man is such an unexpected outcome of the process."[95] Progress was the intended outcome, but it was the result of an indirect process of development, not a directly implanted urge. Barnes even included a section on "the sternness of God" in which he stressed that punishment was always the penalty for failure.[96] In effect, genetics seemed to support the old idea of predestination, and dysgenic mutations must be eliminated from the human species as they were eliminated naturally from wild ones.

In his conclusion, Barnes again tried to strike a balance between the belief that progress is the intended outcome of creation and the Darwinian view that it is not inevitable. He attacked the theory of emergent evolution, saying that it explained nothing, and argued that the emergence of mind in nature represented a gap that could be bridged only by the Creative Mind of God.[97] In the end, Barnes retained the view that mind was an additional creative force implanted in the universe and actively directing evolution through the purposeful activity of animals. His opposition to a mechanistic view of life thus led him to retain a feature that had once been characteristic of Lamarckism. Yet he was more willing than most religious thinkers to come to terms with the prospect

92. *Scientific Theory and Religion*, chap. 14, esp. 459.
93. Ibid., 491–92.
94. Ibid., 510–11.
95. Ibid., 522.
96. Ibid., 523.
97. Ibid., 590–91.

that progress was not inevitable in the short term, and in particular, with the view that the human species was not necessarily the final goal of the process.

Barnes's position on this last point can be judged more accurately from his extended correspondence with Robert Broom, a long-standing opponent of Darwinism and advocate of the view that humankind is the final goal of evolution. In 1930–1931 there was an exchange of letters in which Broom himself made the point that natural selection "seemed to me like the Calvinistic doctrine of Foreordination." He rejected the view that mutations were the source of evolution, praised the doctrine of vitalism in biology, and expressed support for Smuts's views on holism.[98] Broom then sent Barnes several newspaper cuttings with articles on his anti-Darwinian views and his new discoveries of fossil hominids. Barnes replied, criticizing Broom's goal-directed model of evolution, although he admitted that reading Fisher's book made him aware that we still have much to learn about the origin of species. Barnes agreed with Broom that there was purpose and plan in evolution, but suggested that the plan was manifest more in the changing environment than in the raw material of mutations: "It may well be that man represents the final achievement of the mammals. But, if he kills off all the rest and is finally himself extinguished by his environment, I should expect that some other biological type would develop and finally surpass man in mental power as much as man surpasses the reptiles of the Jurassic."[99]

Although Barnes supported the Modernist position, his views had foundations quite different from those of most other members of the movement. Far from being an exponent of immanence and idealism, he was a realist who saw God's designing hand at work in the rigid operation of the laws of nature and in the active power of mind added to those laws. He had no interest in the new idealism in physics, nor in the various philosophies of holism, emergence, and the like used to promote a nonmechanistic biology. One review of *Scientific Theory and Religion* contrasted his worldview with that of Whitehead.[100] Barnes was thus in a position to endorse the Darwinian theory of Fisher, and hence the latter's support for eugenics. Far more sincerely than most Modernists, Barnes wanted Christianity to pay attention to science—in effect, he wanted to modernize Modernism by taking it beyond its uncritical support for outdated biological doctrines.

98. Broom to Barnes, 26 November 1930 and 12 and 21 January 1931, Barnes papers, EWB 11/2/49, 51, 52.

99. Barnes to Broom (carbon copy), 10 April 1931, Barnes papers, EWB 10/4/51. Broom's letters and cuttings are EWB 10/4/50 and 52. On the evolution of nonhuman intelligence, see also Barnes to A. Piney, 7 December 1927, EWB 12/5/32.

100. Lidgett, "Contrasted Cosmologies."

Scientific Theory and Religion was widely seen as a book more about science than religion. A rather equivocal review by F. L. Cross praised the book for not talking down to its readers, but warned that many would have to take page after page of mathematical formulae on trust.[101] The review in *Nature* thought that the book's message was that science had little positive to offer religion—it just ruled many old dogmas out.[102]

Barnes kept open his links to the scientific community, contributing frequently to meetings of the British Association. But within the church he was much more vulnerable. His position on the Eucharist was anathema to the Anglo-Catholics, while both they and the more conservative evangelicals found his views on the Fall unacceptable. Nor did his position have the hoped-for effect of bringing the young back into the fold. By the 1930s they were turning either to outright materialism or to a Christianity that was re-emphasizing the reliance on the supernatural that the Modernists had striven so hard to eliminate. By 1936 Barnes himself was pessimistic about the future of his liberal tradition within the church, although he thought that the rival positions were equally in decline.[103] He saw a spirit of unreason in the air. His final marginalization within the church came with the publication of his *Rise of Christianity* in 1947. Here he attempted to apply his vision of a world without miracles to the story of Christ and the early church, drawing also on the more radical versions of biblical criticism. He was roundly criticized for reducing Christ to a good man revered by His followers for His moral superiority, thereby eliminating the whole idea of salvation. In the resulting outcry he was formally rebuked in the Convocation of Canterbury. One later commentator entitled his section on this episode "The Follies of Dr. Barnes" and quoted *Crockford's Clerical Directory* to the effect that it would have been better if Barnes had remained a Cambridge don.[104] By the 1940s Barnes had become an anachronism within a church that had turned its back on Modernism.

W. R. INGE

Barnes was not the only Modernist to challenge the assumption that evolution is inherently progressive and purposeful. William Ralph Inge (it rhymes, he said, with "sting," not with "whinge") was also a maverick even within the Modernist movement. The deaneries of the Anglican Church have

101. Cross, "Science and Religion in Contemporary Culture."
102. "Scientific Theory and Religion."
103. Barnes, foreword to Harvey, ed., *The Church in the Twentieth Century*, ix–x.
104. Welsby, *History of the Church of England*, 53–56.

W. R. Inge. By Philip De Lazlo, 1934. By courtesy of the National Portrait Gallery, London.

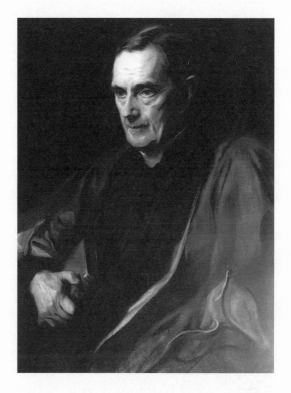

often served as a refuge for those whose brilliance cannot be denied, but whose opinions are so eccentric that they cannot be trusted in the ranks of the bishops. Inge was appointed Dean of St. Paul's in London in 1911 and remained there for the rest of his career. He was a Modernist because, like Barnes, he could not believe in a supernatural realm that interfered on a capricious basis with the operations of the universe. He succeeded Hastings Rashdall as president of the Modern Churchmen's Union in 1924 and served for the next ten years. But where Rashdall, like most Modernists, was suspicious of mysticism, Inge had made his name with his Bampton Lectures, published as *Christian Mysticism* in 1899—for him, personal experience should be the foundation for religious faith, and this included mystical experience. He opposed too rigid an adherence to dogma, seeing the faith as something that must evolve. Like Barnes, he opposed Anglo-Catholic ritualism, although he thought Barnes went too far and did not understand the origins of the tradition he rejected.[105]

105. Inge's *Diary of a Dean* makes frequent references to his contacts with Barnes. See also his autobiographical *Vale* and Fox, *Dean Inge*.

Inge was an instinctive Tory (Conservative), and it was this aspect of his thought—including his support for eugenics—that captured public attention. And Inge certainly had access to the public, writing regular columns at one time or another for the *Sunday Express,* the London *Evening Standard,* and the *Morning Post.* His aphorisms were collected in a little book called *Wit and Wisdom of Dean Inge.* It was the *Daily Mail,* however, that gave him his popular sobriquet, "the gloomy Dean," thanks to his consistently pessimistic analyses of the state of modern society. Inge was convinced that the white races were doomed unless they could prevent the proliferation of the unfit, and shared Barnes's conviction that to purify the race in this way was a Christian duty. He had no scientific training, but worked hard to keep abreast of the latest efforts to popularize scientific developments. His pessimism was justified by appeals not to biology, but to physics and cosmology, with their predictions that the universe was gradually running down toward a "heat-death." He was not an opponent of evolution, but was suspicious of the progressionist worldview created in the nineteenth century and inherited rather uncritically by many Modernists in the church.[106]

Inge was predisposed from the start of his career toward the theology of divine immanence. In a 1902 essay he wrote of God's mind or will animating the universe, insisting that the Idea's incarnation in Christ was part of an ongoing process, not a "catastrophic" break with the past. He was already questioning the uniformity of human progress, suggesting that the art of sculpture, for instance, had reached its peak in the fifth century B.C.[107] From this point on, his writings show an increasing concern for the need to reconcile science and religion. In the preface to the second edition of his *Outspoken Essays* of 1921, he took a characteristically pessimistic view of the church's condition, lamenting, "There is not likely to be any improvement while it is generally believed that a man cannot preach the Gospel without doing violence to his scientific conscience, if he has one. We are so much accustomed to the 'conflict between religion and science' that we have forgotten how unnatural such a warfare is." The reconciliation would be based on one of three alternatives: the Christian Platonism to which he himself inclined, Liberal Protestantism, or Catholic Modernism.[108] Inge explicitly attacked Roman Catholic Modernism, which he thought had no links to the position he wished to adopt. He also criticized Charles Gore for his willingness to take miracles se-

106. For a useful commentary, see Ferrar, "The Gloom of Dean Inge." F. C. O. Beaman's "The Church and Science" contrasts Inge's early interest in mysticism with his later enthusiasm for eugenics.

107. Inge, "The Person of Christ." His essay "The Sacraments" in the same volume, *Contentio Veritatis,* is an attack on superstition and the Anglo-Catholic interpretation of the Eucharist.

108. Inge, *Outspoken Essays,* 18.

riously.[109] An essay on the idea of progress warned that nature neither promises nor denies progress—there may still be some possibility of further advances, but we should not count on too much.[110]

These early essays reveal Inge's support for eugenics, but they also show that he avoided the widely accepted belief in the innate superiority of the white race. His 1919 Galton Lecture ridiculed the notion that there were any pure races left in Europe, although this did not stop him trying to identify the peculiar mental characters of the English (idealism and independence). He argued that for the most part, races are successful only in areas to which they are physically adapted.[111] This relatively relaxed attitude on race contrasted sharply with his growing obsession with the need to limit the breeding of the unfit within the British population, something that he insisted was necessary to prevent the collapse of civilization itself.[112]

In 1925 Inge wrote the concluding summary for Joseph Needham's collection *Science, Religion, and Reality*, expressing the hope that a reconciliation was closer than it had been for fifty years. Here he endorsed the nonmechanistic approach to biology, citing J. Arthur Thomson and J. S. Haldane with approval, and argued that natural selection was no longer seen as the sole mechanism of evolution.[113] The same themes were taken up in a lecture titled "Science and Ultimate Truth," delivered at Guy's Hospital Medical School in 1926. Inge again insisted that on the psychological plane, living things were more than machines, although he accepted that there was no need to invoke nonphysical vital forces. Evolution was purposeful and progressive, at least in the long run, but in the end life itself would be wiped out as the earth became uninhabitable. The only hope of finding purpose in human life was to see traces of eternal values in our experiences, through which God reveals Himself in the world He created.[114]

In his *Christian Ethics and Modern Problems* (1930), Inge saw the church as suffering from the indifference of the young since the Great War, although he insisted that these same young people were anything but indifferent to the is-

109.   Inge, "Roman Catholic Modernism" and "Bishop Gore and the Church of England," *Outspoken Essays*, 137–71 and 106–36.

110.   "The Idea of Progress," in *Outspoken Essays (Second Series)*, 158–83.

111.   "The Future of the English Race," in *Outspoken Essays*, 82–105.

112.   "Eugenics," in *Outspoken Essays (Second Series)*, 254–75; see also another essay under the same title in Inge, *Lay Thoughts of a Dean*, 243–50.

113.   Inge, "Conclusion," 367–71. This piece is reprinted under the title "Science and Theology" with an added reference to Lloyd Morgan's emergent evolutionism in Inge, *The Church in the World*, 141–202; see 172.

114.   Inge, *Science and Ultimate Truth*. The lecture is reprinted in *The Church in the World*, 203–328.

sues addressed by religion. He accepted that the applications of science had led to some forms of social progress, but argued that science itself offered no guarantee of future development:

> The notion that evolution is an automatic machine for bringing on the millennium made the attractiveness of naturalism for the nineteenth century. We can only now view with astonishment the support given to this dream by real men of science, such as Herbert Spencer and, to a modified extent, by Darwin himself. The real attitude of Science is neither an optimism nor a pessimism. There is evolution in some parts of the natural order, involution in other parts. There is no progress in the whole, nor can we assume that the more complex is always 'higher' than the simpler, whatever meaning we may give to 'higher' and 'lower,' words which have no meaning in natural science.[115]

Inge complained about the popular interest in the supernatural that had swept the country since the war. Religion encouraged the belief in a higher kingdom, but did not require higher forces to interfere with the world: "Miracle is the bastard child of faith and reason, which neither parent can afford to own."[116] Science was valuable because it taught the need to respect the truth and expanded our horizons. In his concluding chapters he addressed questions of practical ethics, leading to a reiteration of his support for eugenics. Birth control was repugnant but necessary (a position supported by most Modernists), although Inge was worried that respectable women might become reluctant to bear children. The unfit should be forcibly restricted in their breeding. He was opposed to capital punishment and defended the rights of animals.[117]

Inge's most influential book on science was his *God and the Astronomers* of 1933, written in response to the great public interest in the work of Jeans and Eddington. Like Barnes, he was suspicious of the assumption that the destruction of the old billiard-ball model of the atom was a license to reintroduce spiritual qualities into nature.[118] He disputed the physicists' move toward idealism, again proffering his own Christian Platonism as the only viable philosophy. This was a form of realism that accepted that our perceptions do not give a complete picture of reality, although they are basically trustworthy be-

---

115.   Inge, *Christian Ethics and Modern Problems*, 194.

116.   Ibid., 198.

117.   Ibid., 271–80; see also his essay "Scientific Ethics" (originally an address to the British Science Guild) in *More Lay Thoughts of a Dean*, 257–90.

118.   Inge, *God and the Astronomers*, viii. See also his *Eternal Values*, 38: "Radiation is no more spiritual than a lump of lead."

cause the universe (and hence the human race) is created by a God of love.[119] Jeans's vision of a mathematician God was plainly inadequate, an impoverishment of our experience, because values are also real and yet cannot be expressed in mathematical form.[120] Inge's main purpose, however, was to invoke the directional model of cosmic history implicit in the second law of thermodynamics, and supported by the latest cosmological evidence for an expanding universe, in his campaign against simpleminded progressionism. He expressed amazement that the generation of Darwin and Spencer could have built their faith on the idea of progress, given that the implications of the second law were already well understood at the time: sooner or later all sources of energy in the universe will have been dissipated. We may have a long leasehold on the earth, but we do not have it freehold, and eventually our descendants will all die out in the manner described at the end of H. G. Wells's *The Time Machine*.[121] The existence of life itself was not incompatible with the second law because organic developments were sporadic and localized in a few small corners of the universe.[122] Nature might be creative, but there was no cosmic law of progress. Inge was especially anxious to point out that this fact demolished the philosophies of Bergson and Alexander, in which God was effectively being created through the cosmic process: "Surely a God under sentence of death is no God at all."[123] He was sympathetic to those who saw life and mind as something more than the by-products of matter in motion, but conceded that concepts like "emergence" explained nothing—as he said in a radio talk in 1931, "phrases like 'emergent evolution' only cover up an attempt to assert and deny change in the same breath."[124] Inge thought that these philosophies had become popular because they seemed to challenge determinism, but Alexander's attempt to make God the product of nature pushed this whole approach too far. He acknowledged the work of Whitehead, but claimed to find him so obscure that he could not comment on his thought in detail.[125]

Inge was no opponent of evolution, only of its progressionist manifestations. He knew that the latest work in biology was throwing doubts on the once widely accepted view (which he himself had endorsed) that Darwinism was dead. Natural selection had come back to a greater extent than anyone

119. Inge, *God and the Astronomers*, viii and 6–7.

120. Ibid., 36–46.

121. Ibid., 22–23 and 30.

122. Ibid., 55.

123. Ibid., 10.

124. See Inge's contribution to *Science and Religion: A Symposium*, 143–57, 148.

125. *God and the Astronomers*, 116.

had thought possible, while Lamarckism had been beaten all along the line.[126] These developments only reinforced the view that there was no guarantee of progress: in the later evolution of the human family, progress had been episodic, and in many respects, modern humans were not superior to their ancestors. We might yet follow the dodo to extinction. The human race was not the only goal of creation on earth, and there might be other inhabited planets (here Inge and Barnes were in full agreement).[127] Inge also attacked spiritualism and psychic research as necromancy—although of course he fully endorsed the Christian belief in survival after death.[128]

Perhaps the most surprising aspect of *God and the Astronomers* was Inge's determination not to push his argument for the irreversibility of the cosmic process to the conclusion that many would have deemed obvious. Like Barnes, he found the idea that science could identify the point of creation unsatisfying. Far from looking to what we would now call the "big bang" as evidence that the Creator had started the whole process off, he was unhappy with the conventional view of time and hoped that science would eventually reveal that our cycle of creation was not the only one. He noted that few accepted R. A. Millikan's efforts to demonstrate the existence of a process that would compensate for the expansion of the universe, but found this alternative attractive in principle.[129] For Inge, as for Barnes, it was the orderly and potentially value-laden aspects of the universe that were the best demonstrations of its divine origins. This might be a long way from the old idea of the Creator: "If this is the character of the whole creation, no scientific discoveries can destroy our faith. But there is a thin, bracing air about natural science which blows away a great many cobwebs from the mind and stimulates the imagination to wander through strange seas of thought."[130]

God and the Astronomers was widely read, and in the short term, it may have helped to convince some members of the public that the Modernist trend in the church was coming to terms with the latest trends in science. But within the church, the views expressed by Inge and in Barnes's *Scientific Theory and Religion* (which was published just before *God and the Astronomers*) seemed to herald a turning away from the Christian spirit. What was increasingly wanted was a religion of the heart, not the "thin, bracing air" of science, however

126. Ibid., 137.
127. Ibid., 248.
128. Ibid., 275.
129. Ibid., 64–67 and 248.
130. Ibid., 300–301.

sympathetically interpreted. Inge's own earlier mysticism would have been more in tune with the new attitudes. He retired in 1934 and so was spared the kind of assaults launched against Barnes, although he remained active as an acerbic commentator on public affairs.

CHARLES RAVEN

Charles Raven is remembered by historians of science for his biography of the seventeenth-century naturalist and natural theologian John Ray. But this biography arose from an extended project to deflect the development of science away from its drive toward materialism and back toward a vision of nature that could be reconciled with religion. Like Barnes and Inge, Raven gained some influence in the church during the 1920s, but became isolated by the changing mood of the 1930s. He had a passionate commitment to the unity of God with His creation and undertook a lifelong campaign against those who interpreted Christianity to mean that the world was a mere passive backdrop to the human spiritual drama. His interest in natural history led him to studies of animal behavior, which revealed the active power of living things to take control of their lives. His one exposure to laboratory biology—a course on

Charles Raven. Photograph by Lafayette, 1926. By courtesy of the National Portrait Gallery, London.

genetics with William Bateson—served only to convince him that science was wrong to turn its back on the Lamarckian theory. He remained a passionate anti-Darwinian throughout his life. Where Barnes and Inge tried to absorb the new developments in evolutionary theory, Raven illustrates the tendency for Modernists to identify with exactly those features of early-twentieth-century biology that were being swept aside by the rising tide of materialism.

Raven recalled that there had been little science teaching at his public school, still a typical situation in the 1890s, and one to which he attributed "the inability of the older clergy to appreciate the world of modern thought."[131] He studied divinity at Cambridge under the Modernist theologian Bethune-Baker. This was a difficult time for those students with genuine religious feelings, although Raven recalled with embarrassment the activities of the small cadre of evangelicals. He was already interested in natural history and developed an early enthusiasm for Bergson's vision of creative evolution. When he took his course from Bateson, the result was an immediate distaste for what we would now call genetic determinism and a corresponding preference for the Lamarckian view of evolution, in which behavior can shape the species' future:

> If there is no room at all for use-inheritance in evolution, if our whole physical make-up is strictly conditioned by the immutable germ-plasm, and if therefore our struggles and hard-won virtues have no effect whatsoever upon the course of development, then to speak of the Creator as in any real sense the Father is impossible: a Calvinism of so rigid a sort as to be irreconcilable with Christianity is the only possible theology.[132]

In the 1920s Raven wrote a series of popular books on ornithology, concentrating especially on bird behavior, in the study of which he made a pioneering use of photography. These books reveal his belief that animals are more than mere automata and his hostility to the claim that natural selection can explain all adaptations. In the preface to his *Ramblings of a Bird Lover,* he recorded his debt to Lloyd Morgan's studies of the relationship between instinct and intelligence. Remarking that this was not the place to defend emergent evolutionism, he nevertheless insisted that we need to know more about the evolution of behavior and argued that the amateur observer still had a role

131.   See Raven's autobiographical *Wanderer's Way,* 21. For a modern biography, see Dilliston, *Charles Raven.*

132.   *Wanderer's Way,* 62.

to play in providing basic information.[133] He did not see how natural selection could explain the evolution of nest building, because some element of initiative must intervene to allow birds to transcend the blind compulsions of instinct.[134] Nor would he allow that the beauty of bird song was merely a by-product of territoriality shaped by natural selection.[135]

By this time Raven was already established in the church. He served as a chaplain on the Western Front in the Great War, an experience that prompted him to write "War and the Evolutionary Process," developing the theme that the real conflict was not between the two sides, but between "flesh and blood and the tyranny of blind and impersonal events."[136] After the war he returned to Cambridge, noting that there was now a less hostile attitude toward religion at the university, and became active in the drive to promote a resurgence of the church by adapting its teachings to the realities of modern life and thought. He rejected Anglo-Catholicism as he rejected all forms of authoritarianism—what was needed was not dogma, but a recognition of humanity's role in the universal struggle to achieve God's ends within His creation. He was appointed chaplain to the King, and in 1924 canon of Liverpool, where a new cathedral was consecrated a few weeks after his arrival. Here he played a major role in organizing the 1926 church congress at Southport, which was a key element in the effort to create a more generally favorable climate of opinion toward Modernism. Raven wrote the official account of the congress under the title *The Eternal Spirit* at the same time he was writing his most original contribution to the new natural theology, his *Creator Spirit* of 1927.

Raven's preface to *The Eternal Spirit* reveals his image of how the church should relate to the world. The church should be seen not as the sole beacon of light in an otherwise dark world, but as the focus for the "great dynamo of God's power-house of prayer" that is the universe.[137] While accepting Christ as the "master-spring of the universe," it was vital for Christians to think more positively about the third element of the Trinity, the Holy Spirit, which pervaded the universe and thus balanced the image of a Father who was

133. Raven, *Ramblings of a Bird Lover*, viii and 167–68. Raven wrote to Morgan saying that he had learned more from him than from anyone else; he subsequently asked Morgan to serve on a College of Counsel for Liverpool Cathedral: Raven to Morgan, 8 November 1930 and 13 September 1931, Lloyd Morgan papers, Bristol University Library, DM 128/464 and 510.

134. Raven, *Bird Haunts and Bird Behaviour*, 46–50; see also 84–86.

135. Raven, *Musings and Memories*, 16–17.

136. Ibid., 165–74, see 170.

137. Raven, *Eternal Spirit*, 11–12.

separate from it. Christ Himself used nature for His parables (think of the lilies and the sparrows), and it was thus vital to see the world studied by science as an expression of God's will.

> We have suffered too much from our willingness to relegate to the physical sciences the whole interpretation of nature and from our tendency to encourage bad philosophy and bad religion by contrasting the natural with the supernatural, to be able to afford any neglect of the testimony of creation. No one who knows how many reverent souls are estranged from Christianity by their sense of the indifference and ruthlessness of the universe, will feel it wrong to begin a study of the Spirit with the attempt to remove the notion that the God of religion cannot be the Maker of all things visible and invisible. Unless we can find in the natural order the same values as we are taught to worship in Jesus and to accept from the church, we shall not only leave the breach between science and religion unbridged, but shall be reconstructing a heresy which the Fathers unanimously rejected.[138]

Here was a call for a reconciliation with science—yet it was a call founded on the assumption that science itself must repudiate the old materialism and any theory that seemed to be derived from it. Raven could make common cause only with those scientists who were prepared to see mind and purpose at work in nature. He hailed J. G. Adami's address to the congress as an illustration of the antimaterialistic direction in biology that was required.[139]

The Creator Spirit was Raven's effort to promote the new direction in science, drawing on his own experiences as a naturalist and on the work of sympathetic scientists and philosophers. Morgan, Adami, and Needham are mentioned in the preface, along with Whitehead and Alexander. The book was an "effort to formulate and defend a Christ-centred view of the Universe in such a wise as to heal the breach between science and religion."[140] That breach, he believed, had been disastrous for Western civilization, which would perish "unless the spiritual aspirations of man be redeemed from the bondage of obscurantism and re-united with the best thought of the age, unless science be set free from its association with a deterministic materialism and enabled to serve the highest welfare of humanity." The divorce was not just a consequence of science's growing enthusiasm for materialistic explana-

---

138.   Ibid., 32–33.

139.   Ibid., chap. 4; see 71 and quotations from Adami's address, 77–81.

140.   Raven, *Creator Spirit*, vii.

tions—it was also the fault of a church that had turned its back on God's creation.

> There has been, ever since the dawn of the ascetic movement in the dark ages of the fourth century, a general tendency in Christian thought to regard nature and the natural order, if not as inherently evil, at least as spiritually meaningless, a mere stage on which the divine drama of regeneration was to be played or even a hostile environment from which men were to be set free. The Catholic anchorite, for whom natural beauty was a snare of the devil, and his Puritan brother, for whom the world was at best a vale of tears, join hands here.[141]

Raven repeated his claim that Christians must see the Holy Spirit as the creative force at work within nature, but he now turned to the reformation of science itself. The church's distaste for the study of nature was merely reinforced when science seemed to develop along lines hostile to theism. The conflicts inspired by Darwinism were an obvious example, but Raven insisted that evolutionism itself was not the problem—the real stumbling block was the Darwinian interpretation of it. With the evolutionists' rejection of Lamarckism, "the whole import of the debate is brought to a sharp focus."[142]

> So long as room was left for the efforts of the individual to influence evolution, moral elements could be allowed a place in the scheme of development, and alongside of the mechanism of elimination the Christian could set aspiration and adventure, educability and inventiveness. He was saved from the dominion of chance and necessity; and might still keep his knowledge and his faith running rather jerkily in the same harness.[143]

August Weismann's concept of the germ plasm as totally isolated from the influence of the body had challenged this compromise, and had been reinforced by Bateson's work in genetics. To Raven, Bateson was the high priest of determinism, who had been slow to admit the possibility of any qualification to the rigid laws of heredity—such as those Raven himself had demonstrated in some experiments with moths while still at Cambridge.[144]

141. Ibid., 6. St. Augustine is identified as one source of this negative image of nature.

142. Ibid., 31.

143. Ibid., 34–35.

144. Ibid., 40. On the moth *Leucania brevilinea*, apparently newly appeared on the Norfolk Broads, see 131–35.

As far as Raven was concerned, "if Weismannism is true, not only are the so-cial programmes to which the Church has lately devoted so much attention wholly misdirected, but the conception of God which Jesus proclaimed and his followers accept is demonstrably untenable."[145] The eugenics programs being advocated even within the church by D'Arcy, Barnes, and Inge would be the only way forward, and for Raven, this would be a moral disaster.

Fortunately, Raven claimed, the era of determinism was passing. Al-though biologists had not accepted vitalism, they were willing to admit that mechanism did not tell the whole story. Natural selection was no longer seen as the only driving force of evolution. Raven believed that the work of Adami, MacBride, and the other Lamarckians was of enormous importance; for him, these were the heralds of a new age, not (as they were perceived by most biologists) the survivors of the old. The writings of Bergson, J. S. Hal-dane, Thomson, Smuts, and Morgan were also seen as characteristic of this new nonmechanistic biology. Raven invoked his own observations on the role of spontaneity in bird behavior in support of Morgan's theory of emergent evolution. New psychic properties appeared at key points in evolution, rep-resenting the main stages in the unfolding of the divine purpose.[146]

The later chapters of the book explore the spiritual dimension of this new synthesis. Raven explicitly attacked those church leaders, including D'Arcy and William Temple, who stressed the will of God acting indepen-dently of the world. Like Morgan, he himself could form no concept of God apart from the universe.[147] The universe was a necessary manifestation of God, a self-revelation, best understood by analogy with artistic creation (al-though this did not preclude scientific study of the results). The divine im-pulse ensured a progression toward higher things, but not in the form pro-jected by the old simpleminded optimism: "Man's ancestors have trodden their *via crucis* and left their bones upon it as a testimony. The problem of them, of their suffering and the cruelty that accompanied it, leaves no room for easy talk, as if development were cheap and lightly won."[148] Here Raven launched into the theme of struggle as the driving force of progress: love was not soft-hearted, and all achievements were won at the price of effort and sac-rifice—including the emergence of self-conscious and morally aware per-sonalities. The fact that Raven, opposed as he was to Darwinism, should en-

145.   Ibid., 42.

146.   Ibid., 80–85. Raven wrote to Morgan on the need to reformulate the theory of use-inheritance: 15 June 1930, Lloyd Morgan papers, DM 128/455.

147.   *Creator Spirit*, 105.

148.   Ibid., 116.

dorse the ideology of "progress through struggle" illustrates how loosely that way of thinking was tied to the theory of natural selection. Lamarckians had always seen struggle as the spur to self-development, thereby contributing to the popular image of what is all too loosely called "social Darwinism."

Raven was suspicious of the new psychologies, especially behaviorism, although his experiences in the war told him something of the complexity of the personality. McDougall's concept of the "group mind" interested him, and he was prepared to take seriously the possibility that telepathy kept members of a group thinking along the same lines.[149] Nor would he dismiss mystical experiences, since the purpose of evolution was the creation of higher states of awareness. The book ended with a direct message to Raven's fellow Anglicans, highlighting the kind of church he believed would flourish in the modern world. He warned explicitly against the doctrine of the infallibility of the Scriptures and against the doctrine of transubstantiation, the latter on the grounds that to say that the divine spirit entered into the consecrated wafer was to imply that it did not exist throughout the rest of the universe.[150]

In 1932 Raven was appointed Regius Professor of Divinity at Cambridge and canon of Ely, but from this point on he began to feel increasingly isolated within the church. In a commentary on the 1930 Lambeth Conference he was already lamenting the collapse of Modernists' hopes of reforming the church. There were many, he wrote, who had "believed that the conflict between old and new was virtually over; that the change in our concept of God's nature and activity was now so greatly accepted as to make the reform demanded by the new outlook obvious and acceptable."[151] But the conference had revealed the extent of the continuing opposition to liberalization, while the European scene was presenting a fresh challenge in the form of Karl Barth's theology, with its stress on the otherness of God, his separation from the world and from humanity.[152] Barth and his followers were to become Raven's *bête noir* as their influence waxed through the 1930s. In 1935 Raven used his Riddell Memorial Lectures at Durham as an opportunity to restate his case, arguing again for science as a window onto the wonders of creation and urging Christianity to take note of what it had to offer. Again he warned against simpleminded evolutionary optimism: humanity was the product of an evolutionary process steeped in suffering, and this was an implication that

149. Ibid., 189–94. See also the appendix by Joseph Needham titled "Biochemical and Mental Phenomena," 285–303. McDougall had published his *Group Mind* in 1920, arguing that some species exhibit behavior in groups that cannot be explained in terms of how the individuals behave.

150. *Creator Spirit*, 270–72.

151. Raven, *Looking Forward (Towards 1940)*, 15.

152. Ibid., 25.

only the Christian could understand. "It was from their experience of the Cross that the disciples came to believe that God is love; are we to reject that belief because the Cross-pattern is woven into the whole fabric of our world?"[153] He again presented emergent evolution as the key to the appearance of new levels of reality in the world, with God's incarnation in Christ being the completion of the process, not the intervention of some alien power. He became human so that we could become divine—and this hope needed to be maintained despite the "perversities of Dr Barth's exegesis."[154]

Raven was active in efforts to link the church with left-wing politics, writing a short introduction to *Christianity and the Social Revolution* in 1935.[155] Support for the Left was made all the more imperative by the rise of Fascism. The outbreak of war in 1939 seemed to confirm the hollowness of the progressionism proclaimed all too simplemindedly by many Modernists. Raven had never been such an optimist, but he was shaken by the sequence of events, and he began to focus even more on the materialist trend in science as a factor in the peril facing civilization. His *Science, Religion, and the Future* (1943) used the history of science to trace the rise of materialism and argue once again that the tide must be turned against it. Science had begun with a teleological and holistic view of the world—this was the source of Raven's fascination with the life and work of John Ray, whose writings on natural theology had encapsulated the seventeenth century's belief in a divinely ordered world.[156] Raven argued that a link between science and religion had been maintained until the debates over Darwinism, which were seized upon by militant secularists such as Huxley to promote an image of science as an alternative to religion. A truce had been declared by Lankester and others on the basis of a clear separation into two noninteracting sources of knowledge—but this had left religion powerless to prevent science being hijacked by the materialists.[157] The result was mechanistic biology and behaviorist psychology, both of which reduced humans to the status of automata, thereby fueling the modern world's loss of hope and any sense of responsibility.

In asking if a "New Reformation" could be achieved in science, Raven now appealed first to the physicists. He was a realist rather than an idealist, but

153.   Raven, *Evolution and the Christian Concept of God*, 20.

154.   Ibid., 36.

155.   "Introduction," in Lewis, Polanyi, and Kitchin, eds., *Christianity and the Social Revolution*, 15–27. Raven was also an ardent feminist: see Dilliston, *Charles Raven*, chap. 15.

156.   Raven's biography, *John Ray, Naturalist*, was published in 1942.

157.   Raven, *Science, Religion, and the Future*, chap. 4; on Lankester, see 54.

Eddington and others had shown that the universe was a mysterious place and had undermined the materialists' dogmatism.[158] Analytical psychology, for all its charlatanry, revealed the complexity of personality. But when it came to biology, Raven had little to offer except a repetition of his by now totally outdated claims that Morgan, Smuts, J. S. Haldane, and other anti-mechanists offered the key to a new synthesis. He moved on almost immediately to attack Barth's "neo-Calvinism," with its emphasis on human depravity and the futility of effort, and the new enthusiasm for the "strange and diseased genius of Kierkegaard."[159] The book ended with a series of chapters outlining Raven's vision for a Christianity that would link up with the enthusiasm for social reform now being expressed by younger scientists such as C. H. Waddington. He was particularly scathing about the support for eugenics still expressed by some Christians such as Inge, policies that were little better than those of the Nazism that was threatening to destroy Christianity altogether.[160]

The postwar years were better for Raven. He was made Vice-Chancellor of Cambridge University in 1947 and played a role in having Smuts appointed to the largely ceremonial role of Chancellor. His views were more widely accepted in the church now, thanks to a change in the attitude toward science prompted by the recognition that in the age of the atomic bomb and the Cold War, Christianity could no longer ignore so powerful a force of change. His Gifford Lectures for 1951 and 1952 repeated his call for a theology of immanence and again used the history of science to pinpoint the turning aside toward materialism. There was little new in his view of science, however, which still depended on the antimaterialists of an earlier generation.[161] He still held out the hope that this approach could once again become fruitful in science, and played a major role in promoting the philosophy of theistic evolutionism in Teilhard de Chardin's *The Phenomenon of Man*. Raven indicated that, had he known of Teilhard before his death, he would have been able to make common cause with him, so closely did his evolutionary theology conform to the ideas of Bergson, Morgan, and Smuts.[162] The vogue for Teilhard among religious thinkers in the 1950s suggests that there was renewed enthusiasm for a synthesis with an appropriately nonmaterialistic science from this side of the divide. Raven's career thus ended on a somewhat more

158. Ibid., 64–66.
159. Ibid., 76, 79.
160. Ibid., 83.
161. Raven, *Natural Religion and Christian Theology*.
162. Raven, *Teilhard de Chardin*.

positive note as far as the church was concerned, but his alienation from the majority of scientists was as profound as ever.

Whatever the success of Raven's postwar writings, his fate in the 1930s tells us a great deal about the collapse of the Modernists' efforts to forge a reconciliation between science and religion. His isolation from the scientific community was more than matched by the increasing rejection of his liberal position within the church. Had the situation within the religious community not changed, Modernists might have adapted to the changes in science, because not all were as intransigent as Raven in their rejection of Darwinism. Many remained unaware of the latest developments in biology, thanks to the flood of popular literature on the new natural theology, but that situation was ripe for change as the older generation of biologists became less active. Barnes's efforts to take Fisher's Christian Darwinism into account point the way to what might have happened had the Modernists had the leisure to adjust to the new scientific developments. But the climate of opinion in theology did not stay the same, and both Barnes and Raven found themselves isolated within a church that no longer had much interest in a reconciliation with science, whatever the terms on offer from the scientists. Driven by an increasingly harsh economic and political situation, the churches turned away from liberalism and Modernism, stressing once again humanity's innate sinfulness and need for redemption. There was no room here for a vision of progressive creation; indeed, the new theology of crisis might have welcomed the Darwinian worldview for its very harshness, had it bothered to pay any attention to science at all. It was the theologians, at least as much as the scientists, who turned away from the proposed reconciliation. Just as the New Theology had been squeezed out of the Free Churches a generation earlier, Modernism was eliminated from the Anglican Communion, and along with it, the only party that was seriously interested in making the changes to the faith that would have rendered it more credible to the majority of contemporary scientists.

CHAPTER 9

The Reaction
against Modernism

There were many in the churches who resisted the liberalizing trend that hoped for a synthesis with science. Campbell made few converts to the New Theology among the Congregationalists, and the other Free Churches still had many conservative ministers who saw little to be gained from a compromise with modern thought. To them, Modernism's emphasis on an immanent God deflected attention from the traditional view that humans were a fallen species, in need of redemption from a source of spiritual authority that lay outside the universe. Christ's suffering on the Cross became an embarrassment that the New Theology preferred to ignore. The same feelings were shared by conservative evangelicals within the Anglican Church, but here the situation was complicated by the fact that the rival Anglo-Catholic wing had its own reasons for wanting to retain a belief in the reality of miracles and the redeeming power of Christ's love. The Modernists' rejection of ritualism as a tradition inspired by primitive magic gave the Anglo-Catholics an additional reason to be suspicious of the liberal trend. Charles Gore and his followers had gone as far as they were willing to go toward liberalization in the late nineteenth century; now they dug their heels in against further changes, which they believed would undermine Christianity altogether. The Roman Catholic Church was also opposed to any more than a superficial reconciliation with evolutionism—to them, any attempt to explain the human spirit as the product of natural development was unacceptable. The new natural theology thus came under attack from two separate directions, evangelical and Catholic—yet the weapons used were often the same, reflecting a deep underlying tension between traditional Christian thought and the new vision of God's relationship to the universe and to humankind. The more pessimistic outlook of the 1930s only fueled the suspicions of those who felt that Modernism had betrayed something essential within the Christian faith.

In the Nonconformist churches there was an increasing division between liberals and conservatives, but most conservatives were fairly moderate in their views, and there was only limited support for what became known as Fundamentalism.[1] The views of Barnes and Raven illustrate how far liberal evangelicals could go in the Modernist direction, and few evangelicals took Genesis literally—although there were many who felt uncomfortable with an evolutionism that extended to the development of the mind and the soul. They might go some way with the new ideas, but balked at several elements within the overall Modernist synthesis, especially Barnes's reinterpretation of the Fall.

Many Anglicans, both evangelical and Catholic, were willing to congratulate science on its supposed rejection of materialism, but could not accept the Modernists' efforts to reconfigure the whole structure of the faith. The liberal Anglo-Catholics typified by Charles Gore endorsed the new natural theology and its vision of a purposefully evolving universe. But to them, rejection of miracles and of all supernatural events rested on the denial of a God who transcended the universe we perceive. Those who insisted on God's transcendental nature argued that He who created the world might choose, on occasions of key spiritual significance, to modify the laws He had instituted. They did not believe that the human race was capable of continuing the progressive spiritual development enshrined in the process of evolution. Once given the capacity to choose, we had rebelled against the Creator's plan, obliging Him to become incarnated in the world as Jesus Christ in order to restore the plan and bring us back to our true destiny. Major Anglican theologians such as Gore and William Temple could thus support some aspects of the new natural theology, while at the same time limiting its scope in a way that preserved key aspects of traditional Christian belief.

In the 1930s the rising influence of Karl Barth's reaction against natural theology encouraged a more profound suspicion of science and all its works. The expression of such opinions by popular writers such as C. S. Lewis fueled a growing divide between science and religion that horrified those who

1.   See Bebbington, *Evangelicalism in Modern Britain*, chap. 6, and on science, 207–9.

had once hoped for a reconciliation. Evangelicalism was revived within the Church of England, and it was the conservative rather than the liberal evangelicals who were on the offensive.[2] The result was not so much an attack on specific scientific theories as a growing suspicion of any form of natural theology as irrelevant to salvation. At the same time, the scientists themselves— especially the biologists—had turned against the explicitly antimaterialistic theories preferred by the Modernists and liberal Catholics. The new natural theology was pulled apart from both sides—it wasn't Christian enough for the new generation of believers, and it gave too much away to teleology and vitalism to be acceptable to the younger scientists.

The Roman Catholic Church exerted a discipline that compelled a greater degree of unity than was possible within other denominations. Although still opposed to evolution in principle at the end of the nineteenth century, it had softened its position on the question of biological evolution considerably in the early decades of the twentieth. But it remained implacably hostile to any force that it saw as a threat to the spiritual dimension of human life, and this made Darwinism and the idea that the human mind had evolved from an animal mentality unacceptable. Popular Catholic writers such as Hilaire Belloc took on the secularists in a highly effective manner, giving the church's position a degree of prominence out of proportion to its actual membership.

EVANGELICALS AGAINST EVOLUTION

Conservative evangelicals in the Free Churches and within the Anglican community had always been suspicious of efforts to liberalize the faith in a way that would render it more compatible with science. Their emphasis on the importance of revelation and their belief in the need for personal salvation through Christ made them naturally suspicious of evolutionism. In 1905 James Orr of the Free Church of Scotland proclaimed his opposition to the wholesale liberalization of the faith required to bring it into line with a cosmic progressionism. While quite happy to concede that there was a process of evolution within the animal kingdom driven by divinely ordained forces, Orr drew the line at accepting the human soul as the product of such a trend because this would undermine the doctrine of immortality. Nor was it possible to apply the idea of progress to humanity: the biblical doctrine of sin implied that we had fallen from an original state of integrity.[3] Although quite

2. On the revival of Anglican evangelicalism, see Manwaring, *From Controversy to Co-existence.*
3. Orr, *God's Image in Man;* see 17–19.

liberal in his willingness to see creation as an evolving process, Orr showed that the origin and fall of humanity was still a stumbling block to full acceptance of the scientific worldview.

Another illustration of these concerns can be seen in Charles H. Vine's *The Old Faith and the New Theology*, a collection of sermons and essays published to combat Campbell's New Theology. Vine's preface, while insisting that the book was not a personal attack on Campbell, nevertheless proclaimed that much of his teaching was "confusing, unscriptural and exceedingly dangerous." There was no point restating the truths of Christianity in terms of the modern mind if that meant the destruction of those truths.[4] Albert Goodrich, current chair of the Congregational Union, saw the New Theology as a reaction against the God of Paley's natural theology—the divine artificer who stood quite apart from His creation. Science had now abandoned this view by adopting a new conception of matter in which it was vital rather than inert.[5] But this new evolutionary view made the human race, and the most perfect man, Jesus Christ, no more than the consummation of an age-long world process. Goodrich claimed that the supernatural could be retained in harmony with the doctrine of evolution by noting that many biologists now included the possibility of great leaps in the upward march of life. The origin of life and mind were two such leaps, but the Incarnation was another, even greater step upward.[6] An essay by P. T. Forsyth—who would later emerge as a leading evangelical—compared the New Theology to second-century Gnosticism. It was a "rationalist, speculative, theosophic Doppelganger of Christianity," with "the same etherialized conception of matter, the same amalgam of physics and dreams, the same animus against historic Christianity."[7]

These attacks reveal that the hostility toward the liberalized version of Christianity was based on deep concerns, of which doubts about evolution were only a part. Even so, these issues continued to form an obstacle blocking conservative thinkers' willingness to explore a reconciliation with science. There were few biblical literalists, of course, and outright rejection of evolutionism remained muted in comparison with that in the United States. In America the publication, beginning in 1910, of the pamphlets collectively entitled *The Fundamentals* marked the start of a major revival of emphasis on the word of the Bible. By no means all of the authors of these pamphlets rejected

---

4. Vine, ed., *The Old Faith and the New Theology*, v–vii.
5. Goodrich, "The Immanence of God and the Divinity of Christ," 16.
6. Ibid., 23–24.
7. Forsyth, "Immanence and Incarnation," 58.

evolution outright—but several did, and increasingly the "Fundamentalists," as they were known, became identified with what we now call creationism.[8] While nothing as dramatic as the "monkey trial" of John Thomas Scopes happened in Britain, there was a small antievolution lobby, which attracted a few trained scientists (see chapter 4). It enjoyed little popular support at first, but the formation of the Evolution Protest Movement in 1935 suggests that the changing climate of opinion in the 1930s had a greater influence on events in Britain than did the importation of examples from America.

A number of ministers and laypersons with conservative evangelical views joined scientists such as Ambrose Fleming and Douglas Dewar in attacking evolution. In 1919 the rector of Devises, A. H. T. Clarke, predicted a resurgence of religion and the outbreak of wars between rival faiths. He claimed that the new spiritual awareness would be founded on a rejection of evolution as science undermined the nineteenth century's "discoveries" in this area. Barnes came in for special criticism as an example of a church leader pointing the way in exactly the wrong direction.[9] E. Ray Lankester took the trouble to answer Clarke's fulminations, and the two exchanged a series of increasingly acrimonious "replies" and "rejoinders."[10] Clarke subsequently softened his position in a 1925 article hailing antimechanistic thinkers such as J. S. Haldane and Henri Bergson as evidence of the collapse of nineteenth-century thought. He now felt that neither evolutionism nor creationism could be sustained in their original form, and expressed the need for an image of an "evolutionist Creator."[11]

The Rev. Harold C. Morton published his *Bankruptcy of Evolution* in 1925, insisting that "Darwinism is dead, and will soon be buried without hope of resurrection. But without Darwinism Evolution is the mere empty shell of a venerable speculation."[12] He conceded that evolutionism was a valid reaction against a totally static creationism—clearly there had been a development of the natural world, but the question was how it had been brought about. Darwin's theory of natural selection had swept the board in a generation of thinkers determined to reject any role for the supernatural, and the advocates of relentless struggle in human affairs were the true heirs of this way of

8. On the situation in America, see Numbers, *The Creationists* and *Darwinism Comes to America;* see also Larson, *Summer for the Gods.* On the Fundamentalists who did not reject evolution, see Livingstone, *Darwin's Forgotten Defenders,* chap. 5.

9. A. H. T. Clarke, "The Church of the Future," 126–27.

10. Lankester, "The Church of the Future," "Evolution v. Creation," "A Matter of Fact"; Clarke, "Evolution v. Creation."

11. Clarke, "The Passing of the Nineteenth Century."

12. Morton, *Bankruptcy of Evolution,* 13.

thinking, whatever Darwin himself intended.[13] Surveying the range of theories that had emerged during the "eclipse of Darwinism," Morton asked whether evolution, as distinct from Darwinism, was established—as even anti-Darwinians such as William Bateson maintained. He claimed that the evidence was now going against the theory: the fossil record showed no transitions from one "type" to another, and even within single types, it suggested sudden jumps or saltations from one species to the next. There was a plan running through the whole, but in the absence of any naturalistic explanation, the "mutations" of Hugo De Vries and others offered little more than the old idea of special creation.[14] There was no "missing link" between humans and apes. Morton offered a kind of supernatural saltationism in which God brought about the appearance of new species in accordance with His overall plan. There was a hint that the process might work by the miraculous transformation of existing species, but Morton also warned that the observed anatomical and biochemical relationships between species would exist even if each were a unique divine creation.[15]

Barnes's "gorilla sermons" attracted the anger of those who felt that evolutionism was not an appropriate basis for discarding the tradition of the Fall (see chapter 8). In 1927 George H. Bonner queried whether evolutionism was really as secure as Barnes claimed and expressed surprise that a bishop could preach sermons denying the fundamental creed of his church. Christianity taught the need for salvation, and the evolutionary theory of the origin of the soul was incompatible with this. The materialistic theories of evolution could not explain the emergence of mind, while the pseudo-mystical theories depended on the assumption of a progressive trend that was incompatible with the lack of moral progress in modern humanity. For Bonner, self-consciousness could not be "built up" by evolution—it must have existed for all time, although, like all archetypes, its manifestation in the physical world might date from a particular point in time.[16] There could be no development of ape into human, and to this extent the special creation theory of the Fundamentalists was true. It was Bonner's assertion that no evidence for a human-ape link had stood the test of time that prompted Arthur Keith to link his

13.   Ibid., 69.

14.   Ibid., 103. Hugo De Vries's mutation theory was a popular early-twentieth-century version of the idea of evolution by saltation; see Bowler, *Eclipse of Darwinism*, chap. 8.

15.   Ibid., 148–53, and the appendix on biochemical relationships.

16.   Bonner, "The Case against 'Evolution,'" 589, 593. Other negative responses to Barnes are noted in Bebbington, *Evangelicalism in Modern Britain*, 207–9.

attack with that of Fleming in his response to "Modern Critics of Evolution."[17]

Another attack stressing the confusion among scientists over the causes of evolution was contributed to the *Evangelical Quarterly* in 1929 by Floyd E. Hamilton, an American professor of biblical studies at the Union Christian College at Pyongyang, Korea.[18] Two years later Hamilton published a book in which he claimed to have abandoned with reluctance a long-standing acceptance of "Christian evolutionism" in order to join those normally dismissed as cranks and fanatics.[19] He was now convinced that the scientific evidence itself demanded rejection of evolutionism. The old, confident selectionism proclaimed by the scientists was now replaced with ambiguity and vagueness: "Darwinism is dead, Sir Arthur Keith and the Bishop of Birmingham to the contrary notwithstanding."[20] And with the collapse of Lamarckism and the other non-Darwinian mechanisms, all possible causes of evolution had been investigated and found wanting. American scientists complained about persecution from the Fundamentalists, but it was the scientists themselves who persecuted anyone who dared to point out the hollowness of evolutionary theory. Hamilton accepted that a certain "emaciated type of evolution" could be squared with theism, but insisted that any theory proclaiming the origin of the human soul from animal traits was incompatible with the meaning of Genesis and with the Christian's belief in the need for salvation through Christ.[21]

The most colorful character to attack evolution was Commander Bernard Acworth, retired from the Royal Navy and now a vociferous critic of the growth of air transport. In *This Bondage* (1929), Acworth combined his critique of the air lobby with a bizarre account of how birds fly and attacks on the theories of evolution and relativity. The key to his thinking was a conviction that birds are, in effect, just automata—they have no powers of reasoning or judgment. Bird flight is determined largely by the wind, and because they cannot judge the direction of the wind when flying, they reach their destinations by a spiral curve. Convinced that aircraft are subject to the same limitations, Acworth argued that all hopes of a global air transport network were illusory and were supported only by self-interested industrial and scientific

17. Keith, *Darwinism and What It Implies*, chap. 3.

18. Hamilton, "The Present Status of Evolutionary Faith." On Hamilton, see Livingstone, *Darwin's Forgotten Defenders*, 161.

19. Hamilton, *Basis of Evolutionary Faith*, 8.

20. Ibid., 18.

21. Ibid., 52–53.

groups. The real theory of relativity referred to motion relative to the earth—Einstein's cosmic version, one of evolution's "more reputable concubines," was nonsense because all motion is self-evidently absolute.[22] The significance of all this for evolutionism was that if birds lack mind, so do all other animals, and that means that there is no foundation in the animal kingdom from which the human mind could have emerged. The concept of the "bird mind" was inspired by the evolutionists' blind assumption that animals must have rudiments of the human mental faculties.[23] Conceding that his view of birds was materialistic, Acworth argued that at least he was able to present them as machines perfectly designed by their Creator—the problem with evolution was that it blinds us to the responsibility of free will by reducing us to the same level.[24] The answer was to accept that the Bible was correct when it spoke of divine creation rather than evolution. Acworth also promoted a catastrophist geology that undermined "the monotonous chant of evolutionary fanatics who demand periods varying from one hundred thousand to one thousand million years for the working out of their mutually destructive theories."[25]

Acworth's religious orientation was more apparent in his later book, *This Progress: The Tragedy of Evolution,* one chapter of which was based on a lecture given at the Victoria Institute. Here the range of modern evils attributable to the influence of evolutionism expanded to include feminism, socialism, pacifism (despite its motto of the "struggle for existence"), and unnecessary surgical operations to remove organs deemed no longer useful to humans. Conceding that Fundamentalism (an objectionable American term) was out of favor, Acworth castigated the English churches for turning their backs on the Christian faith by denying the reality of sin and making God Himself responsible for the evil in the world created by human beings—a message that Barnes even had the nerve to preach to the boys of Westminster School.[26] The Church of England and the Free Churches were losing their grip; they were no longer the Church Militant but the Church Tolerant.[27] Evolutionism had become so deeply entrenched that it formed the greatest vested interest in the land. Yet the theory was fundamentally incompatible with Christianity, mocking the perfection of Christ and denying both immortality and free

22. Acworth, *This Bondage,* 54, 57.
23. Ibid., 83.
24. Ibid., 129, in the course of a critique of a sermon by Barnes.
25. Ibid., 163.
26. Acworth, *This Progress,* 65–68.
27. Ibid., 199.

will. Special creation was the only alternative because it fitted all the facts and was in harmony with Genesis. The present state of the earth was clearly the product of some great geological disaster, and the creation of the modern species could fit into a period of days. If simple people wanted to believe in Adam and Eve, there was no harm done, although probably there were many humans created directly as founders of the various races.[28] The Genesis story was absolutely true, and the Bible should be restored to its place of authority in the life of England—only then would the country once again become prosperous, merry, and free.[29]

It would be easy to dismiss Acworth as a member of the lunatic fringe—he certainly had more than the average number of bees in his bonnet—yet his books were based on articles published in a wide number of periodicals, and he claimed to have received a good deal of support for his views.[30] The 1930s saw a minor resurgence of popular doubts about evolution, whatever the developments taking place in scientific biology. Acworth was present, along with Fleming and Dewar, at the meeting that founded the Evolution Protest Movement in 1935.[31] Publications critical of evolution continued, especially in the pages of the *Journal of the Transactions of the Victoria Institute*. The *Evangelical Quarterly* carried an article stressing the bankruptcy of modern scientific theories of evolution in 1933.[32] A Canadian civil engineer, W. Bell Dawson, published a creationist text, *The Bible Confirmed by Science*, in 1936, hailed by both the *Baptist Times* and the *Evangelical Quarterly* as a timely work.[33] Dawson accepted the fossil record at face value, but argued that divine creation was needed to produce new species at the start of each geological period. The Bible was essentially true because the sequence of creations fitted that given in Genesis.

The opposition of conservative evangelicals to evolutionism paralleled that of some Roman Catholic writers (discussed below). A few evangelicals' doubts were based on a literal interpretation of the Bible, but the moderate conservatives were more concerned to preserve the sense of a sinful humanity in need of salvation. Their position coexisted with the more liberal form of evangelical thought, but when the liberal approach went into decline in the 1930s, the conservatives regained any ground they had lost. One prominent

28. Ibid., 277, 292.

29. Ibid., chap. 15 and conclusion, 334.

30. See the preface to *This Bondage*.

31. See the report in the *Times*, 13 February 1935, 10.

32. Knight, "The Evolution Theory Today."

33. Dawson, *Bible Confirmed by Modern Science*, endorsements on the dust jacket of the author's copy.

critic of Campbell's New Theology had been P. T. Forsyth, who went on to write several important books stressing the need for a return to the old Christian principles. It has been argued that Forsyth's influence began to grow in the decade following his death in 1921, suggesting that his traditionalist views were coming back into fashion even outside the Free Churches as the mood of optimism waned.[34] Writers such as Forsyth did not attack evolution directly, but many people now wanted to read more about the Cross and its meaning for the Christian. Such a move in theology encouraged not just hostility to evolutionism, but a more general reaction against the whole idea of natural theology.

LIBERAL CATHOLICISM

In the Church of England, the Anglo-Catholics were by far the more articulate opponents of Modernism. Gore's *Lux Mundi* had pushed this wing of the church as far as it could go in the direction of liberalization—too far for some—by abandoning the need for the miraculous creation of humankind. Few were prepared to move further in the direction of abandoning Christ's divine nature or the miracles that confirmed His divinity. The Anglo-Catholics' rejection of Modernism was based not on a commitment to the word of the Gospel, but on a sense that the church represented a divinely instituted spiritual entity essential for human salvation. Typically, the Anglo-Catholic *Cambridge Theological Essays* (1905) contained chapters defending the power of prayer and the reality of miracles against scoffing materialists such as John Tyndall.[35] In a general attack on materialism in modern life and thought, John Neville Figgis, one of the Cowley Fathers of Oxford, presented the Eucharist as a symbol of God's entry into the world. The church was founded by supernatural events, he argued, and its sacraments made sense only in the light of the complete rejection of a rationalist philosophy.[36] It was more than just a community of like-minded believers—it was God's power at work in the world. The Anglo-Catholic community thus repudiated key aspects of Modernism, yet it was by no means hostile to science or to the new natural theology. After all, Modernism, whose defining feature—its commitment to immanentism—was theological, was only indirectly linked to the new natural theology

34.   Rodgers, *Theology of P. T. Forsyth*; see also R. T. Jones, *Congregationalism in England*, 352.

35.   See A. W. Robinson, "Prayer, in Relation to the Idea of Law," and Murray, "The Spiritual and Historical Evidence for Miracles," in Swete, ed., *Essays on Some Theological Questions of the Day.*

36.   Figgis, *Civilisation at the Cross Roads*, 123–24.

Charles Gore. By John Mansbridge.
By courtesy of the National Portrait
Gallery, London.

proposed by scientists. To the liberal Catholics, nature was indeed a creative
system designed by God, provided that one saw the Creator as transcendent,
existing apart from the universe and able to intervene in its workings directly
(via miracles) and indirectly (through Christ's influence on the church).

Such views were expressed most clearly by Charles Gore, still the central
figure in the Anglo-Catholic movement at the beginning of the new century.
Gore's *The New Theology and the Old Religion* was the most effective critique
of Campbell's position, as Campbell himself conceded. Gore thought that
Campbell's effort to bring theology into line with modern thought had destroyed the traditional interpretation of revelation, although it was a legitimate reaction against the inadequacies of nineteenth-century Protestantism,
especially in the context of the Nonconformists' overreliance on the infallibility of Scripture.[37] The emergence of evolutionism and the breakdown of
the idea of solid matter had stimulated the imaginations of many, and here
Gore introduced Oliver Lodge as the exponent of a vision of divine immanence more or less identical to Campbell's.[38] He quoted extensively from
Lodge, praising his reverent spirit even when he had to criticize his conclusions. Such thinkers were at least in sight of the city of God, although they

37. Gore, *The New Theology and the Old Religion,* 151–52 and 168.
38. Ibid., 5 and 10.

had departed from the mainstream of the church's belief.[39] But science was not the only source of knowledge, and the New Theology had departed from the historical foundations of Christianity by adopting a Platonic and Stoic view of the relationship between God and the world.[40] In reacting against deism it had plunged into pantheism.[41] There was no recognition of the fundamental role played by sin in the human condition—even T. H. Huxley's "Evolution and Ethics" had accepted a natural tendency to evil within society.[42] Gore defended the traditional view of miracles against the attacks of the New Theology (while conceding that Campbell himself retained a belief in the Resurrection). Miracles conformed to the order of the world at a level deeper than that recognized by science, because a rational God would respond to abnormal circumstances in abnormal ways.[43] Campbell and Lodge had gone as far as they could to retain a role for Christ, but their effort to see Him as merely a perfect human being rejected the New Testament idea of the Atonement.[44] Science's faith in progress was misplaced, and whatever the evidence for the evolution of the body, we are all aware that the human race is not what it might have been. We have failed and we know it; the human race is weakened by "an inherited taint or disorder."[45] Any theology that did not take this into account was a betrayal of Christianity.

Gore went on to play a leading role in trying to block the rise of Modernists to positions of influence in the Anglican Church. He resigned the see of Oxford in 1919 at the age of 68 to concentrate on writing. His most important work relating to science was his *Belief in God* of 1921, the first of three works that were republished together in 1926 as *The Reconstruction of Belief.* Gore was still making every effort to keep alive the hope of a reconciliation with science first raised by the *Lux Mundi* authors over thirty years earlier. In a response to comments on *Belief in God,* he insisted "that on the basis of faith, the real conclusions of science can be freely accepted and welcomed."[46] He

39.   Ibid., 18, and for the quotation from Lodge, 14–16. Gore corresponded extensively with Lodge, debating in particular the need for Christianity to be based on miracles: Gore to Lodge, 25 February 1897 and 13 May 1905, Lodge papers, Birmingham University Library, OLJ 1/156/6 and 11.

40.   Gore, *The New Theology and the Old Religion,* chap. 3.

41.   Ibid., 56.

42.   Ibid., 72. Actually, Huxley's "Evolution and Ethics" had attacked social Darwinism by portraying nature as amoral and human moral values as transcending nature.

43.   *The New Theology and the Old Religion,* 112–13.

44.   Ibid., 132.

45.   Ibid., 247–48.

46.   Gore, *Can We Then Believe?* title of chap. 3, 46. On Gore, see Prestige, *Life of Charles Gore;* Carpenter, *Gore: A Study in Liberal Catholic Thought;* Ramsey, *From Gore to Temple,* esp. chap. 7.

insisted that Cardinal Newman and the Roman Catholics were wrong to claim that their church had the authority to decide on matters of scientific fact. At the same time, however, science itself was realizing its limitations, and it was now more widely recognized that science and materialism did not necessarily go hand in hand.[47]

Parts of Gore's books seem quite compatible with the Modernist way of thought: he was not going back on the concession that evolutionism was quite acceptable to the Christian. Darwinism had been seen as overthrowing Christianity because it showed that the Bible was not true, but most Christians had accepted that Genesis need not be taken literally on creation, just as they had accepted that the earth is not the center of the universe.[48] The growth of materialism had been part of a wider revolt against authority and against the injustice implied by the Calvinist view of the human situation. But many scientists had retained or regained their faith—here Gore recalled the experiences of G. J. Romanes, whose posthumous *Thoughts on Religion* he had edited.[49] Science revealed a universe that was orderly, purposeful, and beautiful, and must hence be the product of Mind. Noting Bergson's theory of creative evolution, Gore insisted that the period since Darwin died had confirmed that natural selection was inadequate to explain evolution's purposeful move toward higher things: "It seems impossible to account for progressive evolution of living forms unless some sort of direction, some sort of organic tendency to become this or that, is assumed in nature—which suggests irresistibly a progressive purpose in the order of living things, which has found for the present its culmination and interpretation in man."[50]

So far was Gore prepared to go with the liberals, but what had now been established was only a "higher pantheism" in which science had helped us to see that there was a Mind behind creation. Such a natural religion was unsatisfying to most people because it said nothing about whether God cares about His creatures and wishes them to interact with Him.[51] It said nothing about whether God is personal, or about whether He loves us—let alone whether He might wish to judge us. To answer these questions, so central for Christianity, we have to turn to revelation, and hence we have to believe that God can and does communicate with the world. Gore suggested that there were

47.   Gore, *Can We Then Believe?* 50–51.

48.   Gore, *The Reconstruction of Belief,* 10–16.

49.   Ibid., 41.

50.   Ibid., 59–60. In *Can We Then Believe?* 52, Gore refers to J. N. Shearman's *Natural Theology of Evolution* of 1915, a particularly vague effort to reconcile progressive evolution with teleology that otherwise seems to have gone unnoticed.

51.   *Reconstruction of Belief,* 58–61.

two major reasons why science seemed to raise difficulties with the Christian view of revelation. One was the biblical account of creation, no longer a problem now that we had recognized the abstract nature of Genesis. The other was the concept of free will, so easily denied by the sciences as soon as they tried to apply their methods to human nature. Here Gore was blunt: he claimed freedom as a fact of which we are all conscious and insisted that science could not deal with the human will.[52] Once this was conceded, the possibility of rebellion from the real purpose of God's creation had to be admitted, along with its consequences. Gore noted that E. Ray Lankester— certainly no friend of Christianity—accepted that most diseases were a product of human action, even those affecting animal species.[53] Although suffering was often the stimulus to human progress, the natural world was not the "gladiatorial show" depicted by Huxley and the Darwinians. The way was clear for us to believe that the human race had sinned and was suffering as a result of its action, even though we lived in a divinely planned world that was not created to impose pain on the creatures that evolved within it. God was good, and evil was the product of human sin.

The next question was whether or not the Bible—specifically, the New Testament—gave reliable evidence that God had stepped into the world in an attempt to save humanity from itself. Here Gore challenged the trend in biblical criticism that sought to demystify the events recorded in the New Testament, especially the miracles that proclaimed Christ's supernatural origin. But why was there now such a prejudice against miracles? Here science's notion of a completely self-contained and law-bound nature was indeed a problem. But once it was conceded that God underpins the activity of the whole universe, there was no reason to rule out the possibility that He might find good reason to interfere on rare occasions of extreme spiritual significance, especially given that His scheme of creation had now been perverted by humanity. This would not be a violation of the laws He instituted if we understand those laws as part of a rationally ordered whole with a spiritual purpose. At these points God "bares His arm" in order to make His purpose clear.[54] In the words Gore chose later to clarify his position:

> This world—the human and in part even the non-human world—had become so disturbed and distorted by sin that God the Creator became

52. Ibid., 139–42.

53. Ibid., 158, citing Lankester, *Kingdom of Man*, 33, where it is argued that the relaxation of natural selection in the human population and in domesticated animals has allowed bacteria and parasites to flourish.

54. *Reconstruction of Belief*, 237.

the redeemer, the re-creator, of the world—not to disturb but to restore its true order; and it appears that to manifest the real purpose and order of creation, against the false order of sin, He found it necessary to do exceptional—what we would call miraculous—acts in order to make evident to the faith that even in the physical events of the world His moral purpose is ultimately supreme.[55]

Gore accepted that such occasions would be rare, so that science could get on with its job of investigating the normal operations of the world unhindered. He noted, however, the recent tendency to see miracles occurring frequently in the modern world, citing the cures at Lourdes as an example. He also thought that Lodge and the spiritualists were, in effect, making supernatural occurrences part of everyday life (something that Lodge would have denied, on the grounds that the spirit world was not really supernatural). Gore confessed himself old-fashioned enough to feel an "intense mental repugnance" at the prospect of taking all this seriously.[56] But the virgin birth, the Resurrection, and the Ascension were exactly the kind of divine interventions that would be expected if Christ were indeed the Savior of the human race. The virgin birth was essential to Christianity, and Gore remarked that "I once drew from [T. H.] Huxley the admission that if he believed—what he did not—that Jesus was strictly sinless, he would suppose that that involved a physical as well as a moral miracle."[57]

In his second volume, *Belief in Christ*, Gore explored the traditional notion of the Fall. He contrasted the Christian view that humanity, once created, turned its back on the divine purpose with the Modernist idea that sin was merely the "tiger and the ape" surviving in us.[58] The Christian could believe that creation worked through an evolutionary process that was purposeful, but it was still necessary to accept that once a being with free will appeared, that purpose could be subverted. Genesis does not imply that Adam and Eve were perfect; they were essentially ignorant, capable of achieving perfection or of turning away from it—and they chose the latter course, with disastrous results for their descendants. Gore was never very clear about exactly how the element of sinfulness was transmitted to us, but he noted that progress in society is always blocked, as though we are now incapable of moving toward it. It was to show us a way out of this dead end that Christ came into the world

55. *Can We Then Believe?* 54–55.
56. *Reconstruction of Belief*, 250–51. Gore cited Lodge's *Raymond* to illustrate his point.
57. *Reconstruction of Belief*, 281.
58. *Reconstruction of Belief*, part 2, 565.

and sacrificed Himself for us. The crucial distinction was between sacrifice and punishment—Christ was not punished by a vengeful Father on our behalf; He willingly sacrificed Himself to achieve the divine purpose of redeeming us.[59] Gore thus retained a major role for the Atonement without adopting the Calvinistic doctrine the Modernists so resented.

His final volume, *The Holy Spirit and the Church*, expounded his view of the church as a divinely instituted spiritual body whose development through history revealed its power to transform human life. The sacraments were more than mere symbols: they embodied this spiritual power, and for this reason the consecration of the bread and wine in the Eucharist gave them a unique status worthy of reverence. It has to be said that there is little in Gore's defense of the Catholic view of the sacraments that seems to engage with the kind of assault launched by Barnes and those who took a more Protestant viewpoint. There is no explicit attempt to explore the notion of a spiritual power somehow transferred to or induced in the bread and wine—here at least there seems a genuine incommensurability of viewpoints, with the two sides simply talking at cross purposes. We shall see a good illustration of this below in William Temple's argument that to understand the significance of the consecrated bread and wine, one needed something akin to an artistic rather than a scientific point of view. This was not a rejection of science, but a decision to live in a world transformed by something more than material facts. Gore concluded with an attack on Modernism, insisting that its vision of an immanent God whose purpose we are still working out in human progress was a violation of the central tenets of traditional Christianity, while his own views were a liberalization of the true faith in the light of modern thought.[60]

In his response to comments on these works, Gore argued that science was not, as many now claimed, "bankrupt" if it admitted to the limitations of its powers. On the contrary, it was now one of the "chief glories of our race." Theology owed biology especially a great deal for opening its mind to evolution.[61] Admittedly, many ancient beliefs now had to be reinterpreted, but it was wrong for the Modernists to use these now-abandoned ideas to challenge the whole edifice of belief. They refused to admit the need for a compromise that took some parts of the creed as myths while others were still taken literally. Even the abandoned ideas might still appear in common speech, just as we still talk of the sun rising although we know it is the earth

59. Ibid., 592.
60. *Reconstruction of Belief,* part 3, 952–53.
61. *Can We Then Believe?* 55–56.

that turns. In that sense, Adam and Eve count as "Everyman" when we seek to understand the Fall. Gore recognized the progress of biological evolution, but thought that the origin of humanity's higher faculties was still shrouded in mystery. In these circumstances, given the obvious experiences of the human race that proclaim our imperfection, it was reasonable to suppose that our creation was marked by both progress and catastrophe.[62] Gore accepted and yet also challenged the new natural theology's reliance on progress and purpose in nature, admitting that vision of the world up to the point of human origins, but retaining enough of a discontinuity at the point where free will emerged to allow for the traditional belief that we have somehow become alienated from the divine purpose. More generally, his worldview went beyond anything that could be analyzed in scientific terms by insisting that human spirituality could be shaped by the mystical power that Christ had implanted in the church.

Gore's approach to science was praised by E. O. James in an article in *Theology* in 1922.[63] James had already criticized Barnes's sermons on evolution, arguing that there was a unique turning point marking the origin of the human religious consciousness and insisting that it was reasonable to assume that when we first learned to choose between good and evil, we chose the latter and thus fell away from God's purpose.[64] More substantially, the volume *Essays Catholic and Critical*, edited by E. G. Selwyn (1926), expanded on the liberal Catholic position. Here J. K. Mozley defended the reality of the miracles associated with Christ.[65] E. J. Bricknell attacked Tennant's notion of sin and the Fall, insisting that sin was a moral evil, not a relic of purely biological tendencies, and indicated a deliberate turning aside of early humans from the divine plan.[66] Noting that there were several different races of early humans, Bricknell hinted that these may have been "false starts." If the origin of humanity was a complex process, not a simple progression, it was possible that early humans had started to develop along moral lines and then turned aside. He also referred to an alternative position suggested by the Rev. C. W. Formby in 1923: perhaps the original perversion of the divine plan went deeper into the past than the origin of humankind, affecting the whole of life itself. Stressing the moral character of evil, in opposition to Tennant's view,

---

62. Ibid., 66–67.

63. James, "Science and Theology."

64. James, "Origin and Fall of Man." James returned to these issues again in his "Evolution and the Faith."

65. Mozley, "The Incarnation."

66. Bricknell, "Sin and the Fall."

Formby saw that evil as inherent to the will of all living things, turning the whole creation against God's purpose. This view fitted with the evolutionists' inability to find any clear discontinuity between the earliest humans and their animal ancestors and also explained both the harshness of the struggle for existence and the irregularity of evolutionary progress.[67] Bricknell supported Formby's emphasis on the moral nature of sin, but regarded the suggestion that the life force itself was tainted as pure mythology.

A chapter by Lionel S. Thornton stressed the need to allow for miracles in God's dealings with the universe.[68] A few years later Thornton expanded his ideas into a book, linking the doctrine of the Incarnation to the organic conception of the universe being promoted by Whitehead and others. He was scathing about those who talked of a reconciliation between religion and science based on the two keeping to their own separate territories—a religion based on the belief that God became incarnate within the physical universe could not tolerate such a division.[69] Fortunately science had moved in directions that made a closer connection possible, abandoning its old billiard-ball model of matter and accepting a hierarchy of development in organic and mental evolution. Bergson and the exponents of creative and emergent evolution had paved the way for a vision in which spirit played a central role in the world. The organicist philosophy held that everything is an "organism" and made it clear that human personality, while at a higher level than other forms of organization, could not be seen as fundamentally distinct from those other levels. Humanity is rooted in the cosmic order, and indeed, the human spirit is the "fulfilment of the cosmic series."[70] Much of Thornton's book reads like a Modernist tract, but his emphasis on God becoming incarnate within the universe was linked to the same important qualification of progressionism as that suggested by Gore: The human race, once formed, had not continued to fulfill the purpose of the evolutionary series. We were supposed to link the whole of creation to God, but we have failed to do our job: "It is a failure which frustrates the tendencies of the cosmic series as a whole in its ascent toward eternity and which therefore casts its shadow upon the whole order of creation." The frustration of the divine purpose was the essence of evil—"a withholding of creation from its movement toward God."[71]

67. Formby, *Unveiling of the Fall.*
68. Thornton, "The Christian Concept of God," 146.
69. Thornton, *Incarnate Lord,* xi–xii.
70. Ibid., 46.
71. Ibid., 121.

Like Gore, then, Thornton went along with the liberal trend in theology far enough to adopt the idea of a creative, purposeful universe, but stopped short of the Modernists' belief that the process was still on course as humanity gradually stripped away the relics of its animal past. The Modernists could still retain this sense of purpose because they believed in a God who was totally immanent within the world—if the cosmic order were perverted, it would imply that God Himself had failed. But for the liberal Catholics, God was still transcendent: He sustained the world and made it creative, but He had let it rise to levels of freedom that allowed His intentions to be thwarted. The human race was the embodiment of that frustration of the divine purpose, but a transcendent God could reenter nature from outside, so to speak, and adopt a historically unique way of trying to salvage the wreck through the Incarnation. Science had a role to play in helping us to understand the original purpose of creation, but it could not explain why humanity had turned aside from that purpose, nor could it hope to understand the supernatural events surrounding Christ's mission.

A barometer to measure changing attitudes toward Modernism and toward science itself can be found in the thought of the future archbishop of Canterbury, William Temple. The son of Frederick Temple, who himself became archbishop of Canterbury, Temple was originally trained as a philosopher at Oxford and only gradually realized that he had a vocation in the church. He was at first refused ordination by the bishop of Oxford because he had doubts about the virgin birth and (to a lesser extent) about the Resurrection.[72] His position soon became more orthodox as he sought a synthesis of philosophy and theology around an idealist position in which the divine Mind underpins the whole of reality. He had links with Streeter, Rashdall, and the Modernists, but also with Gore and the liberal Catholics, and seems to have moved gradually further away from the Modernist position. Temple was willing to make some accommodation with science, but he was never very interested in it and wanted to limit its authority both in philosophical thought and in the practical world.

In a series of lectures in 1909, Temple praised Arthur Balfour for stressing the unity and rationality of nature as shown by the latest developments in science, but insisted that this was never enough because what we wanted to know was the purpose of the cosmos. Only religion can answer the question, "Does the Will that lies behind the world know me and care about me?"[73] In

72. See Iremonger, *William Temple,* chap. 7. This was in 1906; Temple was ordained by Randall Davidson three years later.

73. William Temple, *Faith and Modern Thought,* 18–19; on Balfour, see 12.

the 1920s Temple's *Mens Creatrix* and *Christus Veritas* developed the philosophical and theological implications of his thought, propounding the vision of a creative unfolding of the universe culminating in humanity, but stressing Christ as the key to our knowledge of the Creative Mind, especially our recognition of its infinite love. As with Gore, however, there is also a sense that humanity—although the high point of the creative sequence—has taken the wrong course. We arrived at the realization of value by doing what was forbidden, stumbling into things that were the source of evil, yet in so doing we made possible a greater good through our eventual reconciliation with God.[74] Temple's Gifford Lectures for 1932–1934 continued the theme of a universe driven by a Creative Mind: science's efforts to explain the emergence of mind in the course of evolution had failed, but the process of evolution itself suggested that it was planned by mind, so the most logical explanation is that Mind lies at the heart of the world-process.[75] Shortly after the publication of these lectures Temple went to America, where he met Alfred North Whitehead, and the two seem to have got on very well, agreeing to disagree in ways that allowed each to understand the other's position.[76]

There is much in these writings that resembles the kind of idealist progressionism favored by the Modernists, although Temple's Creative Mind is certainly transcendent as well as immanent. There is little about the process of evolution as a scientist would understand it, although Temple was clearly aware of what many scientists were reading into the idea of creative evolution. He was a close friend of Julian Huxley, and Charles Raven later recalled how he had explored with Temple the hope that Christendom would be transformed by the new knowledge of the development of life on earth.[77] Yet Temple was moving steadily further away from the Modernist position, especially as represented by Barnes. He was strongly opposed to the latter's attack on the Anglo-Catholic view of the sacraments, holding that material objects can indeed acquire a spiritual significance. His justification for this view includes an interesting reflection on the limitations of the scientific way of thinking:

> A great many of our difficulties in the past have arisen because we have
> thought in accordance with the habit of mind of science to the exclusion

74. Temple, *Christus Veritas*, 74.

75. Temple, *Nature, Man, and God*, 132.

76. Emmet, "The Philosopher," in Iremonger, *William Temple*, 528–29.

77. Raven, *Teilhard de Chardin*, 9. See also J. S. Huxley, *Memories*, 113, 150—it was Temple who aroused Huxley's interest in religious questions.

of that which is induced by art. Religion is something that binds art, as meaning the outward expression of what is spiritually and emotionally vital, with science, as the impulse to ascertain the truth. Both are there, and we must allow quite as much for all those apprehensions which come to us through the world of art as for those which come through the world of science.[78]

Temple also found Barnes's attacks on the church's attitude toward evolution offensive, and in 1930 he wrote Barnes a blunt letter explaining his objections. He insisted that no one in the church since the time of his father's Bampton Lectures in 1884 had resisted the general idea of evolution. As for Barnes's dismissal of Genesis as a myth, Temple retorted:

> You say: "The story of Adam and Eve is, of course, incompatible with modern knowledge, and serious theologians set it aside." I should have said that the serious theologian never sets anything aside without asking what (if anything) of spiritual value has been faultily expressed here, and taking care to give it better expression. As soon as it is realized that the Garden of Eden is a myth, it is seen to be a very good myth, curiously congruous with evolution, because the Fall is (in the myth itself) a "fall upwards," seeing that by it the knowledge of good and evil was obtained. Similarly Genesis I, as soon as it is taken as a myth, is an overpoweringly good myth.[79]

Here the limitations of the idea of progressive evolution become apparent: Temple still needs the sense of humanity as having turned its back on God, even though he hopes and expects Christ's intervention to be successful in returning us to our rightful place.

In her evaluation of Temple as a philosopher, Dorothy Emmet expressed the opinion that he was never very interested in science, and in particular was indifferent to its empirical methods.[80] He had a generally evolutionary worldview, but this was based more on a kind of Hegelian vision of emerging levels of value than on biology, and his assumption that there were points of discontinuity in the evolutionary advance had more to do with his sense of humanity's uniqueness than with Lloyd Morgan's ideas. The universe was a

78. Temple, speech to the Diocesan Conference, Manchester, 29 April 1927, quoted in Iremonger, *William Temple,* 349.

79. Temple to Barnes, January 1930, quoted in Iremonger, *William Temple,* 491.

80. Emmet, "The Philosopher," in Iremonger, *William Temple,* 535.

rational order, but it was also a moral order, and this was Temple's real concern. His willingness to admit the reality of the Christian miracles marked the boundary beyond which he would not go to accommodate science, but unlike Gore, he seems to have been relatively indifferent to the question of the difficulties this view would pose for those whose main concern was the uniformity of nature. It is also clear that in the course of the 1930s Temple began to reflect the general change in the cultural atmosphere that made the whole concept of progressive development suspect. Emmet quotes a letter to herself from Temple written in 1942 in which he modified the view expressed in his Gifford Lectures to accept that the world at this point in time is not a rational whole—its order will be revealed only at the end (like a drama, not a painting), so it may well appear unintelligible to us now.[81] A slightly earlier letter to Gregory Dix offered an overview of his career in which his earlier efforts as a philosopher to build a bridge by which those in the universities could get to a Christ-centered view of the world were seen not as a blind alley, but certainly as starting from a position that no one now accepted. In the 1920s the world was still stable and liberalism still triumphant, and one could take Christian standards for granted. By 1939 this was no longer possible, and a theology of redemption had become more necessary (although Temple could never go along with Barth's complete rejection of the relevance of natural theology).[82]

THE MENACE OF THE NEW PSYCHOLOGY

If there was limited agreement between Modernists and liberal Anglo-Catholics over the need for a natural theology that would reveal the creativity of the universe, the two sides could also concur in their response to another manifestation of scientific materialism: the new psychological theories developed in the early decades of the twentieth century (see chapter 5). By the 1920s both Freudian analytical psychology and behaviorism had become powerful forces attempting to transform our vision of human nature. The older psychological tradition represented by the work of James Ward had taken free will and moral awareness for granted, thereby sustaining key aspects of the Christian viewpoint. Typical of this tradition is a chapter by W. R. Boyce Gibson in the 1902 volume *Personal Idealism*, distinguishing between an empirical psychology based on inductive methods and essentially

81. Temple to Emmet, 16 July 1942, quoted ibid., 537–38.
82. Temple to Gregory Dix, 1939, quoted in Iremonger, *William Temple*, 606.

determinist in its viewpoint, and a "science of free agency" based on intro-spection and wedded to freedom as its only possible working postulate.[83] Coupled with the claims of emergent evolutionism, such a psychology could be seen as fully compatible with the new natural theology. The new psy-chologies proclaimed a darker vision of human nature, which was enslaved either by learned habits imposed through conditioning or by deep-seated unconscious impulses. In their fully developed (and, of course, mutually in-compatible) forms, both theories were unacceptable to the Christian, and strenuous efforts were made to combat them or to undermine their more dangerous aspects.

Behaviorism challenged the traditional notion of human nature by elim-inating the category of mind from psychological analysis and treating both humans and animals as learning machines that can be programmed with habits to determine their behavior. No Christian could accept such a view of human nature since it made the concept of moral responsibility completely redundant. Analytical psychology had more complex implications. Freud proclaimed that religion was a myth to be explained away in terms of the ten-sions between the conscious and unconscious components of the mind. His view that the central feature of the unconscious was its sexuality was also shocking to many traditionalists. Yet the basic idea of the unconscious was not in itself incompatible with Christian belief; indeed, it offered a way of understanding the less savory aspects of human behavior that any moralist or religious thinker had to take seriously. For the Modernist, the unconscious could be the seat of those ancestral animal urges that we are destined to over-come as our spiritual evolution progresses. For the traditionalist, the uncon-scious was the seat of those sinful impulses that have marred humankind since its first creation.

By the 1920s most intellectuals would have been aware of the new theo-ries of human nature, and accounts of the "new psychology" by religious thinkers were starting to appear. Yet many of the early efforts in this direc-tion were less than fully engaged with the more radical theories, trying instead to reassure their readers by arguing that psychology was still able to help the Christian understand the complexities of human nature. Examples include (in chronological order) books by T. W. Pym (1921), F. R. Barry (1923), H. Crichton-Miller (1924), J. Cyril Fowler (1924), W. Fearon Halliday (1929), and the Rev. E. S. Waterhouse (1930, based on a series of radio talks). Barry briefly rejected Freud's claim to have "explained away" religion, Crichton-

83. Gibson, "The Problem of Freedom in Its Relation to Psychology."

Miller belittled his notion of a "death wish," and Halliday conceded that the emphasis on sexuality might explain some diseased personalities.[84] None gave any real sense that the new psychologies threatened to undermine the Christian view of human nature, and indeed, both Crichton-Miller and Fowler adopted a progressionist vision of biological and social evolution fully in tune with the outlook of the new natural theology.

An explanation for this relative lack of concern was offered by the noted Modernist theologian B. H. Streeter in his Hale Memorial Sermon, *The Church and Modern Psychology*, delivered in 1934. Streeter noted that British religious thinkers knew far less about behaviorism than their American counterparts, although there had been a "boom" in interest in analytical psychology after the Great War because of the recognition that the new psychological techniques offered a way of treating shell shock and similar traumas. Many British psychologists were not hostile to religion, and religious thinkers had taken their ideas into account—Streeter's own edited volume *The Spirit* (1919) had made numerous references to the more complex view of the human personality now emerging from psychology.[85] The new psychology had been denounced more in the press than in the pulpit. The effect of psychology on the value of religious experience was less in Britain than in America because, outside Nonconformist circles, the churches stressed the ethical message of religion rather than its emotional impact. Streeter saw behaviorism as a useful challenge because it had pushed the mechanistic approach to its ultimate extreme and thereby exposed its limitations—in effect, it was a *reductio ad absurdum* of materialism. Christians should not, however, belittle the value of a close link between the mind and the body, nor should they deny the role of instinct in shaping our behavior. Streeter thought that William McDougall's *Introduction to Social Psychology* of 1908 was a major British contribution to the new psychology because it stressed the influence of social instincts. Freud's theory was presented as an extreme form of the same way of thinking, with animal instincts forming the basis for what Christians had traditionally seen as original sin.[86] This was certainly the Modernist position on the Fall, but to present Freud as merely continuing McDougall's project was to misunderstand the radical challenge that the analytical viewpoint offered to the pro-

---

84.   Barry, *Christianity and Psychology*, 166–67; Crichton-Miller, *New Psychology and the Preacher*, 82–83; Halliday, *Psychology and Religious Experience*, 24–25. The other works cited are Pym, *Psychology and the Christian Life*, Fowler, *Psychological Studies of Religious Questions*, and Waterhouse, *Psychology and Religion*.

85.   Streeter, *The Church and Modern Psychology*, 10–11.

86.   Ibid., 19 and 25.

gressionist view of human nature. Streeter also saw the concept of the libido as merely a reformulation of the old idea of a life force. His lecture shows how Modernist theologians were able to assimilate the new psychology into their own way of thinking by refusing to acknowledge how profoundly Freud had broken with the tradition of human rationality.

A few commentators realized the true nature of the threat, however, and the proportion of critical studies increased through time. By the 1930s Streeter's optimism would have been widely seen as misplaced. Theologians no longer tried to pretend that psychology was a friend to religion; rather, it was now portrayed as another example of the dangerous trends within modern thought and science that must be resisted if Christianity was to survive. One of the first critical comments was Henry Balmforth's *Is Christian Experience an Illusion?* This work arose from a paper given to the Joint Conference of Anglican and Free Church Fellowships in 1922 and appeared with an introduction by the bishop of Manchester lamenting the lack of belief among the urban working class. Balmforth linked Freud with anthropologists such as Frazer who had tried to dismiss religious belief as a "racial daydream" created by the human mind.[87] He insisted that no attempt to explain the origin of such beliefs could undermine the Christian's sense of their value, which was based on personal experience.

W. B. Selbie, principal of Mansfield College, Oxford, gave the Wilde Lectures at Oxford on the new psychology, and had the bishop of Gloucester comment on his text before it was published as *The Psychology of Religion* in 1924. Like Balmforth, he stressed the positive value of religious faith to the believer, but also argued that Freud's efforts to explain why we believe in God said nothing about the question of whether or not there really was a God who was the object of that belief. The psychologists confused God with the idea of God, and thought that by explaining the origin of the latter, they had shown the nonexistence of God Himself.[88] Selbie thought the new techniques merely confirmed the fundamental role played by faith in our psychological makeup, and he argued that we should trust our faculties when they informed us of the existence of a spiritual world.

In 1939 Selbie returned to the subject with his more comprehensive *Christianity and the New Psychology*. In his preface he noted that many saw the latest developments in psychology as the greatest challenge faced by the Christian religion. Yet if psychology were truly scientific, it could not conflict with religion, and his goal was to demonstrate its points of value to religion while

87. Balmforth, *Is Christian Experience an Illusion?* 24.

88. Selbie, *Psychology of Religion*, 296–97.

exposing where the psychologists had overextended themselves. All the rival schools of thought in psychology adopted a mechanistic approach, which limited their ability to appreciate the value of human experience. He dismissed behaviorism quite curtly with the insistence that our experience of freedom and value is real.[89] As in his earlier book, Selbie focused mainly on Freud and his followers, paying special attention to the work of Carl Jung. Both Freud and Jung thought that religion was an illusion, but Selbie repeated his earlier point that explaining the idea of God as a projection from the mind did not justify the assertion that God does not exist.[90] The psychologists did not understand the value of faith to the healthy personality. Freud had spent so long studying diseased personalities that he saw sexual perversion everywhere. Jung's views were in some respects the greater threat, since they were a more direct challenge to Christianity. He did not understand the religious experience because he saw no difference between Christianity and more primitive religions.[91] Like the behaviorists, but for different reasons, the analytical psychologists were determinists, belittling our efforts to transcend the internal or external constraints on our conduct. But in fact we can resist our unconscious desires, and sin still has meaning in the sense that we often fail to do this even though we have the power.[92] The psychologists' efforts to destroy religion foundered on the obvious fact that faith was vital to the well-being of the personality, and this biological value suggested that there must be some objective truth in religious beliefs.

The assessment of psychology as the greatest current threat to Christianity was endorsed by the bishop of Southwark, writing in the introduction to a collection called *Psychology and the Church* in 1925.[93] One of the chapters in this collection, written by W. R. Matthews, the Modernist author of *God and Evolution,* repeated the claim that an explanation of belief in psychological terms did not disprove the existence of God.[94] The Rev. L. W. Grensted's *Psychology and God* (1930) was based on his Bampton Lectures, and began with an assault on behaviorism for its denial of our sense of freedom derived from introspection. Freud did not deny the value of introspection or the reality of consciousness, but his view of the mind was equally mechanistic in its denial

89.   Selbie, *Christianity and the New Psychology,* 19–20.

90.   Ibid., 43, 61.

91.   Ibid., 24, 55.

92.   Ibid., 75–79.

93.   Introduction to O. Hardman, ed., *Psychology and the Church,* ix–xiv.

94.   Matthews, "The Psychological Standpoint and Its Limitations."

of free will.[95] Grensted thought that both William McDougall and William James had unwittingly delivered religion into the hands of its critics, the former by refusing to address the question of value and the latter by promoting the philosophy of pragmatism.[96] The bulk of Grensted's study was an effort to defend the value of religious faith to psychological health and to build on this the claim that the objects of faith—God and Christ—must therefore have objective existence. The psychologists' efforts to dismiss sin as a meaningless concept concealed their lack of any true sense of values.

A more complex evaluation was David Yellowlees' *Psychology's Defence of the Faith*, based on a presentation at a summer conference of the Student Christian Movement in 1929. Despite its title, the book sought to engage with the writings of Freud and Jung in some detail, although, like most religious commentators, Yellowlees insisted that psychology's efforts to explain the origin of belief were irrelevant to the question of whether or not God actually existed. He accepted the reality of the unconscious on the basis of his own experiences with disturbed personalities: we could not dismiss the whole of the Freudian system just because we thought its author went too far in linking everything to sex. Some mental illness might well arise because people had been exposed to repressive religious ideas at an early stage in their lives.[97] Yet sin was not the same as guilt, and many real blackguards were not neurotic— they were quite comfortable with the evil they perpetrated.[98] At the opposite end of the spectrum, there was an underlying uniformity in the experiences of those whose lives had been touched by Christ, suggesting that this form of "suggestion" was of real value.[99] Yellowlees thus indirectly joined the ranks of those who thought that the new psychology had a positive message for religion, although he saw the need for many of its more extravagant claims to be exposed.

The rather ambiguous assessment of science in F. Leslie Cross's *Religion and the Reign of Science* (1930) included a chapter on psychology, pointing out the baleful effects of behaviorism's denial of consciousness. Cross paralleled Yellowlees' belief that Freud's idea of the unconscious had some value, provided the excesses of his reliance on the sexual impulses was allowed for. He also hinted that this aspect of mental life might explain much of what was at-

95. Grensted, *Psychology and God*, 22–23.

96. Ibid., 29–20 and 44–45.

97. Yellowlees, *Psychology's Defence of the Faith*, 68–69 and 111–13.

98. Ibid., 62, 65.

99. Ibid., 140.

tributed to the paranormal.[100] Others were even more suspicious, and as late as 1939, the Rev. J. C. Conn could title his book *The Menace of the New Psychology*. This book was based on material originally published in the *Evangelical Quarterly*, and the author—a trained psychologist with a Ph.D. from Glasgow—pulled no punches in dismissing the scientific credentials of the new psychology, if only for the vast range of differing interpretations of the human mind on offer. Behaviorism turned us into automata by insisting that consciousness is irrelevant. It reduced humans to the level of beasts and made a mockery of the hope of immortality.[101] In one sense at least, Conn agreed with Streeter—behaviorism was materialism gone mad. He quoted McDougall and Lodge on the absurdity of its claims. Analytical psychology was a greater challenge because it explicitly claimed to have shown that religion was an illusion. Freud claimed that all mysticism was founded on sexuality, and Conn conceded that some mystical experiences did have a sexual content—but this was clearly irrelevant in the case of Jesus.[102] The psychologists' discovery of the mechanisms prompting belief did not disprove the existence of God, and the sense of sin lingered in the mind even after the analyst had produced a "cure." Infantile sexuality was Freud's greatest absurdity, explaining all the vices of the human mind but none of the virtues. McDougall and many others had shown that we do have the capacity to control our baser instincts, and such self-conquest was essential for Christianity to work.

Conn's accusation that the new psychology was a threat to religion illustrates both the difference between evangelical and Modernist perspectives and the growing realization even among those interested in science that Freudianism was not just a modification of the old nonmaterialistic psychologies. Where Streeter saw Freud as merely extending the insights of McDougall, Conn recognized that he had completely overturned the idea that human nature had faculties capable of transcending our animal ancestry. Yet Conn still held to McDougall as the foundation for a psychology that Christianity could feel comfortable with, evidently unaware that his almost vitalistic view of life was no longer taken seriously in most scientific quarters. Recognizing the perils of the new psychology left religious thinkers who wished to defend a reconciliation with science dangerously exposed, because in this case, the outdated foundations of their alternative worldview were all too apparent.

100.   Cross, *Religion and the Reign of Science*, 64–67.

101.   Conn, *Menace of the New Psychology*, chap. 2.

102.   Ibid., 39–40.

SCIENCE AND MODERN LIFE

Outside the ranks of the Modernists, many Christians had doubts about whether science could offer any contribution to a truly moral perspective. Gore and Temple did not reject the new natural theology but expressed a more basic feeling that its model of creative evolution confirmed only the rationality of God, not His goodness or His saving grace. Conservative thinkers were more generally suspicious, questioning the nineteenth century's assumption that human ingenuity presented the opportunity for unlimited moral and material progress. They were less than enthusiastic about the applications of scientific knowledge that some Modernists accepted as necessary for the furtherance of the divine plan. In the simplest interpretation, science gave us the knowledge we needed to control nature for the good of humanity. But the Great War and the increasingly obvious inequalities of the new industrialized society made many church leaders unhappy about the pace, if not the direction, of change. The enthusiasm of some Modernists for birth control and eugenics revealed the ambiguities of the appeal to progress: was it really justified to interfere with human reproduction on the assumption that scientists know best how to conduct social policy? The belief that moral guidance comes from a supernatural source, not from our own evaluations, led theologians to question the social and economic developments associated with science and technology. More generally, the shallow materialism of much of modern life was seen as a threat, and science was all too easily identified as the source of this way of thinking. Church leaders were thus inclined to challenge the rationalists' claim that science itself was a source of moral values, especially when science was presented as a viable alternative to the traditional forms of education.

Temple's views show how a churchman who was not hostile to the anti-materialistic interpretation of science could deny that science was relevant to moral and spiritual values. There were major limitations on how far he was prepared to go in allowing a role for science in practical affairs. Anna-Katherina Mayer has noted Temple's contribution to the "neglect of science" debate of 1916 in the form of an address to the Education Section of the British Association. Here he noted the efficiency of Germany's use of resources, coupled with that country's moral failure, and expressed a preference for the British system of education in which the classics were used to promote ethical values.[103] This attitude did not arise from political conservatism; indeed, Temple was a prominent Christian Socialist and chaired the Conference

103.   Mayer, "Moralizing Science."

on Christian Politics, Economics, and Citizenship (COPEC) when it was created in 1924. Julian Huxley thought he supported eugenics, although there is little evidence of him engaging actively in the movement. On the practical side, Temple was aware of the growing power of science and technology, but he believed that the ethical values that should determine how this power was applied should be derived not from science itself, but from philosophy and religion.

Temple was by no means the only church leader to express concern about the consequences of the unrestrained application of science and technology. The most widely publicized effort in this direction was the appeal for a "scientific holiday" by Arthur Burroughs, bishop of Ripon, at the 1927 meeting of the British Association. Burroughs had gained some notoriety in 1915 by writing a letter to the *Times* calling for the nation to admit it had sinned as part of the effort to ensure that the war against Germany became a spiritual conflict. One of the many responses was a letter from E. Ray Lankester, asking him what he meant by sin in this context.[104] Burroughs's biographer, H. G. Mulliner, claims that he was not hostile to science, but failed to realize that those exposed to its influence found it difficult to see the universe as designed for humanity, and were thus suspicious of facile talk about a "heavenly Father."[105] Curiously, this biography says nothing about the 1927 sermon, but as Mayer observes, there is indirect evidence that Burroughs's attitude toward science in general may have been triggered by his distaste for Arthur Keith's materialistic Darwinism and a preference for J. S. Haldane's efforts to build a scientific idealism in which personality once again became central to our worldview.[106] The sermon itself noted Haldane's objections to science's normal tendency to eliminate personality from consideration, but focused mainly on the material consequences of scientific advance. It quoted Bergson on how humanity's control of material nature had expanded while its moral vision remained static. Burroughs then introduced his notion of the scientific holiday:

> Dare I even suggest, at the risk of being lynched by some of my hearers, that the sum of human happiness outside scientific circles would not necessarily be reduced if for ten years every physical and chemical laboratory were closed and the patient and resourceful energy displayed in them

104.   Mulliner, *Arthur Burroughs: A Memoir*, 33.
105.   Ibid., 99.
106.   Mayer, "'A Combative Sense of Duty.'"

transferred to recovering the lost art of getting on together and finding the formula for making both ends meet in the scale of human life.[107]

The result might be a loss of further technological advances, but ordinary people would gain time to catch up with all the things they had scarcely been able to assimilate, so fast had the pace of progress become. The implication was clear: science was a dehumanizing force, both through its elimination of personality from our worldview and because it had become the tool of an industrial complex that was undermining traditional values in the search for profit. Both Temple and Burroughs, in their different ways, warned that science's threat to religion might arise not just from its conceptual challenge to traditional elements of the faith, but also from its role as part of the general trend toward a more materialistic way of life. These fears could only intensify as the social and political tensions of the 1930s highlighted the bankruptcy of the old progressionist assumptions.

THEOLOGY IN THE THIRTIES

These suspicions of science drew on fears arising from the social disruption caused by industrialization and the role played by military technology in the Great War. It was all too easy to see science as an impersonal quest for practical knowledge that could be misused by anyone powerful enough to pay scientists to do their bidding. Philosophers such as Bertrand Russell and literary figures such as Aldous Huxley harped on the same theme. But for European culture as a whole, the 1930s marked a final departure from the optimistic liberalism of the late nineteenth century, which had been prolonged into the 1920s in part by the attempted synthesis of religion and antimechanistic science. The decade that saw the depression bring home the moral failure of industrial capitalism and witnessed the rise of brutal totalitarian regimes in Italy and Germany could no longer pretend to itself that all was going to come out for the best in accordance with some inevitable divine plan. The result was a return to religious faith among a significant proportion of the intellectual community and a reinvigoration of a more traditional Christianity based on a recognition of human sinfulness and the need for supernatural redemption. Those who moved in this direction were inclined to become even more suspicious of science's practical applications and its link to

107.   From the report in the *Times*, 5 September 1927, 15; see also *Daily Telegraph*, 5 September 1927, 9–10, and *Daily Mail*, 5 September 1927, 7.

the quest for the domination of nature. As far as claims for an intellectual synthesis between nonmechanistic science and religion were concerned, they were indifferent rather than actively hostile. Nature once again became a neutral backdrop against which the human drama of sin and redemption would be played out, and science became irrelevant to the moral and spiritual questions of the day.

Both Catholic and Protestant churches benefited from the new atmosphere.[108] In France, there was a revival of Roman Catholic philosophy based on Thomism, which exerted some influence across the English Channel. A typical product of this trend was Hubert S. Box's *God and the Modern Mind* (1937), which surveyed all the popular reinterpretations of the previous decades from Bergson and emergent evolutionism to the ideas of H. G. Wells and Julian Huxley, dismissing them all and calling for a revival of the traditional view that God was transcendent, existing apart from the universe yet capable of exerting influence on it when He chose.[109] Anglo-Catholicism was stimulated by this reinvigoration of its traditional philosophical base and lost interest in the links with science that liberals such as Gore had tried to maintain.

Those with more Protestant inclinations came under the influence of Karl Barth and Reinhold Niebuhr, who were forging a new theological orthodoxy that turned its back on the liberalizing trend. The key event in Britain was the publication of Sir Edwyn Hoskyns's translation of Barth's commentary on the Epistle to the Romans in 1933. Here was a new sense of crisis in human affairs and a sense that we are all alienated from God and depend on His mercy for salvation. Barth was hostile to natural theology, arguing that the God it hoped to uncover was not the God of the Christian revelation. Charles Raven returned to Cambridge just as Hoskyns's translation appeared, and for many the two became rival symbols of the old liberalism and the new orthodoxy. Raven was horrified to see his hopes dashed as even the Student Christian Movement, once firmly committed to liberalism, converted to the more traditional view of God. In the 1940s he would become embroiled in several bitter controversies with Anglican theologians as he attempted to defend natural theology against those who dismissed it as irrelevant to the true faith.[110] The revival of interest in the theology of P. T.

---

108. For a survey of the link between cultural and theological developments in the 1930s see Hastings, *History of English Christianity*, chap. 17.

109. H. S. Box, *God and the Modern Mind*, see chaps. 1–3 for a critique of popular ideas.

110. See Dilliston, *Charles Raven*, 205–10 and 286–90. On Barth and natural theology, see Matczak, *Karl Barth on God*, 253–69 and 291.

Forsyth illustrates the same trend.[111] The onetime liberal Nathaniel Micklem published his *What Is the Faith?* in 1936, insisting that science could say nothing about the Fall and salvation.[112]

What is most striking about the new orthodoxy is the power it held to inspire the writers who returned to the fold. Some proclaimed the Christian message in forms that reached a wide audience, giving the church a sense that it was achieving a true revival within British culture. The most effective of these converts in the context of the relations between science and religion was Clive Staples Lewis. Raised in the Church of Ireland and with strong links to Ulster Protestantism, Lewis turned his back on religion while he built his career as a scholar. He regained a general faith in God in 1929 and became a Christian two years later, gradually forging deeper and deeper links with the Catholic wing of the Anglican Church.[113] At Oxford, he built a close relationship with J. R. R. Tolkien, who would write the popular *Lord of the Rings* saga. Lewis himself wrote widely on theology and made radio broadcasts airing his Christian views. His science fiction trilogy, *Out of the Silent Planet* (1938), *Perelandra* (1943), and *That Hideous Strength* (1945), gave voice to his misgivings about science as an agency by which humans try to dominate the world and each other, imagining a cosmic conflict between the forces of good and evil in which science is often exploited by the wrong side.[114] The same point was made at a more academic level in his *Abolition of Man*, which refers to eugenics and other forms of social control to make the point that, far from conquering nature, we may eventually end up enslaved by it as we gain the power to put our baser desires into practice.[115] Lewis was not opposed to the idea of animal evolution, but like many earlier critics, he insisted that the origin of the human soul required a supernatural input.[116] When a dinner-party guest, Helen Gardner, expressed the opinion that Adam would have been an apelike or Neanderthal figure, he sneered at her "Darwinism" and never invited her again.[117]

111. See Rodgers, *Theology of P. T. Forsyth.*

112. Micklem, *What Is the Faith?* 156–57.

113. Lewis's conversion is described in the concluding pages of his autobiographical *Surprised by Joy;* see also A. N. Wilson, *C. S. Lewis: A Biography.* Wilson notes that Lewis's modern disciples, at least in America, are split between those who stress his Protestant and his Catholic leanings. The later episodes of Lewis's life were dramatized in William Nicholson's play *Shadowlands,* later filmed with Anthony Hopkins as Lewis.

114. Collected as C. S. Lewis, *Cosmic Trilogy.* These novels are discussed briefly in chap. 11 below.

115. Lewis, *Abolition of Man,* 29–34 and 38–39.

116. Lewis, *Problem of Pain,* 60 and 65. Lewis speculates on a Fall before the Fall of Man that had perverted the whole of creation; see 121–22.

117. Wilson, *C. S. Lewis,* 210.

In his later career he opened a correspondence with Bernard Acworth and moved closer to the antievolutionary camp.[118] Lewis defended the plausibility of miracles, although he did not apply his argument to the miraculous creation of species.[119]

Lewis's hostility to a naturalistic account of human origins was shared by another literary exponent of the Anglican position, Dorothy L. Sayers. The two corresponded regularly after 1943, by which time Sayers had already gained prominence through her radio play about the life of Christ, *The Man Born to Be King*. Sayers was not indifferent to science, having taken a close interest in Eddington's efforts to forge an idealist philosophy consistent with both science and religion.[120] As a member of the Church Action Group organized to challenge anti-Christian views expressed in the press, she attacked Julian Huxley in 1939 for claiming that if God existed outside the universe, we could have no knowledge of Him.[121] Sayers was happy to explore the idea of God as Creator, but her approach was that of the artist, not the scientist: her *Mind of the Maker* of 1941 focused on the parallels between divine and artistic creativity, dismissing the scientists' efforts to disentangle words from their metaphorical meaning as being against human nature.[122] In a 1942 address titled "Creative Mind," she developed this theme, using the futility of the debate between scientists and theologians over transubstantiation to show that the term "substance" had many layers of meaning. At best, physicists such as Eddington could only pick out elements of reality to construct their abstract view of the world.[123] Sayers accepted that the Genesis story of creation did not have to be taken literally, but suggested that if "God had at some moment or other created the universe complete with all the vestiges of an imaginary past," it would make no difference to what we observe or think.[124] A good dramatist often gives his or her characters a "background" that is sketched in during the play or story without ever being covered explicitly. This comparison was offered not as a serious hypothesis, but as an illustration of how irrelevant the evolutionary past was to the real spiri-

118.   See Ferngren and Numbers, "C. S. Lewis on Creation and Evolution," which reprints Lewis's letters to Acworth.

119.   Lewis, *Miracles;* see 42, where Genesis is described as poetry or a folk tale.

120.   See the letter to her son, 10 August 1938, quoted in *Letters of Dorothy L. Sayers,* vol. 2, 86–87.

121.   Letter to *John O'London's Weekly,* 2 January 1939, quoted in *Letters of Dorothy L. Sayers,* vol. 2, 107.

122.   Sayers, *Mind of the Maker,* 33.

123.   Sayers, "Creative Mind," reprinted in her *Unpopular Opinions,* 43–58; see 47–49. On transubstantiation, see also the comment on the debate between Arnold Lunn and J. B. S. Haldane in Sayers's letter to J. C. Heenan, 31 August 1940, quoted in *Letters of Dorothy L. Sayers,* vol. 2, 179–80.

124.   *Unpopular Opinions,* 54.

tual drama of humanity. Sayers took the New Testament miracles absolutely literally, wondering what it would have been like to live with someone who could turn water into wine and multiply loaves and fishes.[125]

Lewis and Sayers were both confident that the younger generation had turned its back on liberalism's attempt to demystify religion in order to placate science. For them, science was either a danger, because of its potential for misuse, or an irrelevance, because it focused on questions that were not really important. The success of their writings shows how different the atmosphere of the late 1930s and early 1940s was from that of the earlier generation, in which undergraduates had pleaded with Barnes to make the case for evolution against Genesis. A significant proportion of the literate public had absorbed something of the new theological orthodoxy that had affected the churches. Modernism had collapsed under the weight of changing religious and cultural values, and those who sought to popularize religion now felt that the best way to do it was to emphasize exactly those aspects of the Christian faith that had been abandoned by those earlier theologians who wanted to use the "new" natural theology to build a synthesis with science. The fact that science itself had changed—that Darwinism, for instance, was now becoming the basis for a major reshaping of evolutionism—passed these writers by. They were simply not interested enough to find out what was going on in the sciences, although had they taken the trouble to investigate, they would have probably felt that their more general fears about science were confirmed. The isolation of Charles Raven within this new climate of opinion symbolizes just how much things had changed, both in the churches and more generally, since the heyday of Modernism.

ROMAN CATHOLICISM

Like many others, Lewis had been inspired by the writings of G. K. Chesterton, who converted to Roman Catholicism in 1922. Along with Hilaire Belloc, Chesterton struggled throughout the 1920s to combat the rising tide of secularism in British culture and liberal theology in the churches, dismissing the latter as a mere capitulation to materialism. He was one of a small number of highly articulate members and converts who gave the Catholic Church a significance in the nation's cultural life out of all proportion to its numerical strength in the population. While Modernism flourished in the Anglican and even in some Free Churches, the strength of authority and tra-

---

125. Sayers to James Welch, 17 September 1941, quoted in *Letters of Dorothy L. Sayers*, vol. 2, 293–96.

dition that held the Roman Catholic Church together stood as a bulwark. Having stamped out its own Modernist movement at the turn of the century, it could focus on defending rather than modifying the traditional articles of the faith. When the tide of opinion turned, it could claim some credit for keeping the faith alive, and played a significant role in the new theological climate of the 1930s, both by continuing its existing campaign and by attracting converts who—when confronted with the stark choice between materialism and belief—decided that the most traditional church of all was the best foundation for a renewal of faith.

The church was not a static entity, although its hierarchy was able to exert a degree of control over the expression of opinions that prevented the emergence of the kind of party divisions that nearly tore the Anglican community apart in the 1920s. Behind-the-scenes lobbying was always going on to allow new opinions to come up for consideration, although the church refused to be rushed into changing its official position. It had never been opposed to science, and indeed, the Pontifical Academy of Sciences was reconstituted in 1937. But from time to time certain radically new theories with broader implications found themselves under suspicion, and this was certainly the case with evolutionism. Partly because of the stridency of the secularist writers of the late nineteenth century, the church had condemned liberalism along with theories that were seen as its manifestations, including Darwinism. The early twentieth century, however, saw a loosening of opinion on this question. While remaining adamantly opposed to the selection theory and to the idea that the human soul could have evolved from a brute mentality, the church did allow discussion of the general question of bodily transformation. It was always the issue of the soul, not the literal word of Genesis, that motivated Catholic suspicion of evolutionism in the twentieth century. Like the Anglo-Catholics, then, the Roman Church could go some way with the new natural theology's effort to found a nonmaterialistic view of nature, provided always that certain clear boundaries were marked around the territory within which the idea of creation by law could be applied. In the heat of debate, though, it is difficult to be sure whether popularizers such as Belloc believed in evolution at all, and there is no doubt that many Catholics remained opposed to the theory in even its most non-Darwinian forms. At the same time, Roman Catholic scientists had to make sure that they did not transgress the limits of the church's teachings on such contentious matters (see chapter 1).

In the 1920s Catholics saw themselves as defending the only remaining bastion against the rising tide of materialism. For this reason, they provided some of the most articulate efforts to probe the assumptions being taken for granted by secularists and even by liberal theologians in other churches. Belloc and Chesterton were by no means the only writers to offer fundamental critiques of modern thought and culture, and typically the converts (including Chesterton himself) were the most critical. W. E. Orchard began as a Congregationalist minister supporting Campbell's New Theology, but later converted to Rome after unsuccessful attempts to build a rapport with the Church of England in the early 1930s. Shortly before he converted, Orchard published his *Present Crisis in Religion*, bewailing secularism in general and identifying the evolutionists' denial of the Fall as a key factor that—when adopted by Modernist theologians—denatured Christianity to such an extent that it no longer had any emotional appeal.[126] Orchard thought that other areas of science, especially the new physics, had moved away from philosophical materialism, which made it all the more frustrating that the general public were proving so hard to wean away from the hedonistic materialism that had accompanied the Victorian loss of faith.

Orchard was not the only Catholic writer to take an interest in the idealist physics of Jeans and Eddington, but the general attitude was more cautious. The Catholic periodical *The Month* carried two articles on physics in 1931, both skeptical of the press coverage suggesting that science had now found room for religion. Reginald Dingle noted that it was not long since popular science had promoted the materialism of Haeckel. He cautioned Catholics to be aware of the strength of the church's intellectual tradition going back to St. Thomas and to think carefully about any new and probably superficial philosophies being offered by the younger generation of scientists.[127] An equally suspicious comment by the Jesuit H. V. Gill concluded by noting that for all the talk of science now supporting religion, nothing was said about revelation, and until science could accommodate itself to the teachings of Christ, it had little to offer.[128] A more positive response came from the Oxford Jesuit Martin D'Arcy in 1934, although he preferred Whitehead's organic philosophy because it not only demolished materialism but did so by denying the authority of mathematics.[129] D'Arcy subsequently had the unenviable job of countering the influence of A. J. Ayer's *Language, Truth, and Logic*, published in 1936, at Ayer's own university, Oxford. D'Arcy certainly impressed many younger academics, one of whom remembered him later call-

126. Orchard, *Present Crisis in Religion*, 50–51. On Orchard's move to Rome, see his *From Faith to Faith*. For a study of those who converted in this period, see Pearce, *Literary Converts*.

127. R. J. Dingle, "What of the New Physics?"

128. Gill, "Science Gets Religion!"

129. M. C. D'Arcy, "Science and Theology."

ing Ayer an embodiment of the devil on a radio "Brains Trust" program.[130] Another Jesuit, C. W. O'Hara, contributed to a 1930 series of radio talks on science and religion, insisting that the latest developments in physics had undermined materialism in a way that not only made the universe seem a divine creation, but allowed for the possibility of both miracles and revelation.[131]

If Catholic responses to the new physics were mixed, evolutionism formed an even more critical focus for debate. Here the main developments were theological rather than scientific, as most commentators remained unaware of the resurgence of Darwinism among biologists in the 1920s and 1930s. In the late nineteenth century the church had temporarily closed ranks against evolutionism, and the works of several Catholic writers on the subject were put on the index of prohibited books. St. George Jackson Mivart, the morphologist who had created a non-Darwinian version of evolutionism, had been supported by the church in the 1870s, but by the end of the century his claim that the theory (in this non-Darwinian form) was consistent with the teachings of the Church Fathers was rejected. Mivart was excommunicated shortly before his death in 1900, although not for his defense of evolutionism.[132] The situation with respect to evolution began to change in the new century, and in 1909 the Pontifical Biblical Commission formally removed the problem created by the apparent conflict with Genesis by proclaiming that the text need not be taken literally.

By the 1920s the argument that a basic concept of biological evolution was not incompatible with church teachings was becoming more widely accepted. A key figure in this trend was Canon Henri de Dorlodot, the Belgian scholar whose *Darwinism and Catholic Thought* was translated from the French in 1925. Dorlodot sidestepped the issue of Darwinian natural selection, insisting that the actual mechanism of evolution was outside the scope of theological debate.[133] For him, the crucial point was that Darwin had conceded divine creation for the initial origin of the first living things. Opposed to this view stood Absolute Evolutionism, which substituted spontaneous generation for the miraculous origin of life, and Creationism or Fixism, which required a miracle for the appearance of each species. He argued that there was nothing in Holy Scripture, as interpreted by the tradition of Catholic exegesis, opposed to the theory of natural evolution, nor even to Absolute Evolu-

---

130. Longford, "A. J. Ayer's Tour de Force, *Language, Truth and Logic*." D'Arcy wrote a critical review of Ayer's book in T. S. Eliot's journal *Criterion*; see Rogers, *A. J. Ayer*, 123–24.

131. C. W. O'Hara, in *Science and Religion: A Symposium*, 107–16.

132. See J. R. Moore, *Post-Darwinian Controversies*, 120–21, and Gruber, *Conscience in Conflict*.

133. Dorlodot, *Darwinism and Catholic Thought*, 4.

tion. Many of the Church Fathers had taught the idea of spontaneous generation, and none said anything that denied the possibility of evolution. Dorlodot himself rejected Absolute Evolution, but only on scientific grounds—
there was simply no evidence of a process that could form living cells from
inorganic matter.[134]

Dorlodot's position was expanded in Ernest Messenger's *Evolution and
Theology* of 1931, although its subtitle, *The Problem of Man's Origin*, shows that the
author intended to address an issue that the Belgian scholar had carefully
avoided. The book appeared with laudatory prefaces by two Catholic theologians from opposite sides of the Atlantic, both endorsing its claims, but
for very different reasons. Father Cuthbert Lattey thought that the book's
value lay in the extent to which it allowed the whole question to be put aside
as irrelevant to theology.[135] The Very Reverend Dr. Souvay took a harsher
line, castigating hard-liners on both sides of the debate and insisting that the
Catholic tradition was now the only defense against materialism, the Modernist theologians having more or less given up and joined the enemy.[136] Messenger followed Dorlodot in showing that the Church Fathers seem to have
taught spontaneous generation and allowed evolution. But in the third section of his book, he turned to the question of human origins and made it
clear that the theory of evolution could not be extended to include the emergence of the human soul. There was nothing to rule out the involvement of
at least some secondary causes in the formation of Adam's body, although
Messenger was inclined to think that some supernatural agencies had also in-
tervened. He was quite prepared to accept that "Adam" was merely a term
used to denote the first true humans, not a single individual. But however
many persons were involved, the origin of their human souls certainly required miraculous intervention. There was a fundamental discontinuity at
the origin of humanity, and yet this step had occurred as the culmination of
a long process of creation that had been destined from the start to lead up to
humankind.[137] Messenger thus provided the foundation for what would become the position of the Catholic Church on this issue in the later twentieth
century: the evolution of the body is acceptable, but not the evolution of the
soul.

Messenger did not discuss the adequacy of the selection theory as an account of the secondary causes that produce evolution. But his account

134.   Ibid., 126.

135.   Lattey, preface to Messenger, *Evolution and Theology*, xiv.

136.   Souvay, introduction to Messenger, *Evolution and Theology*, xvi–xviii.

137.   Messenger, *Evolution and Theology*, e.g., 91, 116, 224, 275–76.

strongly implied that the whole process was the unfolding of a divine plan, reinforcing the feeling that something more purposeful than the Darwinian theory was required. To the extent that Catholic writers were prepared to accept evolution, they would thus be inclined to go along with the supporters of the new natural theology in their emphasis on progress and teleology in evolution. But where the Modernists wanted to extend progress to include the appearance of the human species, the Catholic Church drew a line at this point. Progress was necessary to show that evolution was under divine control, not to show that it could extend to humans. The insistence that a discontinuity was required for the origin of humanity also explains why Catholic writers were inclined to promote the idea of evolution by saltations or sudden jumps, the last of which would be responsible for those natural (as opposed to supernatural) transformations that gave rise to the first humans.

There was already some Roman Catholic support for evolution in the 1920s. The *Dublin Review* even carried an article endorsing the Darwinian position adopted in H. G. Wells's *Outline of History*, although it was critical of his misrepresentations of the church.[138] Support for the idea of progressive or teleological evolution came from the Jesuit J. Ashton, writing in the same journal in 1927. But after surveying the theories of J. Y. Simpson, J. A. Thomson, Lloyd Morgan, and others, Ashton made it clear that natural causes by themselves were not enough: the mystery of the emergence of new properties required us to believe in a transcendent Power that superintended the whole process.[139] In the following year the same periodical carried a review of books by J. B. S. Haldane, Joseph Needham, and Julian Huxley, stressing the alleged opposition of most scientists to the mechanistic approach. Praising Dorlodot's "masterly discussion," the article hailed his vision of organic development as the unfolding of divinely implanted potentials as the key to a reconciliation with the idea of evolution.[140] An article in *The Month* in 1934 also praised the nonmechanistic wing of biology and its application to evolutionism, but coupled this with an evaluation of science's impact on society. Noting that modern agriculture and technology now provided material comfort for all, the article asked if this was necessarily a good thing, considering how people wasted their time on frivolous pursuits.[141]

138.   Burke, "Mr. Wells and Modern Science."

139.   Ashton, "Evolution as a Theory of Ascent," 72.

140.   Watkin, "Professor Haldane on Science and Religion," 126.

141.   M. Taylor, "Catholics and Biology." This is probably the Monica Taylor who edited the memoir *Sir Bertram Windle*.

There were many Roman Catholic writers who remained suspicious of evolutionism even in its nonmechanistic forms. If Burke was prepared to go along with Wells's Darwinism, others were not, and *The Month* carried a series of articles by Richard Downey ridiculing the hominid fossils Wells had cited as links between apes and modern humans and his gullibility in accepting the now discredited selection theory.[142] In 1926 an article in the same journal anticipated a speech by the American paleontologist Henry Fairfield Osborn at the British Association meeting by dismissing him as a propagandist for a theory that still lacked real evidence.[143] Arthur Keith's address in the following year generated a predictable critique insisting that his confidence in the Darwinian theory of human origins was misplaced; an even more withering attack on Keith appeared in the *Catholic Gazette* in 1930.[144] Two years later J. B. S. Haldane's support for Darwinism in his *Causes of Evolution* came under fire.[145]

All of these writers conceded that the real backbone of Catholic opposition to Darwinism was the work of Hilaire Belloc. It was Belloc who took on rationalist writers such as H. G. Wells to defend the foundations of Christian civilization during the heyday of liberalism and skepticism. Along with G. K. Chesterton, he attacked the materialistic view of human nature in a series of popular works, defending the church's view that human spirituality was a divine creation and appearing at times to reject the theory of evolution altogether. It was Belloc and Chesterton, far more than the Evolution Protest Movement, who sustained the popular myth that Darwinism was dead even within science. Like the evangelical antievolutionists, these Catholic writers wanted to preserve a role for the supernatural in the creation of humanity and for the concept of the Fall. Their opposition to evolution was based not on a literal interpretation of Genesis, but on a respect for the church as a bastion of civilization and the only source of spiritual power capable of resisting the rise of materialism. In this sense, their views paralleled those of Gore and the Anglo-Catholics, but they were articulated in a far more popular format.

In the 1930s the Catholic position began to seem more attractive to some intellectuals as the failure of liberalism and rationalism became all too apparent. Writers such as Belloc and Chesterton became fashionable again

142. Collected in Downey, *Some Errors of H. G. Wells.*
143. Broderick, "Evolution Still an Hypothesis."
144. Keating, "Where Does Adam Come In?" and Blyton, "Three Current Kinds of 'Dope.'"
145. Brinkworth, "Can Darwinism Be Revived?"

among those who thought that the best way out of the crisis was a return to the traditional emphasis on sin and the need for redemption. By the end of the 1930s the Protestant churches, too, were reviving their traditional values. Many Nonconformists had retained a more conservative evangelical outlook, and now there was a revival of similar values even in the more intellectually self-conscious ranks of the Church of England, as the marginalization of Barnes and Raven illustrates. The Modernists' attempt to force a reconciliation with science was increasingly out of favor as Anglicans demanded that their religion reconstitute its traditional vision of humanity's need for redemption. If this meant denying some of the supposed truths of science, so be it. But the new mood was more one of indifference than hostility—Belloc and Chesterton might still proclaim the death of Darwinism, but the new generation of theologians largely ignored biology as irrelevant to their vision of the world. Few even noticed that Darwinism was in fact enjoying a resurrection within scientific biology. Had they been aware of the emergence of the genetical theory of natural selection, it would simply have confirmed the distaste they had conceived even for the more teleological evolutionism of the Modernists.

The new orthodoxy was marked not so much by a theological assault on particular scientific theories as by a refusal to allow that natural theology was relevant to salvation. The only exception to this view was the growing chorus of opposition to the "new" psychology, in which the Modernists' efforts to soften the potential danger were unmasked. Here science did challenge the sense of human responsibility quite directly and thus came into conflict with the new orthodoxy on its own territory—yet the theories of Freud were hardly "new" by the 1930s. At best this was a catching-up exercise rather than a genuine effort to grapple with the latest research. Nevertheless, it fueled the suspicion that science was a source of danger to human values rather than the basis for a way forward. The sense that Western culture was in crisis had undermined the Modernists' faith in progress, and it was this wider transformation of attitudes that seems to have triggered the rise of a more conservative Christianity. The breakdown of the proposed reconciliation between science and religion came about more because religious thinkers lost interest in science than because they were forced to confront the reemergence of materialism in biology.

PART THREE

The Wider Debate

CHAPTER 10

Science and Secularism

M any scientists and religious writers attempted to shape the views of the country's intellectual elite, but some realized that there was a much broader audience waiting to be influenced. Eddington and Huxley, like Inge and Lewis, wanted to carry their message to a wider readership. They were not necessarily trying to convert people to a particular church, but wanted to influence their thinking on broader issues and thereby affect the general cultural life of the nation. This appeal to a wider readership had become all the more important as the middle class expanded, became ever more secularized, and was increasingly influenced by the popular press and broadcasting. Secularization, however, did not necessarily mean a loss of interest in matters spiritual, even if it emptied the churches. There were many still willing to consider the possibility that there is more to life than material pleasure, and those with strong views realized that it was vital to make the effort to shape this more diffuse spiritual and moral awareness.

The background to this expansion of popular writing was the ongoing decline in church attendance that produced a steadily increasing number of educated people who had no formal religious affiliation (see chapter 6). The secularized Sunday of sport and leisure was accepted as normal even by the middle classes, now also being wooed by the mass-circulation newspapers with their emphasis on sensation and scandal. Church leaders and others concerned with the moral health of the nation were naturally worried by these trends, although the militant secularism of the Rationalist Press Association was by no means increasing its influence. Indifference to formal religion, rather than outright hostility, seems to have become the prevailing attitude among the masses, although many still looked to nondenominational and even non-Christian forms of spiritual activity to give some semblance of meaning to their lives.

As Alan D. Gilbert remarks, evaluation of the situation depends on how one chooses to define "religion": if the broadest sense of the term is adopted, reducing it to a vague sense of ultimate concern, then religion never declined, but merely changed its form of expression.[1] The introduction to the *Daily Telegraph*'s 1904 collection of correspondence, *Do We Believe?* noted the decline of formal religion but insisted that people were not really skeptical:

> On the contrary, Science may have won its victories in an intellectual sphere, but the human heart is not satisfied with deductions which seem to leave it without any satisfaction for its intimate and half-avowed tendencies. Dogmatic Christianity may indeed have decayed, but those instincts to which, as a form of religion, it has always appealed, are as fresh and indomitable as ever.[2]

After the trauma of World War I, those who proposed new "spiritual foundations of reconstruction" urged that the burden of spiritual and moral education be shifted away from formal religion, suggesting that history, literature, and even science be used as means to expanding pupils' horizons.[3] Whether by this kind of deliberate replacement or as a result of growing indifference, formal religion certainly came to play a less obvious role in education. The survey *How Heathen Is Britain?* taken in 1946, emphasized that there was by then a massive ignorance of Christian belief, and C. S. Lewis's preface demanded popular education in Christian principles to replace what was evidently lacking in the schools.[4] But there was no equivalent effort to chart the survival of belief in less orthodox forms. There had in fact been an explosion of interest in various forms of mystical or nonmaterialistic beliefs, including astrology, Theosophy, and spiritualism. Aldous Huxley's *Crome Yellow* of 1921 ridicules a clergyman's "end of the world" sermon, but also the society lady

1.   Gilbert, *Making of Post-Christian Britain*, 3–7.
2.   Courtney, ed., *Do We Believe?* 4.
3.   Hayward and Freeman, *Spiritual Foundations of Reconstruction*.
4.   C. S. Lewis, preface to Sandhurst, *How Heathen Is Britain?*

who hopes to win on the horses by casting their horoscopes.[5] E. Ray Lankester agonized over the revival of superstition in the *RPA Annual* for 1922 and conducted campaigns against telepathy and spiritualism. He preserved a clipping from the *Westminster Gazette* in 1925, which proclaimed that "the old certainties of science have given way to the recognition of a sense of mystery."[6] Here was a broader debate than that within the churches, but it covered many of the same issues and invoked the same basic positions. The more traditional Christian writers also lamented the fuzzy-minded mysticism of the time, which they saw as deflecting interest away from the serious question of humanity's spiritual crisis. Yet it was precisely this climate that could foster an interest in the vaguely defined kind of theism represented by the Modernism of the Anglican Church or Oliver Lodge's philosophy of spiritual progressionism.

The Modernism of the churches was not the modernism of the avant-garde in the artistic, academic, and literary worlds. This distinction is crucial to understanding the cultural climate of the early twentieth century. Modernism in religion was a last-ditch attempt to bring the churches into line with the thinking of the late nineteenth century, reflecting the more positive aspects of the Victorian effort to come to terms with the idea of evolution through faith in progress. This is very different from the modernism promoted by the artists and scientists who are seen as the leading figures redefining the direction of twentieth-century thought. For these modernists, the idea of gradual progress was out of date; what mattered was the recognition of a break with the past necessitated by a growing awareness of the uncertainties built into the quest for knowledge. Samuel Hynes's survey of the Edwardian period stresses that there was already a strong reaction against Victorian values among the elite.[7] Virginia Woolf thought that everything changed in 1910 with the first exhibition of post-impressionist art; others saw 1913 as the turning point with Stravinsky's *Rite of Spring*.[8] The artistic and academic elites who appreciated these new insights were mostly contemptuous of Christianity, seeing its certainties as even more ludicrous than those of the secular progressionists. They rejoiced in the demolition job performed by Lytton Strachey's *Eminent Victorians*. The idealist philosophy that had dominated Oxford and, to a lesser extent, Cambridge was replaced by the new

5. A. Huxley, *Crome Yellow*, chaps. 9 and 10–11.

6. Lankester, "Is There a Revival of Superstition?" Lankester's papers include the clipping of "Science at a Full Stop" from the *Westminster Gazette*, 1 May 1925.

7. Hynes, *Edwardian Turn of Mind*.

8. Annan, *Our Age*, chap. 4; see Everdell, *The First Moderns*.

realism of G. E. Moore and Bertrand Russell, and later by the logical positivism of A. J. Ayer. The sense of desolation perceived in T. S. Eliot's *The
Waste Land* expressed the new modernists' vision of a universe devoid of
meaning and morality. The scientific worldview certainly played a role here,
as I. A. Richards proclaimed in 1926: if one took away the artificial confidence
in progress of the Victorian evolutionists, what was left was a universe driven
by impersonal forces and on its way to inevitable extinction via a temporary
fluctuation that created ephemeral sentient beings.[9] The image of a "lost generation" whose hopes were destroyed by the Great War is another powerful
component of our sense that the 1920s was an age of despair rather than one
of continued optimism.[10]

There is a tension here, however, between the image of an elite culture
dominated by hopelessness and the evidence presented above that the postwar years were by no means so different from the old world of optimistic progressionism. In fact, even the elite were not uniform in their opinions. There
was as yet little sign of what C. P. Snow would later call the split between the
"two cultures." Scientists such as Huxley and Haldane (father and son) were
members of an intellectual circle defined by family networks and educational
background in the public schools and the ancient universities of Oxford and
Cambridge. They still interacted with the artistic and literary elite, if only
through family connections such as Julian's brother, the novelist Aldous Huxley—himself fascinated by science. Nor was there unanimity among the literary figures. The Bloomsbury set was disliked by the Oxford wits, and Evelyn Waugh emerged from the latter group to challenge many of the attitudes
of what Noel Annan calls "Our Age."[11] Eliot turned to Anglicanism in 1927,
horrifying his literary colleagues but pointing the way to a flood of conversions to Anglicanism and Catholicism in the 1930s.[12] Many were ambivalent
about the more radical modernism because, in the end, they could not stomach that movement's total rejection of nineteenth-century values.

Outside the elite, social class also generated massive differences in cultural values.[13] The inner circles of atheism were narrowly circumscribed
within Oxbridge and Bloomsbury, and most of their protagonists would have
dismissed with contempt anyone who read the "middle-class" novels of
H. G. Wells or J. B. Priestley. But to obtain an overview of national culture,

9.   I. A. Richards, *Science and Poetry*, chap. 5, "The Neutralization of Nature."

10.   Eksteins, *Rites of Spring*; Wohl, *Generation of 1914*.

11.   Annan, *Our Age*.

12.   Hastings, *History of English Christianity*, chaps. 12 and 17.

13.   On class differences in this period, see McKibbin, *Class and Culture*.

it is vital to survey this "second-rate" literature. The run-of-the-mill may be less than interesting if one is seeking the origins of the next wave of artistic creativity, but it gives a far better reflection of what the majority is reading and thinking. Nor were all members of the intellectual elite indifferent to the wider public: both Huxley and Bertrand Russell stand out for their efforts to disseminate the latest ways of thinking to ordinary readers.

The first few decades of the twentieth century thus presented a more challenging climate than the rationalists had expected for continuing their offensive against religion. Some members of the aesthetic and academic elite had converted to exactly the kind of worldview their fathers and mothers had feared would emerge as a consequence of science's meddling with hitherto unchallenged assumptions. They made their way where only the boldest Victorian agnostic had dared to go, ejecting purpose and progress from the evolutionary perspective and welcoming the hedonism that now seemed the only possible form of morality. But if idealism was banished from Bloomsbury and Oxbridge, it still flourished both in the churches and in the writings of a host of lesser intellectuals and literary figures. Outside the circle of the elite there was still a feeling that things might be retrieved from the debacle of the war and progress put back on course. How to achieve this was the chief point of debate. The Modernists in the churches made common cause with the more generalized theism becoming popular in an age that sought meaning without dogma. Secularists rejected even this relic of the past and insisted that progress would resume only once the shackles of superstition had been thrown off for good. As Alan Gilbert notes, both participated in what George Bernard Shaw called the ideology of "meliorism," the conviction that we can make things better, perhaps even perfect, in this world—they disagreed only over whether or not this was God's purpose for humanity or our own.[14] In Shaw's dichotomy, the opposing ideology is that of "salvationism"—the conviction that only by appealing to forces outside this world can we be saved—and it was this ideology that was in temporary eclipse during the first three decades of the century. The salvationist viewpoint persisted in both the evangelical and Catholic churches, of course, and was preserved in the popular mind by a few writers such as Belloc and Chesterton.

Only in the 1930s did the salvationist ideology begin to acquire widespread intellectual respectability again, as a steady stream of converts began to flow into Anglican and Roman Catholic churches. Here again we encounter the difficulty of defining the culture and ethos of a generation divided by class and other loyalties, let alone changes from one generation to

14. Gilbert, *Making of Post-Christian Britain*, 54–55.

the next. Do generations or decades have an identifiable ethos or culture? It is easy to suppose that they cannot, since in one sense all historical change is continuous. But it would be foolish to deny that episodes such as the Great War or the Wall Street crash can act as punctuation marks. Noel Annan thinks that generations, at least, do have a reality, if only because we perceive them as units by hindsight, but a study of the 1930s by John Baxendale and Chris Pawling warns us of the ways in which the popular image of that decade has been constructed and manipulated.[15] We may be able to take the decades or generations as convenient markers, but only if we make ourselves aware of the multiple layers of culture concealed by the icons of the popular imagination. Just as the twenties cannot be reliably interpreted through the myth of postwar pessimism, our vision of the thirties must be freed from the myth of betrayed idealism—another product of the literary elite (and one that is incompatible with the projection of a similar pessimism onto the previous generation).[16] The same decade saw a reinvigoration of concern for social democracy and the rise of the Marxist alternative to Fascism. Meliorism still fought in its own corner, and for the Marxists it took on the messianic overtones once characteristic of religion. The rise of Christian orthodoxy was also real enough—it finished off Modernism in the Church of England—but was only one facet of a complex response to ever more stressful national and international problems.

For those who wanted to reach the general public, there were many new avenues available, but they required new techniques adapted to the age of sensationalism. The vast number of books sold by writers such as Eddington and Jeans shows that those who got it right could influence tens or even hundreds of thousands of readers, besides making a small fortune in the process. The RPA's influence stemmed in part from C. A. Watts's understanding of how the mass market for books and pamphlets worked. In addition, there were the popular newspapers and magazines, always on the lookout for a story. The press was industrialized in the decades spanning the turn of the century, so that our period was dominated by press barons such as Lord Northcliffe and Lord Rothermere, whose ability to manipulate public opinion conferred, in Stanley Baldwin's phrase, "power without responsibility."[17] The sensationalism and trivialization of the press that we complain about to-

---

15.   Annan, *Our Age*, 15; Baxendale and Pawling, *Narrating the Thirties*.

16.   Baxendale and Pauling, *Narrating the Thirties*; see Hynes, *The Auden Generation*.

17.   Curran and Seaton, *Power without Responsibility*; see 46 for Baldwin's remark. See also S. J. Taylor, *The Great Outsiders*.

day began here. The newspapers offered both opportunities and dangers for those intellectuals willing to write for them. In addition to commissioning articles, they sometimes canvassed the opinions of the famous on a variety of serious questions. There was a fad for "symposia" on various topics in the 1920s, parodied in Ronald Knox's *Caliban in Grub Street*, and several of these involved religious beliefs and reactions to science. Some scientists and theologians got the chance to participate, and hence to project their views to a wide audience—at the risk of seeing them trivialized and countered by what appeared to be equally authoritative statements by famous names who knew nothing at all about the subject. The better newspapers carried detailed reports of events such as the annual meetings of the British Association and the various church congresses. Even the mass-circulation papers would pick up on a scientist such as Keith apparently attacking religion or on controversial sermons such as those of Bishop Barnes on evolution. Popular weekly and monthly magazines would do the same—*Tit-Bits* was not above pirating the odd report from the pages of *Nature*, and it sold over two million copies a week.[18] Everything was adapted not only to the short attention span of the reading public, but also to the prevailing social ethos—the newspapers and magazines, while critical of individual members of the ruling class, were devoted to maintaining the status quo and would adapt most things to serve the purpose of keeping the empire intact. Only a few publications with a self-consciously radical ideology attained any degree of popularity, Robert Blatchford's socialist *Clarion* being one (it also carried better than average coverage of scientific matters).

An even newer medium through which the masses could be reached was radio, which in Britain meant the national noncommercial service, the BBC.[19] There was little explicit coverage of science, and religious broadcasting was at first slow to get off the ground, although Lord Reith was determined that the service would maintain a respectful attitude toward the churches. Atheistic and materialistic views were banned from the airwaves, and only in 1928 did a more liberal attitude permit some questioning voices to be aired, so long as they maintained a respectful tone. Bertrand Russell and the RPA were still excluded. Even so, one of the first products of the new freedom was a series of talks on science and religion, with contributions from Julian Huxley,

18. See Broks, *Media Science before the Great War*, 14 and 33.

19. Asa Briggs's monumental survey of the BBC's history is summarized in his *The BBC: The First Fifty Years.* See also Pegg, *Broadcasting and Society, 1918–1939;* Scannell and Cardiff, *Social History of British Broadcasting;* more specifically, see K. M. Wolfe, *The Churches and the British Broadcasting Corporation, 1922–1956.*

J. Arthur Thomson, J. S. Haldane, Barnes, Eddington, and Inge.[20] Along with the philosopher C. E. M. Joad, Huxley became a national figure in the 1940s through his contribution to the "Brains Trust" program, although most of the questions discussed did not concern biology or philosophy.

Our survey begins with the forces arguing for a more vigorous repudiation of past values. These include the literary and philosophical elites, some of whom at least tried to articulate the new sense of pessimism in a manner that would influence the wider debate. Other members of that elite, Huxley being an example, sought common cause with popular writers such as H. G. Wells in trying to retain a sense of the mystery of life without appealing to a personal God. At the same time, these exponents of "religion without revelation" maintained an uneasy alliance with another decidedly nonelite movement, the rationalists who continued the Victorian program for the elimination of religion altogether. Their mantle would eventually be taken up by the Marxists of the 1930s, anxious to incorporate any hint of a greater purpose in life into their own vision of humanity's ultimate destiny.

AGAINST IDEALISM

The late Victorian period was far from being dominated by the pessimistic vision of nature articulated by the exponents of scientific naturalism. Idealism flourished in the British universities, and it was this nineteenth-century worldview that was preserved by the Modernist movement within the churches. Only around 1910 did the rival form of modernism begin to take hold of the artistic imagination and the worlds of science and philosophy. Now the fears of the Victorian era were realized: nature was fragmented and rendered meaningless, while the human observer was exposed as powerless to gain anything more than a purely local perspective on reality. Truth and beauty were both banished; the individual could only seek his or her own way of coming to terms with an ultimately meaningless existence. In the arts, this was expressed by a revolution in which the old search for harmony was replaced by a desire to shock everyone into a realization of the human situation. In academic philosophy, idealism was replaced by a new realism and a determination to expose the limitations of knowledge. Outside the ranks of the self-consciously innovative elites, however, the old, more optimistic worldview continued to reign. The Modernists of the Anglican Church had

20.   Wolfe, *The Churches and the British Broadcasting Corporation*, 29; the talks were published as *Science and Religion: A Symposium.*

been trained by the very idealists whose worldview was now repudiated in the name of the new modernism.

In the literary world of Edwardian Britain, it was the Bloomsbury group that perceived itself as the avant-garde, the articulators of the new pessimism. For writers such as Virginia Woolf, religion was dead and the search for new sources of meaning difficult. The possibility that the new developments in science played a role in creating this new and harsher image of the human situation was clearly expressed in I. A. Richards's *Science and Poetry* of 1926. Richards was a Cambridge don who had written on the principles of literary criticism. Like many supporters of the latest innovations, he took Eliot's *The Waste Land* as a symbol of the new approach and sought to explain why the new literature expressed such a pessimistic view of the world. (In fact, Eliot publicly dissociated himself from Richards's interpretation because he thought the sense of desolation read into the poem was as much an act of belief as any other.)[21] According to Richards, poetry gives order and meaning to experience, but this is more difficult in a world that has been stripped of the old human values. Where once nature was seen as alive and purposeful, so that human life was in tune with the universe itself, the world was now stripped of meaning, leaving the poet to turn inward upon himself or herself. Richards singled out theories such as emergent evolutionism as attempts to smuggle purpose back into nature and dismissed them with the following words: "but alas! the reasons for suggesting them have become too clear and conscious. They are there to meet a demand, not to make one; they do not do the work for which they were invented."[22] Science had at last exposed the values that had been smuggled into the old forms of knowledge and had thereby revealed the insecurity of our position in the world—even if it did offer ways of reducing the odds against us. The new models of the mind offered by the psychologists were largely responsible: "A sense of desolation, of uncertainty, of futility, of the groundlessness of aspirations, of the vanity of endeavour, and a thirst for the life-giving water which seems suddenly to have failed, are the signs in consciousness of this necessary reorganization of our lives."[23] Richards offered no hope that the latest developments in science would reverse the trend. On the contrary, he thought that the expansion of scientific knowledge would ultimately destroy all the old traditions and

21.   On this point, see Stephen Medcalf's review of Joseph Pearce's *Literary Converts,* "Cries in the Wilderness."

22.   I. A. Richards, *Science and Poetry,* 50.

23.   Ibid., 64.

sources of meaning. Poetry was vital because it provided the meaning from within ourselves, replacing what we now know does not exist in the world itself.

Richards's analysis suggests that the revolutions in science from Galileo onward had now reached their ultimate conclusion by denying all purpose in the world. The full implications of materialism and Darwinism would have to be faced. Small wonder that those who felt themselves the inheritors of such an intellectual legacy should reject religion as an illusion and see efforts to read mind and purpose back into evolution as a false hope. Philosophy, too, was turning its back on the idealism of Edward Caird, T. H. Green, and F. H. Bradley, which had dominated the field in the British universities during the late nineteenth century. Idealism was essentially an attempt to read purpose into the world of the senses, and the generation of thinkers trained by these philosophers continued to express this aspiration into the twentieth century. But as these pupils moved out into the wider world of politics, religion, and the professions, the tide within the universities turned as academic philosophers began to participate in the modernist revolution. G. E. Moore and Bertrand Russell were the two most obvious symbols of the new direction in philosophy, and for Russell, at least, science was an integral part of the quest for a new foundation for knowledge.[24] His technical work on the foundations of mathematics with Alfred North Whitehead was of interest to many scientists, although the two then moved in very different directions. Russell remained true to a humanism that denied all sources of value in the external world. What distinguished him from most academic philosophers was his determination that the implications of this viewpoint should be passed on to as wide a circle of readers as possible.

Russell had lost any faith he ever had in the existence of God even before he went to university, and he incorporated his unbelief into all his mature writings. His essay "A Free Man's Worship," first published in 1903, articulated the implications of the scientific worldview:

> That Man is the product of causes which had no prevision of the end they were achieving; that his origin, his growth, his hopes and fears, his loves and his beliefs, are but the outcome of accidental collocations of atoms; that no fire, no heroism, no intensity of thought and feeling, can preserve an individual life beyond the grave; that all the labours of the

24.   A useful survey is Passmore, *A Hundred Years of Philosophy*; see chap. 9 on Moore and Russell. Recent studies of Russell include Moorehead, *Bertrand Russell: A Life*, and Monk, *Bertrand Russell*. See also Berman, *History of Atheism in Britain*, 202–3, in which Russell is depicted as a popular but not really original advocate.

ages, all the devotion, all the inspiration, all the noonday brightness of
human genius, are destined to extinction in the vast death of the solar
system, and that the whole temple of Man's achievement must inevitably
be buried beneath the debris of the universe in ruins—all these things, if
not quite beyond dispute, are yet so nearly certain, that no philosophy
which rejects them can hope to stand. Only within the scaffolding of
these truths, only on the firm foundation of unyielding despair, can the
soul's habitation henceforth be safely built.[25]

For Russell, the quest for meaning was to be found not in poetry, but in the
search for knowledge and the desire to help everyone live as rich a life as possible under these limited circumstances. All efforts to make the world seem
more purposeful were false, including Henri Bergson's attempt to see evolution as creative, if not goal-directed. Bergson's philosophy was attractive because it depicted humanity riding with the cavalry charge of life itself, but it
made critical thought impossible, and when we ask what reasons there are for
accepting such a worldview, we find there are none.[26] Russell conceded, however, that faith in progress remained high in the English-speaking world—to
this extent, it was hard work to convey the pessimism so characteristic of the
intellectual elite to a still trusting public.[27]

Russell's assault on religion continued in the postwar decades. His lecture to the National Secular Society in 1927, titled "Why I Am Not a Christian," rejected the argument from design on the grounds that the universe was
full of suffering (but conceded that most people didn't worry about the end
of the universe if it was to be postponed for millions of years).[28] Some of
Christ's moral teachings were attractive, but by no means all—His belief in
the reality of hell, for instance. The chapter on science and religion in Russell's *Scientific Outlook* of 1931 resisted what he saw as the new trend by physicists such as Eddington and Jeans to seek a reconciliation with religion. Their
statements did not go as far as many religious thinkers hoped, and were in any
case an expression of their own (very conservative) views rather than an extension of their actual science. There was no justification for the assumption

25.   Russell, "A Free Man's Worship," originally published in the *Independent Review*, December
1903; reprinted in Russell, *Philosophical Essays*, 59–70, in his *Mysticism and Logic*, 46–57, and in his *Basic Writings of Bertrand Russell*, 66–72. Quotation from *Philosophical Essays*, 60–61; *Basic Writings*, 67.

26.   Russell, *Philosophy of Bergson*, 11–12 (originally published in *The Monist*).

27.   Russell's opinion on this was noted in the Introduction above; see his "Eastern and Western Ideals of Happiness," reprinted in *Sceptical Essays*, 100, and in *Basic Writings*, 555.

28.   Russell, "Why I Am Not a Christian," reprinted in *Basic Writings*, 585–97; see 589–90. This
essay was originally published by Watts in 1927.

that the indeterminate character of atomic events left room for the freedom of the will—indeed, Russell (like Einstein) thought that determinism would ultimately be restored when we knew more about the laws of atomic behavior.[29] Jeans's mathematician God was ridiculed as a projection of the human mind and was of little real use to theologians:

> Theologians have grown grateful for small mercies, and they do not much care what sort of God the man of science gives them, so long as he gives them one at all. Sir James Jeans's God, like Plato's, is one who has a passion for doing sums, but being a pure mathematician, he is quite indifferent to what the sums are about.[30]

The inference from the second law of thermodynamics that the universe must have had a beginning might be provisionally accepted, but this did not mean that it had a Creator. Russell knew that old-fashioned materialism had been destroyed by the new physics, but argued that materialism itself was now stronger than ever because the reign of law was being extended to ever-wider areas of biology and psychology. He was particularly scathing about the attempt to salvage a vestige of the argument from design by supposing that evolution is purposeful. If that purpose was to produce humanity, then our meanness and cruelty, to say nothing of our capacity for inventing new weapons, made us a somewhat lackluster crown of creation.[31] Most biological functions were now being explained in materialistic terms, and the mental functions would soon respond to the same techniques.

Lloyd Morgan's emergent evolutionism was singled out for particular criticism: "It would be easier to deal with this view if any reasons were advanced in its favour, but so far as I have been able to discover from Professor Lloyd Morgan's pages, he considers that the doctrine is its own recommendation and does not need to be demonstrated by appeals to the mere understanding."[32] Perhaps there was a divine plan, but if this included children dying of meningitis and older people of cancer, Russell could not see the point of it. Science had effectively replaced religion, and even though it was frequently misused for immoral purposes, there was no going back. The goal must now be to make sure that the new knowledge was used for the benefit of humanity rather than for evil. Here Russell echoed the negative view of ap-

29.   Russell, *Scientific Outlook,* 112.

30.   Ibid., 115.

31.   Ibid., 127.

32.   Ibid., 135.

plied science in his *Icarus* of 1924, but offered a more hopeful view of its long-term applications. The new realism offered no automatic assurance that knowledge would be applied to good ends, but the rationalist had to hope that knowledge was itself a good thing.

The critique of natural theology was extended in Russell's *Religion and Science* of 1935, a contribution to the popular Home University Library series. Here he noted the new consensus emerging among biologists in favor of the theory of natural selection.[33] He repeated his doubts about the value of the human race as the goal of evolution: "From evolution, so far as our present knowledge shows, no ultimately optimistic philosophy can be validly inferred."[34] Bishop Barnes's hope that there would turn out to be some ultimate purpose after all was dismissed, along with the vitalistic and teleological theories of J. S. Haldane and Lloyd Morgan.[35] Eddington and Jeans were attacked once again, and Russell went out of his way to criticize those who now maintained that mysticism offered a valid source of insights that could complement science. In his conclusion he stressed that scientists were always tentative when proposing new interpretations—although the practical discoveries that resulted were of permanent value. This willingness to test and modify its conclusions made science the natural enemy of dogmatism. But there were signs of the times in the message Russell drew from this point. He conceded that most religious thinkers now accepted the beneficial nature of the purification of their beliefs required by science—the real dogmatism of the modern world came from the rival regimes in Germany and Russia. Fascism and Communism were the new challenges to intellectual freedom, and all true scientists should protest against the threat of a new dark age.[36]

Russell was by no means the only philosopher to pillory the idealism of Jeans and Eddington. Other critics included G. Dawes Hicks, H. W. B. Joseph, and Leonard Russell,[37] while in 1937 L. Susan Stebbing published her *Philosophy and the Physicists* in an attempt to stem the tide of popular enthusiasm for their ideas. Stebbing was particularly worried that so many ordinary people, to say nothing of religious leaders, were assuming that the views of these distinguished figures represented the typical opinions of the scientific community. In fact there was no warrant for their idealism, although this did not mean that science still upheld an old-fashioned materialism. In Stebbing's

33. Russell, *Religion and Science*, 71–72.

34. Ibid., 81.

35. Ibid., 191–92.

36. Ibid., 248–52.

37. Hicks, "Professor Eddington's Philosophy of Nature"; Joseph, "Professor Eddington on 'The Nature of the Physical World'"; L. Russell, "A. S. Eddington."

view, Jeans and Eddington "approach their task through an emotional fog; they present their views with an amount of personification and metaphor that reduces them to the level of revivalist preachers." They ought to avoid cheap emotionalism, but "of this obligation Sir James Jeans seems to be totally unaware, whilst Sir Arthur Eddington, in his desire to be entertaining, befools the reader into a state of serious mental confusion."[38] Their effort to give the impression that the new cosmology made us all feel insignificant was nonsensical—to apply human values such as "indifference" or "hostility" to the cosmos was to misuse the terms completely. Kant had found the expanded vision of the heavens a source of awe, not fear.[39] Stebbing was especially harsh in her critique of Jeans's mathematician God (she accepted that Eddington's position was more subtle), arguing that his projection of mathematical models onto the mind of a Creator gave his readers a completely false impression of the real implications of the new physics. Like Russell, she also pointed out that such a rational God bore little resemblance to the Father figure required by most religions.[40] While admitting that the new physics challenged the old determinism, Stebbing insisted that its theories did not rescue the freedom of the will in the way that Eddington assumed. Macroscopic bodies may be built of atoms, but they do not obey the laws of the atomic world, and physicists continue to use determinism successfully whenever they predict the behavior of any macroscopic body. Stebbing saw an element of comedy in the way in which the theologians had now thrust the mantle of the prophets upon the scientists.[41] It was odd how the supposedly "mysterious" nature of matter in the new physics was taken as a sign of hope by many Christians, who apparently thought that it allowed mind to be put back into nature. The world was in a sorry state, whether it was the product of God or mind, and if that state was the consequence of human greed, it was our responsibility to put things right.

Stebbing's and Russell's books were aimed at a general audience. But for most British readers (and certainly for most radio listeners), the most prominent philosopher of the interwar years was C. E. M. Joad. An eclectic thinker with a pluralist metaphysics, Joad too attacked Eddington's idealism and Jeans's mathematician God. He conceded in his *Philosophical Aspects of Modern Science* that it should also have contained a chapter on Whitehead, but said he

38.   Stebbing, *Philosophy and the Physicists*, 6–7.

39.   Ibid., 12–13.

40.   Ibid., 21.

41.   Ibid., 143–44.

was so baffled by *Process and Reality* that he had to give up the idea of including a critique of its position.[42] Like most philosophers, Joad had more sympathy with Eddington, noting that his position was epistemological and sought only to undermine materialism by legitimizing other aspects of human experience besides the physical senses. But his argument did not destroy our confidence in the reality of the physical world, nor did it prove that mystical experiences were a genuine link to reality.[43] Jeans, in contrast, was doing metaphysics as badly as an eighteenth-century rationalist, offering an argument for the existence of a rational God that merely portrayed Him in human terms.[44] Together, the two physicists had attempted to depict the universe as friendly to humanity, but they had abandoned their science to do this, and their arguments told us more about their beliefs than about the universe itself.[45] Joad's critique of the new idealism and its religious implications seems to fit squarely into the tradition represented by Russell and Stebbing, yet his pluralist worldview allowed him to adopt a far more positive attitude toward the biological arguments that Russell found unconvincing. His support for emergent evolutionism, along with his eventual conversion to Anglicanism, will be explored in the following chapter.

In the meantime, academic philosophy was taking a new turn with the introduction into Britain of logical positivism, and here the outspoken advocacy of A. J. Ayer added a new component to the assault on religion. Ayer's lively and belligerent *Language, Truth, and Logic* of 1936 linked logical positivism to the British tradition of analyzing the meaning of propositions, redefining philosophy as a critique of the use of language designed to expose the hollowness of metaphysics. For Ayer, propositions that could not be tested empirically were meaningless, and this included the statements of both the moralist and the theist (to be fair, it also included the statements made by agnostics).

> [The theist's] assertions cannot possibly be valid, but they cannot be invalid either. As he says nothing at all about the world, he cannot justly be accused of saying anything false, or anything for which he has insufficient grounds. It is only when the theist claims that in asserting the existence

---

42. Joad, *Philosophical Aspects of Modern Science*, 17. I have to confess to having a great deal of sympathy with Joad's evaluation of Whitehead.

43. Ibid., chap. 1; see also Joad, "Modern Science and Religion."

44. Joad, *Philosophical Aspects of Modern Science*, chap. 2.

45. Ibid., 339.

of a transcendent god he is expressing a genuine proposition that we are entitled to disagree with him.[46]

The existence of the noun "god" gives the illusion that there might be a possible entity corresponding to it, but since it is impossible to frame verifiable propositions including that noun, we must accept that "god" is not a genuine name. Ayer insisted that on the basis of this philosophy there was no logical ground for the antagonism between religion and science:

> As far as the question of truth or falsehood is concerned, there is no opposition between the natural scientist and the theist who believes in a transcendent god. For since the religious utterances of the theist are not genuine propositions at all, they cannot stand in any logical relation to the propositions of science. Such antagonism as there is between religion and science appears to consist in the fact that science takes away one of the motives which makes men religious. For it is acknowledged that one of the ultimate sources of religious feeling lies in the inability of men to determine their own destiny; and science tends to destroy the feeling of awe with which men regard an alien world, by making them believe that they can understand and anticipate the course of natural phenomena, and even to some extent control it.[47]

Noting that some physicists had recently become more sympathetic toward religion, Ayer argued that this revealed their lack of confidence in their own hypotheses, perhaps as a consequence of the crisis through which physics had recently passed. It was also a reaction against the antireligious dogmatism of some nineteenth-century scientists. Ayer's philosophy thus offered little comfort to the new generation of academics and literary figures who were returning to the fold of religion in the mid-1930s. His was a powerful voice in the narrow but influential world of Oxford, checked to some extent by the tireless work of chaplains such as Ronald Knox and Martin D'Arcy. They would struggle for influence over the new generation that would come to power after World War II, and significantly, Ayer abandoned his nihilistic attitude toward morality when the war against Hitler broke out.[48] Curiously, the Marxists came to regard logical positivism as a continuation of the ide-

---

46.   Ayer, *Language, Truth, and Logic,* 116.

47.   Ibid., 117.

48.   For an insight into Ayer's role in the religious debates at Oxford, see Longford, "A. J. Ayer's Tour de Force." More generally, see Rogers, *A. J. Ayer,* esp. chap. 8.

alist tradition, because it too sought to put limits on our ability to describe nature. For Maurice Cornforth, the new philosophy's claim that we cannot truly understand matter merely kept open the door for those who wanted to invoke supernatural agencies—hardly a conclusion that Ayer would have endorsed.[49]

POPULAR RATIONALISM

The academic philosophers may have become skeptics in the early twentieth century, but their elite world was narrowly defined, and only a few shared Russell's ability to write for a wider audience. There were already other movements at work seeking to reduce the role of religion in British culture. In the late nineteenth century, C. A. Watts's Rationalist Press Association had emerged as an active lobby for the promotion of secular values, and it sought to use the scientific naturalism of the late Victorian agnostics as part of its campaign.[50] The books written to promote this campaign continued to be issued in the early twentieth century and were joined by others written by a new generation of active secularists. Scientists such as E. Ray Lankester and Arthur Keith supported the RPA, providing credibility for the work of its popular writers and speakers (see chapter 2). But rationalist literature began to show an increasing stridency as the century progressed, indicative of a noticeable decline in the RPA's fortunes. Although Watts's publications, including his Thinkers Library series, were still issued by the thousands, actual membership in the RPA declined. The organization was unsuccessful in a campaign to persuade Lord Reith to rescind the ban on radio talks opposing religion.

One of the star writers in Watts's stable had been Samuel Laing, whose books were issued in vast numbers in the 1890s and beyond. In his *Modern Science and Modern Thought*, and again in his *Problems of the Future*, Laing drove home the argument that the latest scientific developments had destroyed the credibility of dogmatic religion while leaving the ethical teachings of Christ intact. Evolutionism offered a new vision of human origins and human nature, while the scientific attitude had led many to question the evidence for the miracles on which most religions were founded. Edward Clodd was another popular writer of this period who stressed the transforming power of evolutionary ideas and science's discrediting of the animistic beliefs upon which

49. Cornforth, *Science versus Idealism*, 17–18.

50. On the RPA, see Whyte, *Story of the R.P.A.*; Budd, *Varieties of Unbelief*; Lightman, "Late-Victorian Agnostic Popularizers," and Royle, *Radicals, Secularists, and Republicans*.

all religions were based. Clodd felt it necessary to fulminate against the animistic beliefs that still underpinned many ordinary people's perceptions of the world: "Animists, in the germ, were our pre-human ancestors; animists, to the core, we remain"—as witnessed by newspaper reports of ghosts, the activity of palmists and mediums, and fear of the number 13.[51] These were all survivals of primitive beliefs—but Clodd evidently felt that they were survivals that the rationalists were having a hard time combating.

Clodd's close friend Grant Allen promoted scientific naturalism as expressed in Herbert Spencer's philosophy. He wrote popular science, poetry, and novels, using all these literary forms as a vehicle for promoting an agnostic evolutionism. His best-selling novel *The Woman Who Did*, published in 1895, scandalized society by depicting in sympathetic tones a woman who flouted the social and sexual conventions of the time. Allen's *Evolution of the Idea of God* of 1897 paralleled Clodd's book on animism, but saw religion evolving from the primitive worship of ancestral spirits.[52] Clodd and Allen synthesized Spencerian evolutionism with the skepticism of anthropologists such as J. G. Frazer and transmitted the result to a wide audience. Evolution was also a key theme for C. W. Saleeby, who confessed hero worship for Herbert Spencer as late as 1906. In his eyes, Spencer's "Unknowable" still left room for hope, while removing the threat of dogmatism by showing the absurdity of conferring human characters on something that by definition transcends anything we can really understand.[53]

E. Ray Lankester shared Clodd's fears about the sustained influence of animistic superstitions and campaigned against telepathy, spiritualism, and the like. Lankester was particularly scathing about Oliver Lodge's and Arthur Conan Doyle's enthusiasm for spiritualism. When *Nature* published a relatively supportive review of Doyle's *History of Spiritualism* in 1926, a sharp controversy erupted in which several writers, including Lankester's friend Bryan Donkin, insisted on the dangers of fraud and the difficulties scientists faced in detecting it.[54] Lankester also endorsed the attack on Bergson's philosophy published by Hugh Elliot, who was the editor of the collected letters of John Stuart Mill. Elliot saw Bergson as promoting an impossibly vague idea of the creative life force, which had gained popularity merely because it seemed to

51. Clodd, *Animism: The Seed of Religion*, 97. See also Clodd's *Story of Creation*, 227–28.

52. Grant Allen, *Evolution of the Idea of God*. Clodd subsequently wrote a memoir of Allen.

53. Saleeby, *Evolution: The Master-Key*, chap. 33.

54. The original review is R. J. Tillyard's "Science and Psychical Research"; see also other articles under the same title by J. P. Lotsy, A. A. C. Swinton, and Bryan Donkin.

undermine the old determinism. Much of Bergson's writing was mere ver-
biage with no concrete meaning; about the only clear points emerging were
hostility to materialism and the desire to set up life and mind as antagonistic
to matter. Many of Bergson's criticisms leveled against materialistic biology
and Darwinism were based on simple misunderstandings of science. For Elliot, Bergson's philosophy emerged as the heir to the idealism of Hegel, a new
effort to make mind the key factor in our understanding of the world.[55] As
such, it went completely against the direction taken by science, which had
consistently shown that progress in understanding the world came only from
adopting a materialistic approach. Elliot thought that McDougall's attack on
materialism was more effective, but that he too had failed to provide convincing evidence for the reality of mind as an active force in nature.[56] In his
conclusion, Elliot firmly linked the enthusiasm for Bergson's philosophy to
the forces that sustained religious belief. People had an instinctive tendency
to believe, although what they believed was shaped by their cultural environment. But since this capacity had no survival value, it was almost certainly
growing weaker as the human race evolved.[57] Claims that materialism undermined morality were false—determinism was not the same as fatalism, and
in fact led us to lay even more stress on the need to suppress the darker instincts that could all too easily lead us astray.

Some years later Elliot published his *Modern Science and Materialism*, a
broader defense of the materialist philosophy. He argued that we can judge
the world only through our senses, so even if the universe has qualities that
are not revealed to us in this way, we have to adopt an agnostic position. "Agnosticism is thus a clearance of the mental rubbish handed down to us from
the past, and preserved by the infirmity of our minds."[58] Having demolished
metaphysical pseudo-knowledge in his critique of Bergson, he now wanted to
take a more constructive approach. An overview of modern cosmology,
physics, and biology helped the reader to gain a firm perspective on humanity's place in the world. We are the product of a purposeless evolutionary process, and our descendants will inevitably be wiped out in the heat-death of
the universe—Elliot noted that these views coincided with the pessimistic
stance adopted by Russell.[59] The new physics showed that the basis of mat-

55. Elliot, *Modern Science and the Illusions of Professor Bergson*, 130, 159.

56. Ibid., 186–88.

57. Ibid., 196–97, 200–201.

58. Elliot, *Modern Science and Materialism*, 6–7.

59. Ibid., 39, 68–69, and on Russell, 14.

ter was energy—but energy was not spirit, and explanations that tried to equate energy and spirit were merely an attempt to revive the old animistic viewpoint. In this sense, materialism was still valid. Natural selection might not be the only mechanism of evolution, but it imposed the universal constraint of adaptation to the local environment. Vitalism in biology was a fallacy, useless to science and deriving its support from a desire to reintroduce spiritualistic views.[60] Philosophy had to take the scientific view of the world into account, and hence must abandon cosmic teleology and the assumption that mind was the driving force of nature. Materialism was the only way forward, although Elliot conceded that an idealism based on knowledge as constructed from sense impressions was a valid (but impractical) alternative. This was a long way from what was popularly understood as idealism, however. The book concluded with a critique of the traditional concept of personality, arguing that our minds are epiphenomena, by-products of our bodies with only the level of unity and coherence that the brain's capacity for memory permits.

Elliot's powerful appeals for agnosticism and materialism were echoed by two stalwart exponents of rationalist philosophy, Joseph McCabe and Chapman Cohen. McCabe was one of the leading writers for the RPA. He had published his attack on organized religion, *The Existence of God*, in 1913 and revised it extensively for a second edition in 1933. It was this second edition that dismissed the "Jeans-Eddington outbreak" as a total misrepresentation of science and insisted that only the older generation of scientists still took religion seriously.[61] Like Elliot, McCabe thought that disbelief was steadily gaining ground as more people understood the true implications of the scientific worldview. His *Evolution of Mind* dismissed the popularity of Bergson as ephemeral and presented an orthodox Darwinian account of the gradual evolution of higher mental powers.[62] In 1934, in his *Riddle of the Universe Today*, McCabe harked back to Haeckel's classic materialist text and again dismissed the apparent revival of interest in religion among scientists as an illusion. If G. K. Chesterton thought Darwinism was dead, he should read J. B. S. Haldane's *Causes of Evolution*.[63] The latest fossil discoveries were confirming humanity's origin from an apelike ancestor. Emergent evolutionism was nonsense because even the psychologists had now rejected the mind as a useful concept and serious psychology was becoming difficult to distinguish from

---

60. Ibid., 118.
61. See the introduction to chap. 2 above.
62. McCabe, *Evolution of Mind*; on Bergson, see vii.
63. McCabe, *Riddle of the Universe Today*, 13.

physiology.[64] While Whitehead had praised Bergson's "intuitive grasp of biology," McCabe noted Lancelot Hogben's retort that this was like praising Madame Blavatsky's knowledge of astronomy.[65] The book concluded with two chapters on the new physics, repeating McCabe's dismissal of Jeans's and Eddington's idealism and insisting that the latest theories still upheld a materialist philosophy. Materialism sought to deny spirit, not to define matter, so it was unaffected by any new theories about the nature of matter. The world was still based in space and time and driven by uniform laws, so there was no room for spiritual interference. McCabe ended with a plea for the rational exploitation of science to improve human life.

Chapman Cohen was chairman of the rival Secular Society, founded in 1898 to express an even more fundamental hostility to religion. Watts and the RPA reflected a traditional middle-class ideology that still echoed an evangelical fervor. For them, agnosticism was a replacement for traditional religion, providing an alternative foundation for the same moral values. They had refused to endorse the atheism of Charles Bradlaugh and had thus alienated those who wanted rationalism to strike out on its own in search of a completely new morality. Cohen was determined to demolish the credibility of religion altogether. His *Religion and Sex* of 1919 appealed to the latest developments in psychology to transform the old rationalist argument that religious experiences arose naturally in the primitive mind. Where anthropologists such as Edward B. Tylor and J. G. Frazer had tried to explain religion in terms of how primitive peoples confronted the world, Cohen now claimed that its origins lay in pathological mental states that generated so-called mystical experiences. William James's *Varieties of Religious Experience* was a "remarkable piece of religious yellow-journalism" because it still tried to argue that such experiences offer genuine insights into reality.[66] The Scopes trial in America prompted Cohen to issue a pamphlet called *God and Evolution*, in which he claimed to have more sympathy with the Fundamentalists than with the Modernist Christians who were making a desperate attempt to modify the traditional faith in order to keep up with science. According to Cohen, the vast majority of Christians who believed that it was impossible to reconcile their faith with evolution were quite right: evolution was indeed an atheistic doctrine.[67] Evolution was not Darwinism—there might be more than one way for John Smith to have got from Manchester to London—but

64. Ibid., 40–42.
65. Ibid., 89. Blavatsky was the leading exponent of Theosophy.
66. Cohen, *Religion and Sex*, 10.
67. Cohen, *God and Evolution*, 6.

the fact of change was inescapable, and there was no evidence that the developments moved in any purposeful direction.[68] The authors of the Genesis account clearly meant it to be taken literally, and all efforts to suggest how their words could be reinterpreted were a betrayal of their evident intention to suggest that the supernatural was directly involved in the creation of life. This was why science and religion necessarily came into conflict: the one denied that element of the supernatural that it was the very essence of the other to endorse. Even the efforts to read some overall purpose into the direction of evolution were a violation of what science had discovered. To see a plan in the pattern of progress rather than in the adaptations of individual species to their environments was to ignore the evidently "experimental" nature of evolution and the equally evident imperfections of the human body. The suffering that pervades the animal kingdom makes it plain that the process of evolution cannot have been established by a benevolent God, nor would it justify the process to argue that the suffering of those in the past was necessary to achieve the progress that led to the present.

In 1928 Cohen engaged in a public debate with C. E. M. Joad, who opposed materialism (although at this point he did not accept a role for the supernatural). Joad jokingly referred to the overwhelming personality (as revealed at the dinner table prior to the meeting) that allowed Cohen to browbeat his opponents all too easily.[69] But the fact was that materialism had tied itself too closely to the old atomic model of matter. It was now quite clear that matter was something more complex, and it was much less easy for us to feel that we had any real knowledge of its fundamental nature. Under these circumstances, a dogmatic materialism was untenable, and the reality of mind had to be accepted. Cohen replied by insisting that this effort to establish a "halfway house" was itself flawed. Unwittingly, at least, Joad was championing an old-fashioned vitalism and the idea of God (a prophetic claim in view of Joad's subsequent intellectual development).[70] Although by no means happy with Eddington's philosophy, Joad did at least accept that his arguments against rigid determinism were a problem for the materialist. Cohen was unimpressed, and in his God and the Universe of 1931, he joined in the chorus of rationalist opposition to Jeans and Eddington. He claimed that science had always advanced by adopting the materialist position, and hence by displacing religion. But a few scientists who, for lack of courage or clear thinking, were afraid to make a clean break with the past had offered a

---

68.   Ibid., 12–13, 15.

69.   Cohen and Joad, *Materialism: Has It Been Exploded?* 19–20.

70.   Ibid., 33–34.

compromise that religious thinkers had clung to with desperation. The vast majority of scientists were "openly disdainful of all religious subjects"— witness even Joseph Needham's admission that ninety percent of all biologists were mechanists.[71] Cohen's book included a whole chapter attacking Eddington, with the latter's reply printed in the following chapter.[72] Other chapters denounced Jeans's mathematician God and Einstein's claim to believe in Spinoza's rational God—which to Cohen was of no use to anyone who was genuinely religious (as Einstein himself admitted).

The fulminations of Cohen and McCabe against the new idealism seem to reflect a certain desperation, a sense that the rationalists' hoped-for transformation of society was evidently not coming about as expected. McCabe's appeal to Haeckel's *Riddle of the Universe,* that icon of turn-of-the-century materialism, suggests a mindset still shaped by the discourse of the previous generation. These were materialists desperately worried by the growing perception that science had turned in a new direction that was no longer so amenable to their interpretation of nature. Not that there was anything wrong with their analyses of the scientists' own opinions—Eddington and Jeans were exceptions among the physicists, and most biologists had indeed abandoned the once promising neovitalism. The religious thinkers who took idealism and neovitalism as typical of modern science were being misled, and in that sense, the much-vaunted synthesis of science and religion was built on shaky foundations.

But if the secularists' analysis of science was more accurate than their opponents', they were less secure in their understanding of contemporary culture. Outside the scientific community, the situation in the 1930s was changing—not because the new natural theology had really taken hold, but because more people were having doubts about the assumption that a rational analysis of the human situation would lead to intellectual and social progress. The RPA and the Secular Society were never very strong on politics—their supporters merely hoped that the sweeping away of superstition would free the world to move forward. They had ideas about how moral values might be transformed, but little interest in social policies or political activity. There were many who shared their suspicion of religion but who doubted that more knowledge of how the world worked would improve the human situation. Russell's *Icarus* had expressed profound doubts about the application of science to human affairs advocated by J. B. S. Haldane and J. D. Bernal, as did Aldous Huxley's *Brave New World* of 1932. One could reject religion without

---

71. Cohen, *God and the Universe,* 13.
72. Ibid., chaps. 2 and 3. Chap. 4 is Cohen's response to Eddington's reply.

being sure that science, or even rational thinking, could fill the moral vacuum left behind, and by the mid-thirties it was becoming ever more clear that the new technology was not an unmixed blessing. The way forward involved social reconstruction of a kind that would ensure the application of science for the benefit of humankind, rather than for private profit or military advantage. The destruction of organized religion would be a by-product of these changes, not their cause. Those scientists who became Marxists were attracted to exactly such a program, although they were by no means the first twentieth-century thinkers to explore the possibility of constructing a new social order on foundations that no longer included the Christian churches.

THE SOCIAL REFORMERS

The most widely read advocate of the need for a complete transformation of society was H. G. Wells. His views included hostility to organized religion, support for materialistic theories in biology, and a belief in science as a vital component in the rationalization of the social order. In the early phase of his career, Wells called himself a socialist and worked with the Fabian Society, but he soon went off in his own direction, stressing social control at the expense of any benefit to the masses. But there were other socialists who were also influential in pointing the way to a new order not based on religion. Almost as well known as Wells himself was Robert Blatchford, who published the socialist newspaper *The Clarion* and wrote in it himself under the name "Nunquam." Blatchford's socialism arose from his idealism, and his principles were such that he got on even less well with the Fabians than did Wells—he even disliked the policies of the first Labour government.[73] He was also hostile to Wells's enthusiasm for compulsion as the only way to a rational society, holding out the old liberal hope that eventually common sense and decency would prevail. His defense of socialism, published first in the *Clarion* and then as a book, *Merrie England* (1893), eventually sold over three-quarters of a million copies.

In his early life, Blatchford was a secularist to the core. He used a review of Haeckel's *Riddle of the Universe* to attack religion, drawing an impassioned response from the Congregationalist minister T. Rhondda Williams.[74] In 1902–1903 he threw open the pages of the *Clarion* to the defenders of Christianity—even Chesterton thought this a generous offer, especially as no hos-

---

73. See L. Thompson, *Robert Blatchford*. Blatchford and the *Clarion* feature prominently in Peter Broks's study *Media Science before the Great War*. On Blatchford's views on evolution and religion, see Kent, *From Darwin to Blatchford*.

74. Williams is discussed in chap. 7 above; see his *How I Found My Faith*, 67–68.

tile replies were permitted. But Blatchford was only biding his time, and soon weighed in with his *God and My Neighbour,* a forthright statement of his own reasons for rejecting Christianity. Those reasons included many based on science, and Blatchford recommended the RPA's series of reprints of the works of Huxley, Tyndall, and Laing. It has to be said that his grasp of the history of science was abysmal, drawing as it did on all the classic images of the church's hostility to new ideas. According to Blatchford, Copernicus was excommunicated for heresy, Bruno burnt at the stake for teaching the new cosmology, and Galileo tortured and imprisoned.[75] Blatchford believed that the church had responded with the same pattern of hostility to the new geology and evolutionary theory, and it is clear that the latter represented an important component of his hostility toward religion. The *Clarion* offered much better coverage of science than did many other popular papers—Blatchford had already reviewed Kropotkin's *Mutual Aid* enthusiastically, and would later praise Lankester's defense of Darwinism.[76] He now insisted that the Bible was full of fables, evolutionism especially contradicting the idea of a Fall from an original state of perfection.[77] Most of Blatchford's book was given over to a critique of the whole idea that the Bible was a divinely inspired revelation, much of it based on his conviction that miracles were totally implausible. If God had created the whole universe, including ourselves, then He could hardly hold us responsible for defying Him, so the idea of salvation was also nonsense. In response to the Rev. R. F. Horton's protestations that there was still a need for Christ, Blatchford said that such hypocrisy, in a world filled with suffering tolerated by the churches, made him feel ill.[78] Blatchford referred to Kropotkin again to argue that evolution was steadily developing the cooperative instincts, and this became the basis for his hope that the human race would move steadily toward a society that did not tolerate inequalities. Since we are all products of our heredity and our environments, he felt that no one should use his or her superior endowments to gain advantage over others. We should aim for a world free of poverty and ignorance, a goal that would never be reached so long as Britain was a Christian nation.

Not surprisingly, *God and My Neighbour* attracted a flood of rebuttals from the ministers of a wide variety of churches. At first, Blatchford kept up his assault in the *Clarion,* but his own opinions changed dramatically over the fol-

75. Blatchford, *God and My Neighbour,* 4.

76. See Broks, *Media Science,* 36, and Lester, *E. Ray Lankester,* 189.

77. Blatchford, *God and My Neighbour,* 10 and 124.

78. Ibid., 192.

lowing decades. By 1920 he had become aware that the physicists had now shown the atom to be composed of electric charges, and unlike many traditional materialists, he conceded that this did make a difference. This was "conduct unbecoming a material entity," and he began to wonder whether, if the material universe was composed of mysterious qualities, the human spirit might also be real.[79] His wife died in 1921, and like many who had lost a loved one, he found it difficult to believe that she was really dead. Perhaps the spiritualists, whom he had long ridiculed, were right after all—and after receiving what he regarded as a convincing communication from his wife's spirit, he came out openly in support of them in his *More Things in Heaven and Earth* (1925). He did not, however, give up his opposition to orthodox Christianity, maintaining to the end that the superstitions he had campaigned against in *God and My Neighbour* were both false and increasingly recognized as false.[80]

If Blatchford's socialism was idealistic, H. G. Wells's commitment came from an enthusiasm for social control by an elite. In this respect he shared the viewpoint of the Fabian Society run by Sidney and Beatrice Webb, although they thought he went much too far, and Wells soon broke with them. This was a socialism far removed from the grassroots sympathy for the working man that was building the basis of real political power for the Labour Party, but it was characteristic of the intellectual world of early-twentieth-century Britain, with its confidence in progress and its conviction that this would be achieved only by rational control. Wells's success was built on his science fiction stories of the 1890s, but from the start of the new century he began to sketch out his plans for the future of the human race. In his *Anticipations* (1901), he already foresaw a new rational ruling class that would take charge of Western society, eliminating the unfit through eugenics and dominating the rest of the human race. *A Modern Utopia* (1905) identified five classes, the highest being the Samurai, the rational, scientifically trained leaders who imposed their will on the rest. Reactions to such books were mixed: many people saw them as inspiring visions of future possibilities, but others were horrified by the brutality they implied in their dismissal of the unfit.[81]

In the interwar years Wells became increasingly certain that the future rationalization of the social order would not come about until the old structure of society had been destroyed by war, leaving the elite to emerge triumphant out of the resulting chaos. This was the vision of *The Shape of Things to Come*

79.   Blatchford, *My Eighty Years,* 262.

80.   Ibid., 201.

81.   The darker side of Wells's vision, and popular reactions to it, are stressed in Coren, *Invisible Man;* other studies include Batchelor, *H. G. Wells,* Mackenzie and Mackenzie, *Time Traveller,* D.C. Smith, *H. G. Wells,* and West, *H. G. Wells.*

H. G. Wells. By Sir William
Rothenstein, 1912. By courtesy of the
National Portrait Gallery, London.

(1933), filmed spectacularly two years later by Alexander Korda, with Wells's cooperation on the script. The book and the film provide a graphic portrayal of a world destroyed by modern weapons of war and of the emergence of an Air Dictatorship as the foundation for the new world state (Wells was always fascinated with transportation and its ability to transform society). By the end, a united humanity at last begins to explore the universe properly and is beginning to meld itself into a single organism.[82] This vision is translated in the film into the final scenes surrounding the "space gun" that will send the first people to the moon. But the film concedes the darker side of this ambition—many are dismayed by the arrogance of the scientists and seek to destroy the project, and the conclusion echoes to the words "Where will it end?"

Needless to say, there was no role for organized religion in this project. Wells began from an agnostic position that allowed him to propose human reason as the guide to all future conduct. That was his position in the end, too—but in the meantime, he passed through a phase in which he saw the possibility of transforming religion into something more compatible with the needs of struggling humanity. The first evidence of this came in his novel *Mr. Britling Sees It Through*, a surprisingly sensitive treatment of the social con-

82. H. G. Wells, *Shape of Things to Come*, 430–32. For Wells's own outline of the development of his views on the world state, see his *Experiment in Autobiography*, vol. 2, chap. 9.

sequences of the Great War. Toward the end of the book, faced with the claim that a God who permits this carnage is not worthy of worship, Mr. Britling argues that the theologians have got it wrong. After all, Christ was mocked and nailed to a cross—so perhaps God is finite, a struggling God trying to shape the world through us. In effect God becomes a personification of human effort and aspiration.[83] This point was developed a year later in *God the Invisible King*, which contrasted the "Veiled Being" of the ultimate mystery of the universe (something like Spencer's "Unknowable") with God the redeemer, the finite, personal God seen by Mr. Britling. The Veiled Being is amoral and is represented by the life force that drives progress yet causes endless suffering.[84] Only the personalized, struggling God was of interest to humanity. Wells insisted that this God was plausible within the scientific worldview; atheistic critics such as Elie Metchnikoff were attacking an outdated religion based on priestcraft, but science could see itself as part of the struggle for progress in this world that was the basis of the real God's activity.[85] There is some similarity between Wells's position and Samuel Alexander's vision of a God who emerges from the evolutionary process, but it was in no sense a version of Christianity, and Wells insisted that his God needed no church. Not surprisingly, the new religion got short shrift from those committed to the old.[86] Wells's agnostic friends were worried by his new tendency to personify the collective human spirit, and it has to be said that he soon abandoned the idea, treating it as only a passing phase in his intellectual development before he returned to the agnostic fold.[87]

Science had been an important part of Wells's worldview since the year he spent studying under T. H. Huxley at the Normal School of Science in 1884–1885. From Huxley's pessimistic view of cosmic evolution (and also from Lankester's little book *Degeneration*) he learned the Darwinian lesson that there is no inevitability about progress and that the material universe must end in the stagnation of heat-death. He also learned about the promise of the new knowledge and its potential to transform human life. These themes run throughout the science fiction stories that made him famous in the 1890s, especially "The Time Machine." His plan to transform the world along rational lines was built on the assumption that the human race must take charge of its own destiny in an uncertain and often hostile universe.

83.   Wells, *Mr. Britling Sees It Through*, 396–99.

84.   Wells, *God the Invisible King*, 18–20.

85.   Ibid., 83, 102.

86.   See, for instance, Lacey, *Mr. Britling's Finite God.*

87.   See E. Ray Lankester's letter to Wells, 21 September 1916, Wells papers, University of Illinois; for Wells's later assessment, see his *Experiment in Autobiography*, vol. 2, 665–74.

When Wells conceived of writing a survey of history to counter the pessimism engendered by the Great War, he decided to begin with humanity's
evolutionary origins in order to show both how far we have risen and how
little we can trust the Darwinian process to guarantee further development.
Lankester joined the advisory panel to offer guidance on evolutionary biology. and it was his version of Darwinism (which made no concessions to the
new-fangled genetics) that Wells adopted and was forced to defend once *The
Outline of History* was published.[88] Hilaire Belloc seized on this point to proclaim that Wells's science was out of date, although it is clear that Belloc's real
target was his secularized interpretation of European history, which marginalized the church by treating it as an obstacle to progress. Belloc had a reputation as a savage controversialist, and there is no doubt that Wells was stung
by his criticisms, especially as he had to issue a privately printed booklet in
which to make his reply.[89] When a new edition of the *Outline* was published
in 1926, Belloc went on the offensive again with a series of articles that were
widely reprinted. Watts's publishing house now issued a more substantial rebuttal, in which Wells dismissed Belloc's claim to have access to a superior
European science of which the English were ignorant (Vialleton's morphology). He also went through the routine responses to Belloc's equally routine
arguments against natural selection, most of which, Wells believed, had come
half digested from Samuel Butler.[90] Wells certainly felt the strain of the controversy, and there were many literary figures who thought that Belloc was the
outright winner.[91] Yet *The Outline of History* sold in vast numbers and was regularly updated through its many editions. If the literary world thought that
Belloc's evaluation of Darwinism reflected the best modern science, there
were many ordinary readers who got the opposite impression from Wells.
Anyone in touch with science knew that Darwinism was being transformed
and was coming to dominate biological thinking. An article by H. J. Randall
published in the *Edinburgh Review* in 1929 portrayed the anti-Darwinians as
fighting a losing battle against a theory that had transformed our view of the
world.[92] Wells himself was stimulated to gather a new team, this time including Julian Huxley, to begin work on a serious popularization of modern

88.   On natural selection, see Wells, *Outline of History,* chap. 2, esp. 15–17; chaps. 2 and 3 offer an
overview of evolution up to the emergence of humanity.

89.   Wells, *New Teaching of History.*

90.   Wells, *Mr. Belloc Objects to* The Outline of History, 24.

91.   See Waugh, *Life of Ronald Knox,* 232; for an account of Wells's reaction, see Coren, *Invisible
Man,* chap. 7.

92.   Randall, "Intellectual Revolution in Nineteenth-Century England," 49.

biology, *The Science of Life* (1931).[93] Although Belloc and Chesterton continued to pretend that Darwinism was dead, the message about its revival among working biologists was at last being offered to the public.

The economic troubles of the 1930s forced many to rethink their attitude toward Wells's vision of a planned society. Even Bertrand Russell and Aldous Huxley, both well aware of the dangers of regimentation, became more sympathetic. Joad organized a Federation of Progressive Societies and Individuals in 1932, which included Russell, both Huxleys, and Wells.[94] Attitudes toward the Soviet regime in Russia began to change. Wells distrusted Marxism and had initially been hostile to the Bolshevik revolution, but following a visit to Moscow in 1934 he adopted a more favorable stance toward Stalin.[95] But Communism held only limited appeal for British intellectuals in the 1930s; even students were not very enthusiastic. It was the politically active scientists who became absorbed in the Marxist vision of a planned society that would be built on allegedly "scientific" principles (their story has been told in chapter 2). Yet at Cambridge, where there was by far the most active left-wing group in the early 1930s, barely five percent of the scientific community was involved.[96]

Although never very strong politically, the Communist Party cannot be ignored because its members made a point of trying to reach out to the masses. Where Wells and the advocates of the planned economy had nothing but contempt for the working class, the Communists went out into the world organizing demonstrations and exhibitions and helping to run labor unions. They clashed, often violently, with Oswald Mosley's Fascists. Opposing organized religion was seen by many Communists as a duty, whatever the feelings of the few who wanted to heal the breach between the churches and Marxism. In some respects, their task was made easier when Belloc and the Catholics came out openly in support of Franco in the Spanish Civil War. It was by writing for the lay reader that left-wing intellectuals could most actively influence society at large. J. D. Bernal, J. B. S. Haldane, and Lancelot Hogben certainly took their responsibilities to the masses seriously, writing copiously in the Communist press and for mass-market publishers. Joseph Needham tried to interest the literary elite in his own synthesis of Marxism

93.   The introduction to *The Science of Life*, 3–4, makes it clear that the project was seen as arising from the controversies sparked by *The Outline of History*.

94.   See Bradshaw, *Hidden Huxley*, 31–43.

95.   Wells, *Experiment in Autobiography*: on Marx, vol. 1, 180; on Stalin, vol. 2, 800–808.

96.   Wood, *Communism and British Intellectuals*, 69–70; on Cambridge, see Werskey, *Visible College*, 216–22.

and Christianity. But it was Haldane who emerged as the most prolific science journalist of his time, endlessly promoting the claim that a rational application of the latest findings could be used to help ordinary people live a better life. In 1933 he offered to provide a free scientific news service to the *Daily Herald*, and was turned down. It was the Communist *Daily Worker* that eventually provided him with a platform, and he wrote a regular Thursday article for that paper until it was closed down after the outbreak of war in 1939. These articles were collected in his *Science and Everyday Life* of 1939, reissued two years later as a Pelican paperback. Significantly, this collection contains another article proclaiming the revival of Darwinism (Haldane had earlier responded to Belloc's attack by making the same point). But the article also goes beyond Darwinism, using ecology to argue that there are limits on how rigidly species must be adapted to their environments and praising Lysenko's work in Russia.[97]

In one sense, Haldane was the heir to Lankester, Keith, and the other popular science writers of the RPA, but in another he was a symbol of a new turn in the thinking of the more radical intellectual elite as it responded to the social challenges of the 1930s. Writers like Wells, Huxley, and Haldane all strove to promote a wider awareness that in biology, at least, the sciences were moving in a direction that favored their philosophy. Their message seems to have been lost on most religious thinkers, as we have seen in earlier chapters, but there was clearly a section of the reading public willing to buy their educational works and respond to the message therein. The emergence of Marxism as a significant factor influencing those with left-wing tendencies illustrates the polarization of attitudes generated by the social tensions of the 1930s. While those committed to Christianity became ever more orthodox, those who chose the rationalist philosophy also came under the influence of a cultural force demanding allegiance and commitment as never before.

97. Haldane, "Evolution and Its Products," in his *Science and Everyday Life*, chap. 5.

CHAPTER 11

Religion's Defenders

If there were figures as eminent and diverse as Bertrand Russell and H. G. Wells writing in support of secularism in the early twentieth century, there was an equally prominent and equally diverse group of writers and thinkers still prepared to defend religion in the arena of public debate. Some of these were churchmen themselves, by no means afraid to carry the debate beyond the confines of the theological establishment. Bishop Barnes and Dean Inge were no strangers to controversy and outspoken press commentaries. Professional writers such as Hilaire Belloc and G. K. Chesterton also expressed a specifically Christian viewpoint, as did scholars such as C. S. Lewis who had the knack of writing for a wider readership. An older generation of philosophers was still active in defending the idealist view of knowledge and in some cases drawing out its implications for theism, if not for Christianity. Scientists, too, addressed the public directly in support of a religious perspective. Oliver Lodge, J. Arthur Thomson, and C. Lloyd Morgan wrote innumerable books and articles promoting an evolutionary natural theology, while Arthur Stanley Eddington and James Jeans sold their books on the new idealism by the tens of thousands. By the late 1930s the flood of popular writing by scientists on this theme had begun to dry up as the older generation of biologists became less active—although even the efforts of Wells, Julian Huxley, and J. B. S. Haldane were unable to demolish the belief that "Darwinism is dead" in the minds of those influenced by Belloc and Chesterton.

This list of names, however, reveals the tensions that existed within the antisecularist camp. Some wanted to support a generalized theism in opposition to the materialism that was perceived to have dominated the previous century. Others wanted to defend Christianity itself, but disagreed over whether this was best done by modifying its theology in a more generally theistic direction or by holding fast to the traditional vision of Christ as the savior of a fallen humanity. Modernist Christians and non-Christian theists could form a reasonably united front, since few theists would deny that Christ had been an important teacher. But orthodox Christians tended to view the whole liberalizing trend as a mistake, and pointed to this alliance as proof that Modernism was only a step down the slippery slope toward abandoning the unique Christian perspective altogether. The same threat could be seen in the emergence of a whole range of alternatives to Christianity that could easily become associated with the general defense of theism. Spiritualism, for instance, could be linked to Modernist Christianity—this was Lodge's position—but it certainly had the capacity to set itself up as a rival to the churches, and it was viewed with deep distrust by many orthodox Christians. Theosophy and other forms of pseudo-Eastern religions were specifically opposed to Christianity while generally supportive of the antimaterialist crusade. In the hands of George Bernard Shaw, even creative evolutionism became an alternative religion. All were bidding for public support in a marketplace of rival spiritual products.

For Christians, the crucial issue in the defense of their faith was the extent to which it should be modified to take modern ideas and ways of thought into account. Theologians and religious scientists expressed varying opinions on the questions of evolution, the nature of mind, and the nature and history of the physical world. These issues were central to the public debates; indeed, the concept of evolution formed the foundation of many attempts to produce a modernized form of theism. But the issue of modernization went deeper than this, threatening the very foundations of traditional theology. If religion took science as a model, it might have to abandon its claim to offer certain knowledge based on ancient texts or other sources of inspiration. Was the whole concept of dogmas that must never be challenged or even modified now out of date? The Modernists argued that it was, but the problem with this more scientific methodology was that it might lead to the old foundations on which the church had been built being modified beyond recognition, as more orthodox Christians were constantly warning.

There was no shortage of philosophers, both academic and popular, encouraging religious thinkers to adopt a more flexible attitude that would make theology more "scientific." In 1905 G. Lowes Dickinson published a critique of religion collected from his articles in the *Independent Review.* He accepted that the religious attitude toward life was valuable because it encouraged hope, but believed it was essential that its insights not become matters of faith to be accepted without question. These insights were more akin to music and poetry than to science, but they would have to be consistent with the harsh view of nature revealed by science, and their value for our lives would have to be judged by the test of experience.[1] A few years later M. M. Pattison-Muir explicitly called for theologians to treat their dogmas like scientific theories, subject to revision.[2] In 1929 the philosopher R. Gordon Milburn responded to a question posed in Eddington's *Nature of the Physical World* by insisting that a science of religious belief was possible. Eddington had asked whether the theologians had "any system of inference from mystic experience comparable to the system by which science develops a knowledge of the outside world." Milburn provided a cross-cultural comparison of religious insights based on a hierarchy of five grades of belief, in which the highest was the kind of mysticism reflected in the thoughts of Christ.[3]

Few orthodox Christians were happy with the idea that Christ was merely the source of the highest kind of mysticism. Some regarded the whole attempt to test religious faith against modern experience as a mistake and queried whether the public interest in such questions indicated a genuine concern for spiritual matters. All too often, the Modernists' search for a new kind of faith resulted in a systematic dilution of the Christian message in the hope of finding a level that would still be acceptable to the majority of laypersons. As Ronald Knox put it in his *Some Loose Stones* of 1913—a response to the Modernist volume *Foundations*—it was really a search for "What will Jones swallow?"[4] In other words, it didn't matter whether or not the Christian faith was true—the Modernists were interested only in what a majority could be persuaded to believe. Knox's hypothetical "Jones" was a middle-class man of about sixty—and he noted that the younger generation were now being tempted not so much by materialism as by a range of beliefs including Christian Science and spiritualism.[5] Knox noted that the newspapers

1. Dickinson, *Religion: A Criticism and a Forecast,* 90–93.

2. Pattison-Muir, "The Vain Appeal of Dogma to Science."

3. Milburn, *Logic of Religious Thought,* 34. The question is posed by Eddington in *Nature of the Physical World,* 339.

4. The title of the first chapter of Knox's *Some Loose Stones.*

5. *Some Loose Stones,* 9–12. Knox specifically opposed spiritualism; see the appendix to chap. 10, 195–200.

were full of symposia on questions such as "Is Prayer Answered?" "I Believe in ———," and "Where Are the Dead?" To him they indicated not a real boom in nondenominational religion, but a national fondness for talking about such questions in public.[6]

As a Catholic, and later as a Roman Catholic, Knox felt that Christianity was true and that it was important to speak out about that truth even though many would turn away. To some extent his position was vindicated by the new orthodoxy emerging in the Anglican Church in the later 1930s. But others wondered if such a hard line was appropriate, with even Charles Gore conceding that diluted faiths such as Lodge's evolutionary spiritualism were preferable to no religion at all.[7] A vague theism might be a stepping-stone to real Christian faith, not a temptation to turn away from it. As William Temple, one of those criticized in Knox's attack on *Foundations*, put it in a public letter in reply: "I am not a spiritual doctor trying to see how much Jones can swallow and keep down . . . *I am Jones asking what there is to eat.*"[8] One either made an effort to confront the tensions within modern thought or gave up and accepted that only a few would ever be saved; not an easy choice, and one that would become more difficult in a world sliding toward chaos and war.

FROM IDEALISM TO SPIRITUALISM

The spectrum of writers seeking to resist materialism ran from the most respectable academic philosophers to those trying to publicize popular cults such as Theosophy and spiritualism. In the academic world there were still many advocates of the idealist philosophy that had become popular at the end of the previous century. If no longer at the cutting edge in university departments of philosophy, they were senior figures who had access to publishers and who might be asked to give a series of Gifford Lectures—very much in parallel with the senior scientists who promoted the new natural theology. Their idealism was not easily compatible with Christianity, but most regarded the Absolute as some kind of spiritual power underlying the universe that might be identified with God, even if a somewhat impersonal God. There were also influential thinkers such as the philosopher and statesman Arthur Balfour who, while not actually idealists, had worked hard to stem the

6. Knox, *Caliban in Grub Street*, 2–3; for the list of newspaper symposia, see 17. Knox's attacks on modern pseudoreligions are discussed below.

7. Gore, *The New Theology and the Old Religion*, 13–15.

8. Temple, letter to Knox, quoted in Iremonger, *William Temple*, 161–62. See also Hastings, *History of English Christianity*, chap. 12.

tide of scientific naturalism. Their defense of theism highlighted a theme popular among liberal theologians and the scientists who associated with them: an evolutionary natural theology that saw the development of life as the unfolding of a divine plan. The assumption that evolution was progressive and purposeful was also taken up by a wide variety of popular writers, some of whom went far beyond what even the most liberal Christian could accept. The theme of a blindly creative evolutionary process groping its way upward was central to the thinking of both Henri Bergson and George Bernard Shaw, but their rejection of the claim that humanity was the preordained goal of this process took the debate in a direction that many theists could not accept.

Some opponents of scientific naturalism were so anxious to demolish the materialistic view of the mind that they were tempted to take an interest in the paranormal and spiritualism. A few made their way into the murky world of mysticism and the occult, with its circles of adepts and enthusiasts ever ready to follow the latest guru offering enlightenment. Although most Christians would have regarded this as dangerous territory indeed, the popularity of these more exotic alternatives to traditional theism offers insights into the popular culture of the period.

That there was an association between the idealism of the late nineteenth century and the explosion of interest in mysticism was admitted by Antonio Aliotta, the Italian professor of philosophy who wrote perhaps the best survey of the period, translated as *The Idealistic Reaction against Science* in 1914. Aliotta's purpose was to survey the rise of an idealistic reaction not so much against science itself as against the materialist philosophy so often associated with it. He claimed that Herbert Spencer's concession of the Unknowable had unwittingly paved the way for Bergson's philosophy of the vital force.[9] His survey included the English Hegelians T. H. Green, F. H. Bradley, and John McTaggart as well as the corresponding psychology of values espoused by James Ward, showing the link to theistic evolutionism and the philosophy of spiritualism. Aliotta's conclusion was that the new philosophies and the scientific theories built on them paved the way for a spiritualistic conception of the world based on the existence of an Absolute Consciousness equivalent to God.[10] All this was pure gain, yet he had been forced to concede in his introduction that the reaction against materialism had also opened the floodgates to a surge of irrationalism:

9.   Aliotta, *Idealistic Reaction against Science*, 5–8; on Bergson, see 127–37. See also Aliotta's contribution to Joseph Needham's *Science, Religion, and Reality*, "Science and Religion in the Nineteenth Century."

10.   *Idealistic Reaction against Science*, 460–61.

Once the blind power of impulse was exalted and the sure guidance of the intellect abandoned, the door was opened to every kind of arbitrary speculation; hence the confusion, Byzantinism, and dabbling in philosophy which during the last twenty years have obscured thought and masqueraded under the fine-sounding name of idealism. O unhappy Idealism, how many intellectual follies have been committed in thy name! Theosophy, the speculations of the Kabala, occultism, magic, spiritualism, all the mystic ravings of the neo-Platonists and neo-Pythagoreans, the most antiquated of theories, debris of every kind, heaped haphazard on the foundations of the speculations of the ages—all these have returned to favour in defiance of the dictates of logic and common sense.[11]

Small wonder that some Christian thinkers were worried that the application of idealist philosophy to the defense of religion might not only threaten the faith's basis in historical events, but also leave it open to absorption into this welter of mysticism and the occult.

British idealism did not collapse completely when assaulted by Russell, G. E. Moore, and the realists of the early twentieth century. It continued to function, both because some younger idealists remained active in philosophy and because the students of the first generation were now influential in the wider world.[12] Some of them became leaders of the Modernist movement in the Anglican Church. But—and this was always the problem for the Modernists—idealism promoted a worldview that might be spiritual, but was not necessarily very Christian. It saw each individual mind as a component of the Great Mind or Absolute that underpinned the universe, and thereby threatened the Christian vision of humanity as alienated from God and needing salvation. Several key idealists saw no point in the immortality of the individual soul, since it was already a component of the Absolute. McTaggart denied that Hegelianism could be equated with Christianity and took the idea of reincarnation seriously.[13] He thought the universe itself was a moral agent, but refused to identify his Absolute with God and wrote openly against religious dogma.[14]

11.   Ibid., xv–xvi.

12.   The conventional view that idealism collapsed in the early twentieth century is accepted in P. Robbins, *British Hegelians*, but on its continued influence, see Den Otter, *British Idealism and Social Explanation*, and Hinchcliff, *God and History*, chap. 6. All accept that idealism formed a kind of alternative religion, but on its dangers for Christianity, see Sell, *Philosophical Idealism and Christian Belief*.

13.   McTaggart, *Studies in Hegelian Cosmology*; on reincarnation, see 47–50, and on Christianity, chap. 8.

14.   McTaggart, *Some Dogmas of Religion*.

Other idealists were more positive in their defense of religion. Andrew Seth Pringle-Pattison was a follower of Bernard B. Bosanquet, whose idealism he saw as an explicit endorsement of the idea of God. He contributed a chapter to B. F. Streeter's volume *The Spirit*, in which he proclaimed the material world to be an abstraction governing the relationships between conscious spirits, the whole forming the medium by which God shaped human souls.[15] His Gifford Lectures for 1912–1913, titled *The Idea of God in the Light of Recent Philosophy*, offered a defense of theism linking the idealist perspective to many developments in science. The neovitalism of Hans Driesch and J. S. Haldane had demolished materialism, while evolution had been liberated from naturalistic Darwinism by Thomson and others, who had shown that it could be understood only in terms of cosmic teleology:

> ... and so Darwin may be taken as replacing man in the position from which he was ousted by Copernicus. Man appears, according to the doctrine of evolution, so interpreted, as the goal and crown of nature's long upward effort. The evolution of ever higher forms of life, and ultimately of intelligence, appears as the event to which the whole creation moves; and, accordingly, man is once more, as in pre-Copernican days, set in the heart of the world, somehow centrally involved in any attempt to explain it.[16]

A somewhat more realistic interpretation of teleological evolutionism was proposed by Sir Henry Jones in his Gifford Lectures for 1922. Jones was a student of Edward Caird who had reluctantly abandoned the Methodist ministry for a career in philosophy. Like many Modernists, he saw Christ as revealing the divine essence within all of us. His lectures were titled *A Faith That Enquires*, reinforcing the trend to see theology as something to be tested by experience. Jones saw the universe as a great moral progression, the striving of the divine will, "the perfect in process," which was thus both friendly and helpful toward all our moral aspirations.[17]

The campaign to convince the public that evolution was a purposeful or creative process driven by nonphysical forces was a major feature of the offensive against materialism in the early twentieth century. Other sources of opposition to materialism depended on the assumption that mind was now to

---

15.   Seth Pringle-Pattison, "Immanence and Transcendence."

16.   Seth Pringle-Pattison, *Idea of God in the Light of Recent Philosophy*, 82–83.

17.   H. Jones, *A Faith That Enquires*, 360. Jones was active in social thought and politics; see Boucher and Vincent, *A Radical Hegelian*.

be seen as a real entity influencing the material world. The idealism of Jeans and Eddington was greeted with derision by most academic philosophers but was endorsed by some writers opposed to materialism. The philosopher W. Tudor Jones noted in 1929 that the physicists were returning to idealism and linked this change with neovitalism in biology to argue that science was now endorsing the concept of God. He conceded, however, that science generated only a very basic theism—the real source of religious knowledge was mystical experience.[18] A similar point was made by the popular science writer J. W. N. Sullivan, who realized the limitations of Jeans's argument that the God who sustains the universe must be a mathematician.[19] Others, however, were struck by Jeans's argument that the existence of life on earth required an unlikely combination of circumstances. R. E. D. Clark's *The Universe and God* of 1939 built on Jeans's theory of the origin of the solar system and the antimechanist critique of the spontaneous generation of life to argue that human life was unlikely to have appeared except by the exertion of some designing power. The existence of order in the universe was a sign that something beyond the laws of nature had been at work, and since the collapse of the theory of natural selection, it was impossible for science to explain how more advanced forms of life had evolved. Clark realized that the God revealed by physics was an abstract entity, but appealed to a vision of the living world purged of the "struggle for existence" to argue that the Creator did indeed care for us.[20] Here the new physics was linked not to creative evolutionism, but to something that looked very much like creationism.

There were traditions opposed to scientific naturalism that stopped short of philosophical idealism but remained committed to the belief that the mind was something that stood above material nature. Their adherents, too, favored antimechanistic scientific theories and saw mind as an emergent quality—but some were inclined to look outside the realms of orthodox investigation in search of evidence for the ability of the mind to transcend bodily limitations. They thus formed a natural audience for the claims of the spiritualists and the exponents of even more exotic and mystical philosophies.

The leading opponent of scientific naturalism in the closing years of the nineteenth century had been Arthur Balfour. He was the most eminent product of that last generation of public figures who still took it as their duty to

18. W. T. Jones, *Reality of the Idea of God*, 30–34, 51, 150.

19. Sullivan, *Bases of Modern Science* and *Limitations of Science;* for a summary of Sullivan's position, see chap. 6 of the latter. *Bases of Modern Science* was singled out for criticism in Lancelot Hogben's *Nature of Living Matter.*

20. R. E. D. Clark, *The Universe and God*, chap. 15.

engage in philosophical debate. While active at the highest level in politics (he became Prime Minister in 1902), Balfour was at the center of a network of thinkers determined to resist the scientific naturalism of T. H. Huxley and John Tyndall and equally determined to preserve a role for a social and intellectual elite in the governing of the nation. When Huxley died in 1895, he was in the process of completing his review of Balfour's *Foundations of Belief* and was well aware that there was now a tide running against the philosophy on which he had built his life.[21] Balfour wrote as a liberal Christian, but his chief concern was the philosophical defense of theism.[22] His argument was that science itself was based on faith (in the uniformity of nature) and that belief in a God who created us with the ability to comprehend the universe was by far the most plausible interpretation of the human situation. Balfour was closely linked with the Cambridge physicists who used the ether theory to defend the idea that the physical world constitutes a unified whole designed by God. He shared Lodge's interest in spiritualism and served as president of the Society for Psychical Research in 1893–1894.

Despite his political duties, Balfour remained active in intellectual matters in the early decades of the new century. In 1904 he served as president of the British Association and gave his presidential address on the implications of the new physics, concluding "that as Natural Science grows it leans more, not less, upon a teleological interpretation of the universe."[23] In 1908 he argued that because of these transformations in our worldview, science could be seen as a moral force in the world despite the misuse of its practical applications—those harmful effects were no more indicative of its fundamental implications than were the bigotry and persecutions that sometimes arose from overzealous religious belief.[24] He wrote an extended critique of Bergson's philosophy, seeing it as an interesting attempt to formulate a third viewpoint that was neither naturalistic nor idealistic. But despite being fascinated by Bergson's image of the life force as the essence of freedom in opposition to the determinism of matter, he thought that evolution was best explained

21. Lightman, "'Fighting Even with Death.'" See also Jacyna, "Science and Social Order in the Thought of A. J. Balfour." On Balfour's links with spiritualism, see Oppenheim, *The Other World*, 129–35, and on his links to Lodge and ether physics, see Wynne, "Physics and Psychics."

22. Balfour, *Foundations of Belief*; see the conclusion, 334–39, in which Balfour briefly extends his more fundamental position into an argument for Christianity. He also defends the plausibility of miracles; see 311–15. Note, however, his choice of an inadmissible application of theological principles—the claim of papal infallibility, which in turn is linked to a rejection of sacred texts as a secure foundation for belief, 223–26.

23. Balfour, "Reflections Suggested by the New Theory of Matter," 14.

24. Balfour, "Decadence," in his *Essays Speculative and Political*, 1–52. Originally the Sidgwick Memorial Lecture of 1908.

by assuming that its goal reflected a divine purpose.[25] Balfour's Gifford Lectures of 1914 argued that the debates over whether natural selection destroyed the argument from design were sterile and outdated. Whatever the mechanism of biological evolution, it was clear that with the gradual emergence of mind, something new had been introduced into the world, something capable of affecting the course of events in a manner not predictable by materialistic principles.[26] The real evidence for design came from the higher values that such minds were eventually capable of appreciating. Here he anticipated the concept of emergent evolutionism, but also made it clear that he viewed the mind as an entity independent of the body. He remained committed to the search for evidence of survival of bodily death, eventually accepting as plausible messages purporting to come from a woman he had once loved via automatic writing by a medium.[27] In 1925 he wrote the introduction to Joseph Needham's collection *Science, Religion, and Reality*, still arguing that science did not disprove the plausibility of miracles or explain the freedom of the will.[28]

Balfour's commitment to spiritualism was by no means unusual. The movement experienced a substantial revival following the Great War. Few academics were prepared to lend their support, but an important exception was a lecturer in moral sciences at Cambridge, C. D. Broad. Broad's *The Mind and Its Place in Nature* of 1925 is one of the more substantial philosophical defenses of the theory of emergent evolution. Yet it is also an explicit defense of the use of the paranormal as evidence for the reality of the mind. Broad admitted in his preface that he would be blamed by scientists and by many philosophers for this, but he was quite impenitent. Like Lodge, he insisted that the skeptics were deliberately blinding themselves to the evidence through their commitment to materialism—they "confuse the Author of Nature with the Editor of *Nature*," and it did not follow that because the latter would not accept the reports, they were unworthy of being taken seriously.[29] It was in his final chapter, "Empirical Arguments," that Broad introduced the evidence provided by the Society for Psychical Research, arguing that if any of it was valid, the emergence of some psychic factor in the course of evolution was the most reasonable explanation.[30] Broad had little interest in Christianity, however. In a 1939 article on science and religion—the debate over which had

25. Balfour, "Bergson's Creative Evolution," reprinted in *Essays Speculative and Political*, 103–47.

26. Balfour, *Theism and Humanism*, 39–41.

27. Oppenheim, *The Other World*, 132–33.

28. Balfour, "Introduction," 13–17.

29. Broad, *Mind and Its Place in Nature*, viii. Brian McLaughlin includes a lengthy discussion of Broad's book in his "Rise and Fall of British Emergentism" without mentioning the paranormal.

30. Ibid., 651.

"acquired something of the repulsiveness of half-cold mutton in half-congealed gravy"—he again introduced the evidence for survival after death, but argued that this was a Trojan horse for Christianity. The latest developments in science had made a basic theism more plausible, but if the evidence of the paranormal were accepted as credible, the miracles of Christ lost their unique status. Christ would have to be accepted as merely a gifted psychic, not the son of God.[31] Curiously, this was exactly the opposite conclusion from that reached by Malcolm Grant in 1934, who felt that psychic research would restore Christianity to its former influence because it made both survival and the New Testament miracles more plausible.[32]

Not everyone who saw the mind as a real entity accepted that it survived the death of the body in a personal form. Here the most interesting position was that taken up by the popular "radio philosopher" C. E. M. Joad. In the interwar years Joad was not only an opponent of materialism (we have already encountered his 1928 debate with Chapman Cohen in chapter 10) but also an opponent of Christianity, although he would eventually convert after World War II. In 1932 he debated the truth of Christianity with Arnold Lunn, professing himself amazed at a survey by the *Daily News* showing that seventy percent of the public thought that the first chapter of Genesis was not historical, yet eighty percent thought that the Bible was inspired by God. Joad felt that it was inconsistent to believe in a God who inspired the writers of Genesis and then made them tell lies.[33] This skeptical attitude, however, concealed a very real desire to defend both the reality of the mind against the materialists and the general theistic position that evolution is the working out of a divine purpose. Like Russell, Joad was concerned about the effects of new technology on human life—he complained bitterly about the replacement of the pianola by the gramophone as an example of how people were being reduced to passivity by ever more complex machines.[34]

Joad's *Matter, Life, and Mind* of 1929 was a complex defense of the reality of the mind based on neovitalist physiology and a kind of emergent evolutionism. Like Lloyd Morgan, Joad believed that emergent properties were genuine novelties, but he also insisted that the mental functions existed independently of the body that produced them. He was convinced that for progressive evo-

31.   Broad, "Present Relations of Science and Religion," 138–41. For the "cold mutton" quotation, see 131.

32.   M. Grant, *A New Argument for God and Survival.* This position had been argued much earlier by Alfred Russel Wallace.

33.   Joad in Lunn and Joad, *Is Christianity True?* 168.

34.   Joad, *Under the Fifth Rib,* chap. 7. The pianola plays the notes automatically, but allows some contribution by the operator, who can work the pedals affecting loudness and tone.

lution to be possible, the experiences gained by one generation must be transmitted to the next.

> I see no alternative for those who advocate the doctrine of progressive evolution but to accept [the inheritance of acquired characteristics], and to maintain that, just as the body of the embryo recapitulates in a short period of time all the past history of the physical changes through which the species has passed, so does the mind of the newly-born offspring rapidly ascend through all the levels of feeling and knowledge that have marked the progress of the race in the past, and emerge at a comparatively early stage in possession of the faculties so painfully acquired by its ancestors.[35]

Joad appealed to the controversial experiments by Paul Kammerer and others in support of Lamarckism, but in fact he was convinced that heredity was more than a physical process. He believed that the unconscious part of the mind survived the death of the body and "is reabsorbed . . . into the universal stream of life from which it took its rise, with the result that life as a whole is continually enriched with the acquisitions made by its individualized units when objectified in matter."[36] This process was responsible for progressive evolution, which left Joad very close to the Bergsonian view of life as a creative force struggling toward higher levels of awareness. But Joad's was a more teleological and theistic viewpoint, as expressed in his concluding allegory of a great commander whose plan of campaign is carried out by poor servants who gradually learn what He wants and how to advance more effectively. When they reach the promised land, they will be released from His service and can "gaze for evermore on the beauty that lies at the journey's end."[37]

Joad's view of survival explicitly ruled out the preservation of individual personality so dear to the spiritualists. He acknowledged that many would not like this aspect of his philosophy, but argued that the universe would be trivialized if its purpose required the preservation of individuals rather than the overall stream of life. He was quite right in his assessment that many would prefer individual survival, and the surge in the popularity of spiritualism after the Great War reveals how great was the public desire to believe in something more than the material world. Along with Lodge, one of the most influential exponents of spiritualism was Sir Arthur Conan Doyle, who pub-

35. Joad, *Matter, Life, and Mind*, 175.
36. Ibid., 157.
37. Ibid., 413.

lished and lectured widely on the topic and engaged in a debate over the reality of psychic phenomena with Harry Houdini. Doyle was trained in medicine and originally absorbed the materialism characteristic of many members of that profession. In 1887 he began to take an interest in psychic phenomena. He dabbled for a year or two in Theosophy, but it was only in the new century that he became convinced that the spiritualists were genuine. He saw the movement as the basis for a new religion that would be not so much an alternative to Christianity as a modernization of its beliefs in the light of new knowledge. The religious significance of spiritualism came not only from its proof of survival, but also from the spirits' teachings about the afterlife. Doyle entitled his first detailed account *The New Revelation*, explaining how we could now be sure that we would all undergo a process of spiritual development in the next world. He believed that the Christian idea of the Fall had to be abandoned in the light of both the biological and spiritualist evidence for progressive evolution—a position very close to that advocated by Lodge.[38] The fact that Doyle subsequently went on to endorse photographs of fairies produced by two girls in 1917 reveals the extent to which he became committed to the reality of the other world—and to the skeptics, the depths of his credulity. His *Coming of the Fairies* argued that the photographs were either the "most elaborate and ingenious hoax ever played on the public" or an epoch-making discovery. He was convinced of the latter, although he admitted that he had been unable to convert Lodge and had responded to a good deal of skepticism in the popular press. The fairies were not spirits, just elemental expressions of the life force, but if their existence were accepted, people would be more willing to accept the truths of spiritualism.[39]

Spiritualism made an impact at all levels of society. Among the upper classes there was continued enthusiasm for various kinds of mystical philosophies with decidedly nonmaterialistic worldviews. Perhaps, as Aliotta claimed, this enthusiasm was a by-product of idealism, although few of those involved would have been exposed to that philosophy at an academic level. Theosophy was still popular; Annie Besant became president of the Theosophical Society in 1907 and gained support—and funds—from many wealthy people through the 1920s, although the movement declined into a crank sect in the following decade.[40] More secretive were occult organizations

38. Doyle, *New Revelation*, 72. For an account of Doyle's conversion, see chap. 1 and also Doyle's *Memories and Adventures*, 99–10, 103–4, 138, and 172–75.

39. Doyle, *Coming of the Fairies*, 41; on Lodge, see 18, and for the "hoax" quotation, 2. See also Brandon, *The Spiritualists*, on some of these later episodes.

40. On the development of Theosophy, see Washington, *Madame Blavatsky's Baboon*. Note that Conan Doyle quotes Theosophist views of "elementals" in his *Coming of the Fairies*, chap. 8.

such as the Order of the Golden Dawn, which attracted not only unsavory characters such as Aleister Crowley but also literary figures as eminent as W. B. Yeats.[41] The influence of these movements in creating a cultural environment inimical to materialism cannot be ignored.

The relationship between these movements and Christianity is complex. Spiritualism was usually presented as a complement, rather than an alternative, to Christianity—at least to a Modernist interpretation of the Christian message. Theosophy was openly hostile to the Christian religion, and its belief in reincarnation was derived explicitly from the East. Yet some Christians were prepared to explore the possibility that reincarnation could serve as the basis for the Modernist belief in the spiritual evolution of the human race. Perhaps our future development took place not in the afterlife, but in successive bodily reincarnations. This view was promoted by the Oxford scholar and later Anglican clergyman Frederick Spencer in his *The Meaning of Christianity* (1912). Spencer anticipated the idea of emergent evolution, writing that "Creation is a great process of unfolding, of emergence of reality, stage by stage, into the state of being with which we are in sensitive communication."[42] But he was convinced that once souls had been created, they were imperishable—hence the need to postulate reincarnation to explain how we would progress to a higher state.[43] Spencer was still arguing in favor of this viewpoint as late as 1935, and again evolution was a key factor:

> Over modern times broods the great conception of evolution, the envisagement of the world as the subject of an immense process of change making for increase of good. Therefore no conception of the future life can be wholly satisfactory that does not set the attainment by individuals of better life in relation to the attainment by humanity of better life, that does not integrate individual development with racial, if not with cosmic, evolution.[44]

Spencer was well aware that the idea of reincarnation was linked to Theosophy, but was willing to use it as part of his own version of Modernist Christianity. Many Christians would have been hostile to such views, but the ap-

41. See, for instance, Howe, *Magicians of the Golden Dawn;* Harper, *Yeats's Golden Dawn,* and Harper, ed., *Yeats and the Occult.*

42. Spencer, *Meaning of Christianity,* 81; he later commented on his anticipation of Morgan's ideas in his "Darwinism and Christianity."

43. Spencer, *Meaning of Christianity,* 358–78. He later argued that Christ would also return to help struggling humanity; see his *Revival of Christianity and the Return of Christ.*

44. Spencer, *Future Life,* 245.

pearance of synthetic beliefs such as Spencer's indicates the complex cultural environment of the early twentieth century. Opposition to materialism and support for the idea of a Supreme Being came from many who did not endorse the traditional framework of religion.

CREATIVE AND EMERGENT EVOLUTION

Nowhere is this point more evident than in the wave of popular enthusiasm for the various forms of progressive evolutionism. The spiritualists used evolutionism to endorse the idea of future development for the individual and the race, but for most enthusiasts, the point of creative evolutionism was to explain how we got to our present state. Whether in the form proposed by Bergson or by Shaw, the conviction that evolution was driven by the creative aspirations of mind or spirit seemed to throw off the leaden feeling of despair engendered by the materialism of the previous century. Many fairly orthodox religious thinkers made use of this idea, coupling it with the hope that the end result was preordained by the Creator who set the whole system going in the first place. Some who had no denominational affiliation also welcomed the more general theism implied by the assumption that there was a moral purpose built into the system. Others—and this would certainly include many Bergsonians and Shaw himself—saw creative evolutionism as an alternative to traditional religion. One did not have to believe in a God who created the world to draw comfort from the belief that in contributing to the upward surge of life, the human race was ennobling both itself and the system that gave it birth.

There were two key distinctions within the progressionist worldview. One was between those who wished to retain some form of teleology representing a divine purpose for the universe and those who worshipped a life force that was blindly struggling upward. So long as a sense of transcendental purpose was retained, human values could be seen as the goal of creation, so that moral values became part of nature itself. Even if it had to be conceded that an element of struggle was necessary to push evolution onward in its lower reaches, it could be argued that the human race was predestined to rise above this struggle to a new level of morality. This was the philosophy of emergent evolutionism, and it pinpoints a second key distinction: whether the human spirit was seen as merely a continuation of what had gone before or as a leap into an entirely new world, transcending even the teleology of creative evolution. For the advocates of emergence, life and mind appeared in the course of time and enshrined new values not visible in what went before. The fact that nature was structured so as to permit or even encourage the emergence of these higher values was a sign that it was divinely created. Here progress had breaks or discontinuities that preserved a unique status for humankind. The problem with this philosophy was that the great upward steps were still to some extent mysterious, and a few popular writers still tried to argue that a supernatural force would have to be involved, very much as the more cautiously liberal Christian theologians did.

In the creative evolutionism proposed by Bergson, however, the life force was all-powerful, even though it struggled irregularly upward with no preconceived plan. The human spirit had to be a product of this force, and there could be no sharp division between human values and those of nature itself. Such a position threatened to undermine traditional morality along the lines conventionally attributed to "social Darwinism." One way out of the dilemma was to insist that nature's own values were those we respect in ourselves, thus denying the Darwinian image of evolution altogether. We have seen how Modernist religious thinkers viewed the life force as an expression of God's creativity. In the most extreme cases, not only did evolution become a moral process, but the whole mechanistic worldview was replaced by an organic description of natural processes, as in the philosophy of Whitehead.

A darker side of creative evolutionism was already visible in Shaw's version of the philosophy (discussed below). Shaw despised materialism and Christianity equally, and he was determined to found a new morality consistent with the claim that we are products of nature's blind struggle toward progress. In this view, moral values were no longer distinct; they were extensions of natural forces and had to take account of the harshness that was embodied in those forces. Shaw's Lamarckian form of creative evolution shows that the worship of struggle and the desire to transcend conventional morality were by no means associated solely with the Darwinian theory he despised. Darwinism might appear to eliminate purpose, but in this version of creative evolution, the individual's drive to transcend the limitations imposed on it and so succeed in the race toward progress became the new source of moral values quite independently of the theory of natural selection. The dangers of this Lamarckian worship of struggle would become all too apparent as the century progressed—but the enthusiasm of many liberal Christians for the harsh social policies of the eugenics movement warns us that there is a spectrum of opinion here, not two irreconcilable opposites.

The interpretation of evolution as a necessary progression had been well established by the late-nineteenth-century thinkers who strove to head off the more pessimistic message hidden within the theory of natural selection. This had been Herbert Spencer's purpose, for all that he was pilloried for the naturalistic bent of his philosophy by the next generation. The Hegelianism of the idealist philosophers encouraged the idea of historical progress, but put an entirely different slant on it, one more compatible with those anti-Darwinian theories in biology that stressed the purposefulness of nature. The sociologist L. T. Hobhouse sought to reconcile the Spencerian and idealist visions in his accounts of progressive mental and social evolution, postulating a linear hierarchy of mental stages marking the increased role of mind as the driving force of evolution. Evolution was orthogenetic, moving in a straight line as though toward a goal, irrespective of the adaptations demanded by natural selection. Hobhouse's *Development and Purpose* of 1913 presented this model of evolution as the maturing of a germ for which provision had already been made, and proclaimed, "God is that of which the highest known embodiment is the distinctive spirit of humanity."[45]

By this time the wave of enthusiasm for Bergson had already reached its peak. The original French edition of *Creative Evolution* had appeared in 1907 and had already attracted much attention across the English Channel before the translation was ready in 1911 (this was also the year in which Bergson made a triumphant lecture tour of England). It was widely recognized that the new philosophy of which creative evolution was a part represented a reaction against materialism and an attempt to revitalize a conservative way of thought more in line with religion.[46] Bergson's vision of an irregular but unstoppable ascent of life driven by the *élan vital* reconfigured the argument from design. He attacked the concept of a goal for evolution, but admitted that the vital impulse gave some degree of unity and coherence to the history of life: "If ... the unity of life is to be found solely in the impetus that pushes it along the road of time, the harmony is not in front, but behind. The unity is derived from a *vis a tergo:* it is given at the start as an impulsion, not placed at the end as an attraction."[47] Bergson cited all the lines of evidence used by anti-Darwinian evolutionists to argue that the ascent of life was less open-ended than the selection theory predicted. He also saw the emergence of the human level of mental powers (if not the human physical form) as an inevitable outcome of the vital force's activity in the branch of the animal kingdom where intelligence predominated over instinct.

Many scientists of the period were inspired by Bergson's new vision—even when they realized that it was not, in itself, science—and so were many religious thinkers. Bergson's leading defender in Britain was a philosopher

45.   Hobhouse, *Development and Purpose*, 371. For an account of Hobhouse's intellectual development, see xv–xx; see also Collini, *Liberalism and Sociology*. See also Hobhouse's *Morals in Evolution* of 1906.

46.   See Grogin, *Bergsonian Controversy in France.*

47.   Bergson, *Creative Evolution*, 103.

teaching at the University of London, H. Wildon Carr. He published a popular account of the new philosophy in 1911 and a more substantial version in
1914. These books stressed that the life force was inherently creative, struggling against the limitations of matter to push itself upward. Carr accepted
Bergson's interpretation of the history of life as separating itself into two
great streams, one that developed the power of instinct and culminated in the
insects, and another that developed the power to learn and innovate and
ended in humankind. The new worldview revealed the destiny of humanity
only in the sense that it confirmed our position as the high point of the more
creative trend, not in the sense that it showed us to be preordained. The whole
point of the Bergsonian philosophy was to get away from the idea of nature
working according to predictable laws. We had regained the life of the spirit,
but not God or immortality.[48] This was the conclusion of Carr's 1911 account,
although in the preface to his later study, he admitted that Bergson himself
had lectured to the Society for Psychical Research in favor of survival—Carr
still thought that the notion of a soul that could be separated from the body
made the whole concept of personality too complex.[49] He now implied that
the life force could be understood as a kind of impersonal God acting
through the world. Such a God was not eternal and timeless, however, nor was
it a loving father, and the only message that could be derived from a contemplation of its activity was the value of freedom.[50]

Carr continued to support a Bergsonian viewpoint through the 1920s.
His *Changing Backgrounds in Religion and Ethics* of 1927 focused on the concepts
of individuality and freedom, but included a chapter titled "God and the
Philosophers," in which he insisted that evolutionary theory had demolished
the traditional theodicies and required a reformation of the idea of God that
would eliminate the idea of His personality. God was again proclaimed as the
life force operating within us, and hence neither omniscient nor benevolent.[51]
But Carr was more positive now about the arguments for survival, although
he admitted that the evidence provided by the psychic researchers was weak.[52]
The philosophers were now like the poets in that they praised imagination at
the expense of reason, and on this basis the vision of Jesus was a major contribution. The goal of evolution was a perfect humanity that would be
reached when the poets created the new Jerusalem and philosophers were

48.   Carr, *Henri Bergson*, 87–91.
49.   Carr, *Philosophy of Change*, vii.
50.   Ibid., 187–93.
51.   Carr, *Changing Backgrounds in Religion and Ethics*, 77–78.
52.   Ibid., chap. 9.

kings. This was hardly a view of the future that an orthodox Christian would endorse, and when Carr was asked to give a series of lectures for a religious foundation at the University of South Carolina, he deftly shifted the parameters of his task by noting that the problem of freedom had always been a central theme of Christian thought.[53]

Carr was not the only commentator to note the tension between Bergson's and Morgan's philosophies, thereby pinpointing one of the key distinctions within the evolutionary perspective. Those who followed Bergson in postulating a single life force thrusting itself blindly upward ran the risk of reducing human spiritual qualities to the harsher values of nature's struggle. By insisting on the element of continuity, they committed themselves to a position in which human values had to be interpreted in terms of nature's laws, even though those laws cover forces that seem amoral. The concept of emergent evolution had always taught that life and mind emerged at certain points in the ascent of life, and Lloyd Morgan himself had admitted that these breakthroughs might be abrupt. This element of discontinuity generated philosophical problems, but allowed the higher values to exist uncontaminated by the processes operating lower down the scale. One standard criticism of emergence by the materialists was that it did not really explain the origin of novel properties and was little more than a new name for divine creation. In this sense, emergence was difficult to distinguish clearly from the view (favored by many Christian thinkers) that a distinct intervention by the Creator Spirit was required at these key points in evolution. Morgan had done his best to argue that the emergent novelties, while unpredictable, were expressions of potentialities always latent within nature. To him, the existence of these potentialities was enough to confirm that there was a purpose built into nature by its Creator. In this way, emergence was used to retain a cosmic teleology—it did not postulate a blind life force because it held that all of nature's activities were established to make possible the realization of the Creator's plan. The unique nature of moral values was maintained within a system that was still governed by law. Evolution was creative, but it did not depend on the element of struggle that was so dangerous within both the Darwinian and Lamarckian perspectives.

Some exponents of creative evolutionism stressed its potential to reform theology. In 1918 Edmund H. Reeman published *Do We Need a New Idea of God?* and answered his own question in the affirmative. God could not be both omnipotent and benevolent, so it was better to see Him as the life-spirit of the

53.   Carr, *Unique Status of Man*, preface.

universe, operating through all things, including ourselves. If there was no separation between God and humanity, the ideas of sin and atonement had to be revised—what was needed was a sense of at-one-ment with God in the struggle to progress.[54] Reeman noted the resemblances between this view of God and that proposed by H. G. Wells and in Campbell's New Theology, but he also appealed to Bergson and to Shaw.[55] Edward Douglas Fawcett's *The World as Imagination* (1916) made frequent references to Bergson to support an idealist vision of the whole universe as an evolving system that was becoming the consciousness of a nascent God.[56] The nonadaptive excesses seen in some branches of the history of life on earth indicated the overexuberance of the life force. Fawcett was sympathetic to Alfred Russel Wallace's belief that there might be sentient beings above the human level who helped to supervise the process.[57] Edmond Holmes, writing in 1921, also argued that the idea of evolution had transformed our idea of God in a way that destroyed the old dogmas of the Catholic Church.[58] Thomas Stephenson reviewed Lloyd Morgan's work in the *London Quarterly Review*, noting its author's insistence that there was a divine purpose revealed in the nisus toward the emergence of higher values.[59] A. Wyatt Tilby, commenting on J. C. Smuts's philosophy of holism, linked it to emergence and criticized Bergson for ignoring the need to see a purpose imposed on evolution from the start. He noted that Smuts himself was inclined to accept Bosanquet's view of the universe as a "factory for souls."[60]

The element of cosmic teleology was clear in the model of emergence presented in Samuel Alexander's *Space, Time, and Deity*, originally his Gifford Lectures for 1916–1918. Alexander was professor of philosophy at Manchester. His philosophy makes difficult reading for anyone but a specialist, and his argument that God will be the final emergent entity to appear at the end of evolution was roundly criticized by orthodox religious thinkers. He also opposed the belief that the mind could survive the death of the body, holding that if this were proved, his whole philosophy of emergence would be undermined.[61] Yet he insisted that his philosophy was closer to theism than to

54. Reeman, *Do We Need a New Idea of God?* chap. 7.

55. Ibid., 24 and 69.

56. Fawcett, *World as Imagination*; see, for instance, 141 for a general statement.

57. Ibid., 548–51.

58. Holmes, "The Idea of Evolution and the Idea of God."

59. T. Stephenson, "Emergent Evolution," 68.

60. Tilby, "General Smuts' Philosophy," 245.

61. Alexander, *Space, Time, and Deity*, vol. 2, 424.

pantheism, and that the concept of a nisus toward perfection gave it a distinctly teleological character. Our own desire for righteousness gives us faith in the moral character of the nisus of the universe and the certainty that eventually the highest values will be expressed in God: "God is the whole universe engaged in process towards the emergence of this new quality, and religion is the sentiment in us that we are drawn towards him, and caught in the movement of the world toward a higher level of existence."[62]

One of the most influential developments in the move to see the universe in teleological terms was the philosophy of organism proposed by Alfred North Whitehead in the later part of his career. He became professor of applied mathematics at Imperial College, London, then moved to Harvard as professor of philosophy in 1924 (although he was already in his sixties by then). Whitehead had worked with Russell on the foundations of mathematics. He never shared Russell's unconventional social opinions, but had been a convinced agnostic until his son, Eric, was killed in action in 1918. Russell claimed that Eric's death made Whitehead want to believe in immortality, and that from this point on he was determined to find a way of transforming religion to make it acceptable in the modern world.[63] But to accomplish this, science, too, had to be transformed so that it shook off its commitment to a mechanistic worldview. Whitehead's approach—which followed a prediction made by J. S. Haldane—was to apply essentially organic categories to the whole universe. In this he was inspired by Bergson and by Carr (he had interacted with the latter in the Aristotelian Society), but wanted a far more systematic approach in line with Alexander's philosophy of emergence. He was also impressed by field physics, with its technique of explaining events in terms of the transfer of energy. He became convinced that the universe must be portrayed in terms that made it seem conformable to human values, and he sought to give the whole a purpose by portraying it as a movement toward a preordained spiritual goal. His most detailed statement of his philosophy of organism was *Process and Reality*, based on his Gifford Lectures at Edinburgh in 1927–1928. But both the lectures and the book were highly technical and required new terminology—we have already noted that Joad found it unintelligible. Whitehead was not a gifted lecturer, and by the time his series was finished the audience was down to about half

62.   Ibid., vol. 2, 429.

63.   The best study of this period in Whitehead's career is volume 2 of Victor Lowe's *Alfred North Whitehead*; on Russell's comment, see 188. Other studies of Whitehead's thought include Eisendrath, *The Unifying Moment*; Emmet, *Whitehead's Philosophy of Organism*; and W. Mays, *Philosophy of Whitehead*.

a dozen. The book itself did not sell well in Britain, although the more accessible *Science and the Modern World* of 1926 had been a great popular success.[64]

Science and the Modern World offered an interpretation of the history of science that traced the origins of the mechanistic viewpoint and tried to show that the time was now ripe for a new approach to nature. Whitehead insisted that the future course of history would depend upon the working out of a new relationship between science and religion.[65] Vitalism was a blind alley, since it left the mechanistic view of matter intact and appealed to a separate vital force. The real challenge was to replace mechanism with a new view of the primary entities in nature. If this could be done, the conflict between religion and science would move into a new, more constructive phase: "A clash of doctrines is not a disaster—it is an opportunity."[66] But if science were to be transformed, it was equally necessary for religion to become purified in its search for something beyond the world of the senses.

The latter point was taken up in *Religion in the Making*, also published in 1926, but it was *Process and Reality* that addressed the technical issues involved in the reformulation of science and attempted to relate the new philosophy to Whitehead's own religious views. The technicalities are outside the scope of this study, but Whitehead's broader aims are more directly relevant. He wanted a cosmology based on a system of ideas that would include aesthetic, moral, and religious values as well as the concepts of natural science.[67] Plato's *Timaeus* was held up as the foundation for the kind of vision of cosmic order that was now being articulated by modern physics.[68] The concepts of flux and order were both essential, with flux being necessary to allow for the emergence of novelty within an ordered world. Molecules now had to be treated as social entities, with living organisms as a higher grade of entities with the capacity for thought. In the conclusion, Whitehead charged orthodox philosophy with being too narrow: "Philosophy may not neglect the multifariousness of the world—the fairies dance, and Christ is nailed on the cross."[69] But the idea of God as the aboriginal, transcendent Creator was a "fallacy which has infused tragedy into the histories of Christianity and of Mahometanism."[70] God should be seen as the personification of moral energy;

64. On the reactions, see Lowe, *Alfred North Whitehead*, vol. 2, 250–57.
65. Whitehead, *Science and the Modern World*, 224.
66. Ibid., 230.
67. Whitehead, *Process and Reality*, vi.
68. Ibid., part 2, chap. 3.
69. Ibid., 477.
70. Ibid., 484.

His existence cannot be proved, although a certain rendering of the facts is suggestive. In the end, it is clear that Whitehead's God transcends all the traditional categories, emerging as an expression of the dynamic, creative, and ultimately moral order of the world.

Philosophical reviews of *Process and Reality* were mostly polite, but few faced up to the challenge of creating a new understanding of nature. R. G. Collingwood took the most positive view, linking Bergson and Whitehead as key figures in the emergence of a new, more historically oriented view of reality.[71] Scientists, too, were mostly indifferent, with a few significant exceptions such as Needham and C. H. Waddington. Religious writers, as we have seen, were more enthusiastic—it is doubtful that they understood the details, but they were ready to welcome what they perceived as a new contribution to the assault on the mechanistic view of the world. The theory of emergence and the idea that nature itself was driven by nonmechanistic forces were both key elements in the Modernist version of Christianity and in the more mystical versions of popular writing on spiritual issues. In later decades Whitehead's philosophy was an important stimulus for the founders of process theology. But the distinction between the claim that nature is the unfolding of a divine purpose through the activity of spiritual forces and the Bergsonian concept of the ever-struggling life force is not easy to maintain. As we shall see below, Wyndham Lewis put Whitehead, along with Bergson, in the camp of those who promoted a darker vision of progress through struggle. The whole spectrum of progressive evolutionism had become a murky topic. At first sight, it seemed to offer hope to those seeking to reconcile science and religion. But the modernized version of Christianity required for the reconciliation was impossible for many traditionalists to swallow, and for them, the totally un-Christian "religion" advocated by Shaw was a prime illustration of the dangers of going down this route.

EVOLUTION AND THE HUMAN SPIRIT

Several popular writers promoted a mystical vision of evolution driven by spiritual forces, but stopped short of allowing those forces to create the human soul. Although not explicitly writing within the framework of Christian theology, they sensed the dangers that many Christians saw lurking in the argument that we are nothing but the products of nature. They felt that the intervention of an external or supernatural force was needed to give the human spirit its unique status, even though the rest of nature exhibited creativ-

71.   Collingwood, *Idea of Nature* (published posthumously in 1945 but written in the 1930s).

ity on a lesser scale. One exponent of this position was the "well-known poet, mystic and philosopher" Ronald Campbell MacFie (the hyperbole is from the dust jacket of his *Theology of Evolution*). In fact, MacFie had medical training, served on a medical selection panel during the Great War, and first gained attention through his comments on the dysgenic effects of the war in killing off the best samples of humanity.[72] His *Metanthropos* of 1928 countered the eugenicists' pessimistic assessments of degeneracy and predicted the emergence of a more moral type of humanity, although it conceded that evolution itself was unpredictable precisely because it worked through the emergence of novelty.[73] In his more detailed evaluations of evolutionism, however, MacFie presented it as a purposeful process in which chance played no role. His *Science Rediscovers God* (1930) stressed neovitalistic biology and non-Darwinian evolutionism, appealing to the writings of Morgan, Thomson, and others to support the claim that evolution progressed through an inner drive. The rejection of materialism by the physicists, including Lodge, Jeans, and Eddington, was noted as a further indication that science revealed a purposeful Mind behind the universe.[74] Frances Mason recommended this book to Morgan, and asked him whether MacFie was the sort of writer who should be invited to contribute to her volumes expounding the new natural theology.[75]

All these points were developed in MacFie's most ambitious work, his *Theology of Evolution* of 1933. Here, however, he made it clear that the purposeful force behind evolution did not arise from within the material universe. The origin of life itself, and of the metazoans (multicellular organisms), could not be explained in natural terms.[76] The selection theory was plainly inadequate as an account of progressive evolution, and indeed, the real puzzle was how that theory had managed to "obtain and maintain such a hold not only upon the popular imagination, but on most of the best minds of the latter half of the last century."[77] Now the wave of non-Darwinian thinking had swept it aside to reveal the true purpose underlying evolution. MacFie believed that evolution was driven by dramatic mutations—but he claimed that the really important steps forward could not be explained in materialistic terms. Here again we are "in the presence of transcendental mystery."[78] In the

72. See Crook, *Darwinism, War, and History*, 169–71.

73. MacFie, *Metanthropos*; on evolution, see 8–11.

74. MacFie, *Science Rediscovers God*, 268.

75. Frances Mason to Morgan, 22 February and 19 April 1931, Lloyd Morgan papers, Bristol University Library, DM 128/474 and 490.

76. MacFie, *Theology of Evolution*, chaps. 2 and 3.

77. Ibid., 93.

78. Ibid., 141.

higher reaches of the history of life, however, we can see how the process works—mind itself can affect the material substances of the body and can thus direct how mutations should appear. This was what MacFie called "mento-mutation," although here, too, he subordinated the mental influences to an overarching trend: everything "points to a Prescient Cause *acting independently of the ordinary material causal linkages, and capable of itself arranging new molecular causal colligations and linkages.*"[79] Even so, he could not believe that the normal forces of evolution were capable of producing the great leap to the human mind. As in the case of the origin of life itself, some direct act of the Supreme Mind must have been involved. It was a scandal that schoolchildren were being taught that we are just improved apes, and the Daytonists (American Fundamentalists) and Roman Catholics were to be supported in their opposition to this.[80]

MacFie's theology fused elements of creative evolution, emergent evolution, and outright creationism into what he, at least, thought made a coherent whole. By the time *The Theology of Evolution* appeared, its author was dead, although another posthumous work also proclaimed the *Faiths and Heresies of a Poet and Scientist* on this topic. MacFie had indeed been a poet, and some of his verses expounded his vision of evolution. One poem was dedicated to J. Arthur Thomson, while another, "Chance," proclaimed the necessity for seeing a purpose behind the giddy whirl of nature's activity. Its last stanza must serve as an illustration of MacFie's talents in this area:

> Then though it seem a whirl of Chance
> This ever-changing atom-dance
> Of dust and wind,
> Of brain and mind,
> Of variant and circumstance,
> Yet there must be a Power behind,
> Loving and strong, not deaf and blind.[81]

A different approach was taken by J. Parton Milum in his *Evolution and the Spirit of Man* of 1928, based on a doctoral thesis for the Arts Faculty of the University of London. Milum shared the widespread distrust of Darwinism, which he saw as the end product of a materialistic trend in philosophy. This

79. Ibid., 193, MacFie's italics.

80. Ibid., 286.

81. *The Complete Poems of Ronald Campbell MacFie*, 342; the poem dedicated to Thomson is "Simple Beauty and Nought Else," 330–36.

trend was opposed by Hegelian idealism, but that approach had ignored the study of the natural world until Bergson had shown how to treat life as an active force.[82] Evolutionary theory was now deadlocked between neo-Darwinism and neo-Lamarckism, neither of which was adequate, while emergent evolution was merely a name for the process of "becoming." An acceptable synthesis would have to include a role for life and mind, as suggested by J. S. Haldane, E. S. Russell, and others. The important thing was to distinguish a main line in evolution that led to a steady increase in mental powers without adaptive specialization. The human species was the end product of this line, so evolutionism now "restores to man his unique and transcendent position" in the world.[83] The human mind was an awakening of "the spirit which has lain dormant in living matter through geological ages." Yet, like MacFie, Milum could not bring himself to believe that the soul was a product of normal mental evolution. There must have been a "critical point" when the first humans appeared, and although some might think this suggestion betrayed "a lurking desire to find a place for special creation, and to magnify the uniqueness of man," it was necessary to acknowledge the evidence suggesting that something new had appeared with humanity.[84] Milum rejected the anthropologists' account of the evolution of religious beliefs, suggesting that our earliest ancestors had already acquired the basic capacity needed to recognize the hand of the Creator. Christianity held a germ of truth, even if the figure of Christ was obscured in history, because it was indeed necessary for humanity to be remade to uncover its full potential. Milum's book caught the attention of Lodge, who wrote a favorable review indicating that his own thoughts ran along similar lines.[85] Ten years later Milum published a short article stressing the need for science to take note of human consciousness, in which he placed greater emphasis on the Christian story.[86]

These evolutionary theologies reveal the bewildering variety of ways in which the idea of an inherently creative evolutionary process could be developed. But the reluctance of thinkers such as MacFie and Milum to see the human spirit as merely the last product of evolution's upward striving illustrates a serious difficulty encountered by all who sought to subsume humanity into nature as a whole. Even without the Darwinian selection theory, it

82. Milum, *Evolution and the Spirit of Man*, 25.

83. Ibid., 101. This resembles the position supported by Robert Broom, although his work is not mentioned.

84. Ibid., 101 and 125.

85. Lodge, "Philosophy of Evolution."

86. Milum, "The Science of Man and of God."

was hard to believe that the history of life was the neat, goal-directed process implied by the assumption that the human mind is the goal of a divine plan. Creative evolution itself stressed an element of unpredictability in the process and conceded that many branches of the tree of life were ultimately barren. Worst of all, by portraying the human spirit as a product of nature, the theory made it hard to see how the moral faculties could have emerged from a process that still, in a sense, depended on struggle. That was why some religious thinkers—and not just those committed to a traditional Christian perspective—invoked mysterious jumps or saltations to lift the human spirit at one bound above the morass of evolution. Their fears were in part a reaction against a harsher vision of creative evolutionism, which positively welcomed the element of striving and struggle implied by the theory's rejection of a morally significant goal preordained by the Creator. For all that they despised the trial and error mechanism of natural selection, some creative evolutionists were as committed as any social Darwinist to the belief that progress could be achieved only through struggle—because struggle allowed those whose upward striving was most effective to dominate the whole process. The best-known advocate of this harsher vision was George Bernard Shaw.

Progress through Struggle

Shaw had already begun to stress the creativity of the life force before Bergson published, and he had no compunction in taking over the term "creative evolution" for his own worldview—which he saw as a new religion capable of replacing the old. Bertrand Russell described an encounter in which Shaw told Bergson to his face what he ought to mean by the term.[87] In effect, Shaw revived the old psycho-Lamarckism of Samuel Butler, the claim that living organisms played a creative role in the evolution of their species through their conscious choice of new behavior patterns, the results being reflected in their bodily structure through use-inheritance. As early as 1906 he had begun to attack the soullessness of the Darwinian selection theory, writing that "if it could be proved that the whole universe had been produced by such selection, only fools and rascals could bear to live."[88] In the same year he included comments on evolution in a speech at the City Temple, at which he was introduced by R. J. Campbell, and later gave a talk linking his views to

87. B. Russell, *Portraits from Memory*, 73–74. The four-volume biography by Michael Holroyd, *Bernard Shaw*, says disappointingly little about his evolutionism; see vol. 3, 40–41.

88. Shaw is quoting his own words in the preface to *Back to Methuselah*, liv. He does not indicate the original source, and I have been unable to trace it, although one possibility is a lecture on Darwin to the Fabian Society in 1906, the syllabus of which is listed in Laurence, *Bernard Shaw: A Bibliography*,

George Bernard Shaw. By Harry Furniss.
By courtesy of the National Portrait
Gallery, London.

Campbell's New Theology.[89] The two men were obviously close—each
praised the other's contributions to the transformation of belief, and Shaw
later admitted that Campbell had convinced him that Jesus was a real person.[90] But where Campbell rapidly backed away from the impersonal God of
the New Theology, Shaw continued to use creative evolutionism as the basis
for a new religion. His campaign had already begun in his *Man and Superman*,
which hailed the creativity of the life force and proclaimed humanity as the
vehicle by which that force took conscious direction of its activities. In the
preface to the cheap edition of the play published in 1911, Shaw pointed to

vol. 1, item A74 (the only location given is Laurence's private collection). These words do not occur in
Shaw's "Religion of the British Empire" of 1906.

    89.   Shaw, "Religion of the British Empire," reprinted in his *Religious Speeches*, 1–8; Shaw, "The
New Theology," reprinted in *Religious Speeches*, 9–19.

    90.   Shaw, "Modern Religion I," reprinted in *Religious Speeches*, 38–49, see 41; for other comments
about Campbell, see *Religious Speeches*, 25 and 55.

act three as "a careful attempt to write a new Book of Genesis for the Bible of the Evolutionists," making it clear that this was the new evolutionism that had thrown off the materialism of the selection theory.[91] In the same year he gave a talk on "The Religion of the Future" to the Cambridge Heretics, which attracted widespread and critical comment in the press for its open attack on the Christian view of God and led to a public debate with G. K. Chesterton. Here he proclaimed that "the universe was being driven by a force that might be called the Life Force ... it was performing the miracle of creation [and] had got into the minds of men, as what they called their will." Curiously, although Shaw vilified natural selection for its destruction of the sense of purpose in the world, he accepted that the life force advanced only through trial and error; this made sense because individual purpose was short-sighted and did not always move in the best overall direction. He told his audience "that they could imagine something—the Life Force—beginning in a very blind and feeble way at first, first laboriously, achieving motion, making a little bit of slime to move and then going on through the whole story of evolution building up and up until at last man was reached."[92]

Because the life force effectively replaced God in the new religion, anti-Darwinian evolutionism remained central to Shaw's position. At one time, he claimed, it had been impossible to criticize Darwin without reproof, as Butler had found out, but by the 1920s the anti-Darwinian trend was running full-spate (in the literary world, at least), and Shaw devoted much of the preface of his *Back to Methuselah* of 1921 to a full-scale attack. He described the selection mechanism as a process that opened up a gulf of despair—yet Darwin had been welcomed as a deliverer because he got rid of the arbitrary God depicted by the old argument from design. Science became the true savior of humankind, and only now were we beginning to realize that in the case of Darwinism, it had led us astray.[93] Shaw admitted that neither natural selection nor creative evolution could be disproved, but if someone tells you that you are a product of selection, "you can only tell him out of the depths of your inner conviction that he is a fool and a liar."[94] For Shaw, as, in a sense, for Carr, identifying ourselves with the creative life force became the basis for a new religion because it gave human life a cosmic purpose once again. Cre-

91.   Shaw, "Foreword to the Popular Edition of *Man and Superman*," in *Bodley Head Bernard Shaw*, vol. 2, 531–32. For relevant passages from act 3 of the play, see 662–63 and 684–85 of this edition.

92.   Quotation from the account of the meeting in the *Cambridge Daily News*, reprinted in the pamphlet *The Heretics: Mr. Bernard Shaw*, 5–6; this account is reproduced in Shaw, *Religious Speeches*, 29–37. See Holroyd, *Bernard Shaw*, vol. 2, 215–21.

93.   *Bodley Head Bernard Shaw*, vol. 2, xl–xlii.

94.   Ibid., xlviii.

ative evolutionism would become the new religion of the twentieth century, rebuilt on a scientific basis, with the artists as its prophets.[95] In his play *Too True to Be Good*, published in 1934, Shaw presented Einstein as the agent who had destroyed the deterministic universe of Newton, with Darwinism completing the job of convincing us that we live in a totally capricious universe.[96] Creative evolutionism restored a sense of cosmic purpose, but the fact that the life force worked by trial and error meant that it offered no guidance in decision-making and hence no basis for morality. Shaw was certainly not offering a worship of brute force—he believed that progress would be in the direction of higher morality, and he wanted to ensure that everyone was freed from the restrictions imposed by poverty. But his attacks on the hypocrisy of traditional morality led him to endorse the flouting of convention by the men and women of genius who would point the way to the future, and this could all too easily be identified with the sort of contempt for common humanity associated with the name of Nietzsche.

Shaw exerted immense influence over the generation that came to maturity in the first decades of the twentieth century, as witnessed in a tribute by Joad.[97] He was a member of the Fabian Society and worked closely with Wells when he too was a Fabian (although Shaw was deeply involved when the Society eventually got rid of Wells). Shaw's Lamarckian evolutionism got a certain amount of attention in literary circles, and of course there were still scientists such as E. W. MacBride to whom he could appeal for support. An article in the *Review of Reviews* in 1932 linked Shaw's ideas with those of MacBride and contrasted them with J. B. S. Haldane's hostility to Lamarckism. The article concluded that religion had nothing to fear from a science that, as in the case of Julian Huxley's, sought to explain evolution without destroying the sense of mystery in the world.[98] The *English Review* carried a hostile evaluation of the Darwinism presented in Haldane's *Causes of Evolution*.[99] The religious writer J. C. McKerrow promoted Lamarckism in a 1937 book, which concluded with a mystical evaluation of humanity's role in the cosmic process.[100]

The most bizarre application of the idea of creative evolution came from Sir Francis Younghusband, the eminent soldier and explorer. Here the

95. Ibid., lxxviii–lxxix.
96. Shaw, *Too True to Be Good*, 84–85.
97. Joad, *Shaw*.
98. "Religion and Evolution."
99. Heseltine, "Professor Haldane and Evolution."
100. McKerrow, *Evolution without Natural Selection*.

potentially dangerous implications of the worship of a struggling life force became only too obvious. Younghusband had gained public attention with his best-selling *Heart of a Continent* of 1896 and had led a British military expedition that occupied Tibet in 1904. He retained an interest in Eastern philosophy and Theosophy, eventually dismissing Christianity as a religion fit only for a subject race. After a serious injury in a road accident he rejected faith in a benevolent God and began to see a creative life force as a way of imposing a purpose on the cosmos without denying the reality of struggle and suffering, a view he claimed to have derived from Bergson and the idealist McTaggart.[101] The concluding pages of *The Heart of a Continent* had already talked of cosmic spiritual evolution and the possibility of superior races among the stars. The latter theme was developed in his *Life in the Stars* (1927), which drew on creative and emergent evolution to promote the idea that higher intellects could evolve elsewhere in the universe. It also insisted on the importance of great leaders and the need for *esprit de corps* among their followers, hinting that the leaders of humanity were receiving influences from the higher powers in the stars by something akin to radio.[102] In 1933 Younghusband gave a Hibbert Lecture titled "The Destiny of the Universe" and followed this up with a book titled *The Living Universe.* Here the astronomers' prediction of a heat-death for the universe was contrasted with the biologists' insistence on the creativity of evolution. But evolution was a moral process generating love, and eventually a "world-love" experienced through mysticism. (Younghusband also advocated free love and was abandoned by some of his more respectable friends.) Human spiritual evolution was driven by those more advanced spirits who led the rest along the path toward a true community of minds. The spiritual forces directing evolution were at work throughout the universe, producing life and intelligence on many different planets, and here again there was a role for the leaders. Younghusband believed that in each period of cosmic history one race (currently the one inhabiting the planets of the star Altair) supervises the activity of the world spirit throughout the whole universe, broadcasting its influence like radio waves to be picked up on other planets.[103] The fittest races respond most actively to these progressive forces, and each may hope to evolve eventually into the dominant spiritual force of the universe. Here the idea of a creative life force is transformed into a kind of cos-

101.   Younghusband, *Mutual Influence,* x.; for the story of his accident and subsequent change of heart, see his *Within: Thoughts during Convalescence.* On his career, see French, *Younghusband: The Last Great Imperial Adventurer.*

102.   Younghusband, *Life in the Stars,* chap. 13; on *esprit de corps,* see chap. 10 and epilogue; see also *Heart of a Continent,* 390–91.

103.   Younghusband, *Living Universe,* 212–13.

mic Fascism, ideally suited to the ideology of empire—Baden-Powell, founder of the Boy Scouts movement, liked *Life in the Stars*. Few others took the books seriously, however, and Younghusband's publishers wished he would stick to writing about his travel experiences.[104] Yet he cannot be dismissed as a total crank—he contributed an article to that classic exposition of the new natural theology, Frances Mason's collection *The Great Design*, and gave papers at the Modern Churchmen's Union meetings of 1927 and 1931.[105] He also founded the World Congress of Faiths to promote dialogue between the various religious traditions.

Younghusband's ideas reveal the extreme range of a spectrum centered on the new natural theology and the ideology of progress. The Modernists within the Christian churches and scientists such as Lodge, Thomson, and Morgan had all sought to preserve an image of the human race as both the product and now the supervisor of cosmic spiritual development—almost always with a sense that some were fitter to do the supervising than others. Most of them tried to present the system as a modernized foundation for traditional values, which is why it could be taken as an extension of orthodox Christianity by its less radical exponents. Like Lodge and the Anglican Modernists, they thought that their philosophies restored a sense of the importance of human values while remaining true to the latest teachings of science. But Bergson, Shaw, and Younghusband reveal the harsher side of this way of thinking, as does the enthusiasm for eugenics among many Christian Modernists. Shaw's *Man and Superman* had already raised the prospect of those at the cutting edge of the life force's advance having to take things into their own hands in order to transcend the petty restrictions of bourgeois life. Younghusband merely carried this viewpoint to its logical conclusion.

These implications were not lost on the movement's critics, one of the most perceptive being the avant-garde artist and writer Wyndham Lewis, whose *Time and Western Man* of 1927 launched a blistering attack on the whole effort to see nature in spiritual and organic terms, from Bergson to Whitehead. He saw the ideology of progress as something that necessarily led to a sense of some being superior to others, and having the right to dominate them.

From Darwin to Mussolini . . . is a road without a break. Bergson's 'creative evolution' is as darwinian as the 'will to power' of Nietzsche. It is

---

104. On the reactions, see French, *Younghusband*, 286, 321–25.

105. Younghusband, "Mystery of Nature"; see also A. M. G. Stephenson, *Rise and Decline of English Modernism*, 215–17.

Darwin's law of animal survival by ruthless struggle, and the accompany-
ing picture or the organic shambles through which man reached world-
mastery; broadcast through the civilized democratic world, it has brought
in its wake all the emotional biology and psychology that has resulted in
these values, for which fascism is the latest political model.[106]

Lewis had no time for those who pretended that the drive toward progress had generated the traditional Christian virtues. Many supporters of the
new natural theology would have been horrified by the claim that their enthusiasm for "pushing ahead" could result in an ideology that worshipped
brute force. Yet we have seen enough of the more muscular understanding of
Modernist Christianity to appreciate that the link perceived by Lewis is not
totally fanciful even in the case of ostensibly less radical thinkers. Some certainly seem to have appreciated the dangers, and this may explain the reluctance of Milum and MacFie to accept that the final step up to the human
mind was a typical product of the purposeful evolutionary process. This fear
of subsuming humanity completely within nature may—as Milum admitted—suggest a sympathy for the position of those who rejected evolutionism outright. It was one thing to make nature itself purposeful, quite another
to see humans as merely by-products of the great upward struggle. Those
who sought to keep humanity free from even a progressive vision of evolution were, in one sense, trying to run with the evolutionary hare and hunt with
the traditionalist hounds. But they sensed that to see moral values as nothing
but an application of nature's laws was to threaten a dehumanization that
would end in nightmare. By appealing to a conventional theism in which the
Creator Spirit was both immanent and transcendent, both the driving force
of evolution and something capable of stepping in from the outside to raise
it to new heights, they could keep the unique status of human values safe
from the threat that Lewis uncovered so effectively in his critique. Better to
worship a Creator Spirit who imposed purpose from outside than a life force
that made it impossible to separate human actions from the struggle for existence.

THE CHRISTIAN RESPONSE

Shaw's influence on the intellectual community was immense during the
first three decades of the twentieth century, although it declined in the 1930s.
When G. K. Chesterton entered the debate in 1911 to defend Christianity

---

106. Wyndham Lewis, *Time and Western Man*, 215.

against Shaw's attacks, he was regarded as an anachronism, out of touch with the new skepticism that demanded that religion be rejected or transformed beyond recognition. But a few writers of influence joined Chesterton in try-ing to stem the tide of skepticism, and by the 1930s they were increasing, both in number and in the effectiveness of their appeals for a return to Christian beliefs and values. Some moved closer to the traditional heart of Christian-ity—both Chesterton and Ronald Knox converted from Anglican to Roman Catholicism. Others converted from outright skepticism to the Anglican Church, including T. S. Eliot and C. S. Lewis. Eliot's poetry of the early 1920s was seen as a classic expression of humanity's alienation from a purposeless universe, but he soon began to dissociate himself from this interpretation and finally accepted the Christian message.[107] His *Murder in the Cathedral*, first per-formed in Canterbury in 1935, offered not only a defense of the church but also a veiled warning about the rise of Fascism. Much of the literature pro-duced by these writers reflects indifference to science rather than outright hostility, but in some cases it responds to the use of scientific arguments by the modernizers, and thus forms part of our story.

Hilaire Belloc stood out as the most vocal defender of Christian values. His opposition to H. G. Wells escalated into a personal confrontation that left the latter feeling bitter and humiliated. For Belloc, Wells was the symbol of a provincial, small-minded materialism that was undermining the great tradition of European spirituality represented by the church. His attacks on the Darwinism of the introductory chapters in Wells's *Outline of History* were but a prelude to his critique of the book's shallow misunderstanding of the church's origin and role in the creation of Western civilization. Battle was joined in 1920, with Belloc attacking in both the *London Mercury* and the *Dublin Review*.[108] The former article dismissed Wells's support for the theory of nat-ural selection as a "howler" indicating how out of touch he was with scien-tific opinion. This was to become the litany that Belloc would repeat on many subsequent occasions: convinced that the selection theory had been disposed of, Belloc tried to present Wells as out of touch with what was really going on in science. As he wrote in his *Companion to Mr. Wells' Outline of History*:

> To trot out Natural Selection at this time of day as the chief agent in Evolution is almost like trotting out the old dead theory of immutable and simple elements in a popular chemistry.

107. See Pearce, *Literary Converts*, and for a divergent opinion on Eliot, Medcalf, "Cries in the Wilderness."

108. Belloc, "Mr. Wells' *Outline of History*," and "A Few Words with Mr. Wells." See Speaight, *Life of Hilaire Belloc*, 397–403, and Coren, *Invisible Man*, chap. 7.

When Driesch said, twenty long years ago, "Darwinism is Dead" he was hardly premature.

To quote him now is to repeat a commonplace.[109]

Belloc insisted that Wells had missed the main thrust of modern biology visible in the work of Continental scientists such as the morphologist Louis Vialleton. Significantly, Vialleton was arguing not just against the selection theory but against any form of continuous evolutionism; for him, as for Belloc, the main types of animal life visible both today and in the fossil record were distinct from one another.[110] If evolution occurred, it was by drastic saltations, a point repeated by Belloc in his 1927 controversy with Arthur Keith and again in his *Crisis of Our Civilization* in 1937.[111] Selection was self-evidently not a creative force, and was merely a product of the materialists' desire to eliminate Mind from nature. J. B. S. Haldane subsequently cited Belloc's assertion that Darwinism was dead as the target against which his own defense of the theory was launched (and was criticized by Belloc in turn for his misunderstandings of Catholic doctrine on the Immaculate Conception and the virgin birth).[112] Belloc insisted that he was not against science itself, only against scientists who extended their thinking outside its proper realm. Science was concerned only with the details of nature, not with achieving an overview. For this reason, *"Anyone can, with patience, do scientific work,"* while great thinkers attend to the serious questions of philosophy and theology.[113] When this attitude is coupled with Belloc's failure to understand what was actually going on in biology, one can hardly blame the scientists for thinking that the literary world had turned its back on any attempt to take the study of nature seriously.

Belloc remained active in the 1930s, by which time his warnings about the consequences of Western civilization abandoning its Christian roots were beginning to sound more convincing. In a chapter in the collection *Science in the Changing World* (1933), he insisted on the need to defend the freedom of the

109.   Belloc, *Companion to Mr. Wells' Outline of History*, 10–11.

110.   Ibid., 21–26; see, for instance, Vialleton, *L'origine des êtres vivants*. Readers of my *Eclipse of Darwinism* will find no reference to Vialleton because his name was not mentioned by any of the anti-Darwinian scientists I studied there—perhaps some reflection of the growing gulf between the morphological tradition and the rest of evolutionary biology in the early twentieth century.

111.   Belloc, "Is Darwinism Dead?" and *Crisis of Our Civilization*, 183–84.

112.   Haldane, "Darwinism To-Day," in his *Possible Worlds*, 27–44, see 27; Belloc, "An Article of Mr. Haldane's," in Belloc, *Essays of a Catholic Layman in England*, 267–83. See McOuat and Winsor, "J. B. S. Haldane's Darwinism in Its Religious Context."

113.   Belloc, "Science as an Enemy of the Truth," in *Essays of a Catholic Layman in England*, 195–236; see 226 (Belloc's italics).

G. K. Chesterton (seated, left) and Hilaire Belloc (seated, right). By James Gunn, 1932. The standing figure is the journalist and author Maurice Baring. By courtesy of the National Portrait Gallery, London.

will against governments that used technology as a means of coercion.[114] In his *Crisis of Our Civilization* (1937), he presented Marx and Darwin as twin examples of the consequences of materialism, with the Soviet Union as a clear warning of where their influence was pointing. Unfortunately, his opposition to the Communists led him to support the Franco regime in Spain, hardly a position calculated to inspire respect among those who were becoming even more concerned about the growing influence of Fascism.

G. K. Chesterton had already emerged as a critic of materialism and Darwinism before his conversion to Rome in 1922, and his later assaults on science were seen as a continuation of Belloc's war against Wells and the secularists. Chesterton had originally come under the influence of Gore and the Anglo-Catholics. His essay "Science and Religion" proposed a vision of science almost as simpleminded as Belloc's: "Physical science is like simple addition: it is either infallible or it is false." He went on to attack progressionist evolutionism because it ignored the Fall. Evolution showed examples of physical degeneration, so why could there not be moral Falls? A sense of de-

114.   Belloc, "Man and the Machine."

pravity was, in any case, a product of spiritual conviction, not of physical origins: "If a man feels wicked, I cannot see why he should suddenly feel good because somebody tells him that his ancestors once had tails."[115] His most influential work after his conversion, *The Everlasting Man* (1925), continued this theme. Chesterton believed that Wells and the evolutionists were mistaken in thinking that by making the origin of life and of mind continuous they somehow made them more intelligible: whatever humanity's physical resemblances to the animals, the soul was something new: "Man is not merely an evolution but rather a revolution."[116] There followed a critique of the popular notion of the primitive "cave man" who was still struggling to develop morality and spirituality. In effect, the sciences of paleoanthropology and prehistoric archaeology were swept away as figments of the imagination based on a few scraps of stone and bone. The first humans were already fully human, and already civilized—with the problems of despotism and oppression showing our fallen state. As late as 1935 Chesterton attacked Darwinism in the pages of the *Illustrated London News*, treating it as a product of Victorian extremism and praising Belloc for revealing to the public how scientists had now abandoned the selection theory.[117] He was also a noted opponent of eugenics.[118]

There were other converts to Roman Catholicism who were active in defending Christian values in popular writings. Ronald Knox converted from Anglicanism and became an influential Roman Catholic chaplain at Oxford. His *Caliban in Grub Street* (1930) exposed the shallowness of much popular writing and journalism on vaguely religious topics. Younghusband's cosmic evolutionism was singled out for attack as mere pantheism.[119] Knox professed surprise at how little influence spiritualism had gained, given the effect of the Great War and the popularity of similar fads. He ridiculed Conan Doyle's belief that animals might have souls capable of survival, and asked if he was the only one to feel that it was all too good to be true.[120] The suggestion that there might be a pool of spiritual life into which we were absorbed at death was seen as less attractive—Knox quipped that if this was survival, then he had achieved immortality if his latest articles were pulped and turned up as

115.   Chesterton, "Science and Religion," in his *All Things Considered*, 187–93; see 189–90. On Chesterton's life and work, see Dale, *Outline of Sanity*.

116.   Chesterton, *Everlasting Man*, 23.

117.   Chesterton, "About Darwinism," reprinted in his *As I Was Saying*, 194–99.

118.   Chesterton, *Eugenics and Other Evils*.

119.   Knox, *Caliban in Grub Street*, 66–67, 75–78; see Waugh, *Life of the Right Reverend Ronald Knox*; Morris, *Mgr Ronald Knox*.

120.   Knox, *Caliban in Grub Street*, chap. 11.

next month's *Bradshaw*.[121] Two years later his *Broadcast Minds* attacked a number of writers who had tried to popularize the new materialism, including Bertrand Russell, Julian Huxley, and John Langdon-Davies. If a religion without God came to Britain, said Knox, it would come from the East, not from Huxley & Co.[122] Langdon-Davies had attacked Christianity via Aristotle, and Knox parodied his argument thus:

(i) Euclid was an old Greek, and his views Einstein has proved to be nonsense.

(ii) Aristotle was another old Greek, so his views must be nonsense too.

(iii) Christianity was invented in the Middle Ages, when people thought a lot of Aristotle.

(iv) Therefore Christianity was founded on a false basis and must be untrue.

(v) Therefore one can have as many wives as one likes simultaneously.[123]

There was much more in the same vein, but Knox insisted that he was not attacking science, only the "omniscientists" who tried to build popular materialism on the basis of their own interpretations of science. He acknowledged that there was no necessary conflict between Christianity and science, not even on the question of evolution. The danger was not science itself, but the priests of science who were trying to turn an honorable but limited institution into the basis for a new civilization.[124]

Knox received his friend Arnold Lunn into the Catholic Church in 1933. Originally a lawyer and a prominent Alpine skier (founder of the slalom race), Lunn had already published his *Flight from Reason* in the previous year. This was ostensibly an attack on rationalism, but was in fact a detailed critique of the Darwinian theory and its influence along the lines developed by Belloc and Chesterton. Lunn's *Within That City* offered an accessible defense of the Roman Catholic faith, and suggested that the "flight from reason" was now being checked as people began to realize that the materialistic implications drawn from science were without foundation.[125] Jeans and Eddington had destroyed the materialistic interpretation of physics from within, and

---

121. Ibid., 154. *Bradshaw* was the national railway timetable.

122. Knox, *Broadcast Minds*, 92.

123. Ibid., 190.

124. Ibid., 265–66 and 275. Curiously, the term "omniscientist" was also applied to William Whewell, the early-nineteenth-century polymath who coined the term "scientist"; see M. Fisch and S. Schaffer, *William Whewell: Philosopher of Science*, 87–116.

125. Lunn, *Within That City*, 43.

meanwhile the growing evidence for telepathy and the paranormal was per-
suading an ever wider range of thinkers that materialism had to be wrong.
The Modernists who wanted Christianity to compromise with materialism
were starting to look dated, and the real battle was now between dogmatic
Christianity and dogmatic humanism. Lunn noted that Joad and others
brought up without religious faith confessed that they were now deeply unhappy. There were many who had begun to see that Roman Catholicism was
the only possible form of Christianity that was still acceptable, yet hesitated
to take the final step.[126] Lunn's book was an attempt to convince as many as
possible that the step should be taken.

Alfred Noyes was a popular poet who wrote from an Anglo-Catholic
perspective (although it was widely believed that he had converted to Rome
by the early 1930s). In 1925 he enthused over the implications of modern science and technology in the pages of *Radio Times:* "To those who have any mental or spiritual vision, wireless is, perhaps, the most startlingly vivid scientific
vindication of the belief that this universe is essentially miraculous, essentially a unity, and referable in the last analysis only to the supreme miracle of
that single Reality wherein we live and have our being."[127] But by 1934 Noyes
had joined the ranks of those who thought that modern developments had
led Western culture into a wilderness where no one was certain of anything
any more. His *Unknown God* began with the recognition that many were now
bewildered, groping for a source of moral authority in a world of specialists
who could not see the forest for the trees. Christianity, he felt, had not so
much been abandoned as allowed to lapse.[128] Noyes had himself begun as an
agnostic and had written a long poem with Darwin as its hero, but the death
of a close friend had convinced him that there must be something more to
human personality.[129] He rehearsed the usual story of the nineteenth century
losing its way in the rationalism of Spencer and Haeckel, but unlike Lunn,
found nothing wrong with the basic idea of evolution—only the materialistic explanation of it. As long as a directing Power was recognized as its cause,
evolution held no threat to religion: "The argument is not against evolution,
but against any easy acceptance of happy accidents as an explanation of a vast
and harmonious system of law."[130] Berkeley's philosophy had shown him

126. Ibid.; on Modernism, 45–46, on Joad, 50, and on Catholicism, 58.

127. Noyes, "Radio and the Master-Secret," 550. I thank Anna-Katherina Mayer for bringing
this article to my attention.

128. Noyes, *Unknown God,* 9–11.

129. Ibid., 331. Noyes quotes here from his *Book of Earth,* part of a series on "The Torch
Bearers."

130. Noyes, *Unknown God,* 72.

how to escape from philosophical materialism, and his idealism had now been endorsed by scientists such as Jeans.[131] For all that he refused to attack evolution, Noyes was no Modernist, and one of his principal concerns was to reestablish the human spirit as the central feature of creation, and hence as central to the Creator's purpose: "Not till our world becomes theocentric will the dislocation caused by Copernicus and Darwin be set right."[132] Noyes was impressed by the glory of the heavens, but even more by the thought that it was the human mind that gave the whole experience a meaning. Those such as Barnes who thought that God's purpose could be achieved only if the universe were filled with intelligent beings had ignored the work of Jeans and other astronomers suggesting that the earth might well be the only habitable planet. The human species was unique in its mental abilities, yet those abilities were so limited that the only source of meaning had to come from outside. That was why Christianity could offer comfort, because it was the only religion based on the belief that the Creator became incarnated in the world to share our sufferings.

The Anglican Church, too, had its converts, of course, and the one who wrote most effectively to oppose the modernizing spirit was C. S. Lewis. We have already noted Lewis's contributions to theology in chapter 9; in his humorous *Screwtape Letters* of 1942 and his science fiction books, he tried to present the Christian message in a form that would be accessible to readers not used to theological debate. Lewis always claimed that he was attacking neither science itself nor even the scientific planning of society; it was not that "scientific planning will lead to Hell," but that "in modern conditions any effective invitation to Hell will appear in the guise of scientific planning."[133] He certainly recognized that science in itself need not be opposed to religion—the minor devil, Screwtape, warns his subordinate against trying to lure humans away from Christianity with science. This will only tempt them to think about realities they can't touch and see: "There have been sad cases among the modern physicists."[134] But the idea of progress is a positive force for evil, "hence the encouragement we have given to all those schemes of thought such as Creative Evolution, Scientific Humanism, or Communism, which fix men's affections on the Future, on the very core of temporality."[135]

131. Ibid., 144–46.

132. Ibid., 227.

133. C. S. Lewis, "A Reply to Professor Haldane," in his *Of Other Worlds*, 74–85, see 80. This essay was a response to J. B. S. Haldane's critical review of his science fiction writing, published in 1946.

134. Lewis, *Screwtape Letters*, 14.

135. Ibid., 77.

Screwtape later transforms himself into a large centipede, a natural power understood by a modern writer "with a name like Pshaw."[136] Modernism in the church is thus a potent force for undermining real Christianity, although to be fair, Lewis also ridicules the Anglo-Catholic "Father Spike."[137]

In his science fiction stories, Lewis presented an allegory in which the human race is seen as isolated from the higher beings who inhabit the other planets. In *Out of the Silent Planet* (1938), the hero Ransom is transported to Mars (Malacandra), where he helps the angelic inhabitants thwart the evil intentions of Devine and Weston, who are caricatures of the "progressive" scientists depicted in the futuristic utopias of H. G. Wells—although in a note prefaced to the book Lewis claimed to have enjoyed Wells's own science fiction stories. In fact, the philosophy put into the mouth of Weston reads more like an extension of Shaw's religion of the life force in the direction of Nietzsche.[138] *Perelandra* (1943) is the planet Venus, where Ransom defeats Weston, now transformed into the devil, the Bent One, who is trying to restage the temptation of Eve (the Green Lady). Finally, in *That Hideous Strength* (1945), Lewis pits Ransom against the other evil scientist, Devine (now Lord Feverstone), whose National Institution for Co-ordinated Experiments is trying to create a controlled society by taking over the English town of Edgestow. One of the NICE personnel is a mad clergyman, the Reverend Straik, whose views are a perverted form of Modernism.[139] Significantly, Lewis links the idea of the *élan vital* with magic and the old idea of the Anima Mundi[140]— not to dismiss this nonmaterialist perspective, but to exploit it for his own literary ends. NICE is eventually defeated by a combination of extraterrestrials and the magic of a resurrected Merlin. The novels are imaginative and occasionally effective, although some readers were put off by the all-too-obvious Christian message. Lewis insisted that he wrote them when he realized that there really were many people whose hopes for the future depended on the rationalists' vision of using science to conquer the universe and eventually death itself.[141] He felt that it was essential for the Christian alternative to be presented in a form that such people might at least attempt to read.

136.   Ibid., 114–15.

137.   Ibid., 82–83.

138.   The three novels are collected in Lewis, *The Cosmic Trilogy*; for the philosophy of the life force in *Out of the Silent Planet*, see 121–25.

139.   Ibid., 426–28. Curiously, the acronym NICE is now in use, the British government having established a National Institute for Clinical Excellence.

140.   Ibid., 560.

141.   Lewis, letter to a lady, 9 July 1939, in *Letters of C. S. Lewis*, 166–67.

At the time, Lewis's novels attracted some attention because they represented a highly visible effort to project the new orthodoxy of Anglicanism to a wider audience. Not only had the atmosphere within the church changed with the declining influence of Modernism, but orthodox Christians were no longer afraid to carry their message to a wider public, and to some extent that public was more receptive than it had been for decades. The popularity of Dorothy Sayers's radio play on the life of Christ, *The Man Born to Be King* (1941–1942), is an illustration of this changed attitude. Another indicator is Joad's conversion to Anglicanism in the late 1940s, at which time he produced several books explaining both his dissatisfaction with modern values and his positive experiences with Christianity.[142] Joad, at least, sought to maintain a balance between science and religion, holding that as long as science accepted its limitations and left the world of values alone, it offered no threat. That may have been Lewis's position too, in principle, but in practice he was opposed to many implications of science, especially to evolutionary theory, and suspicious of the scientists' hopes of transforming the world by applying their knowledge. An image we cannot ignore if we wish to understand the changed relationship between science and religion in the Britain of the 1940s is that of the progressive scientist Weston transformed into the devil in Lewis's *Perelandra*.

Wells had reached a wide audience with his efforts to undermine the claim that Christianity was the foundation stone of European civilization. Unlike many rationalists, he sensed that something traumatic would have to happen if the old social and moral order were to be overthrown. But Wells was not the only rationalist whose enthusiasm for right-wing politics must have made many readers increasingly uncomfortable in the late 1930s. Lewis articulated the growing fear that scientific planning would destroy human values by identifying it with the traditional Christian position that humans are inherently sinful and cannot be the source of their own salvation. To the extent that his fears coincided with the popular mood of the country, they suggest that the reaction against Modernism in the churches was part of a general transformation of British cultural life. The only other credible reaction to the rise of Fascism in Europe was to turn to Communism and the materialistic ideology of Marxism. Unlike the old rationalism, Marxism had the advantage of portraying history as discontinuous: if revolution was necessary for progress, the current breakdown of international order need not herald the end of all hope. But for Christians like Lewis, Communism was no better an application of scientific planning than Fascism itself.

142. Joad, *Decadence* and *Recovery of Belief*; see esp. chap. 5 of the latter on science and religion.

The literary debates of the early twentieth century thus show a substantial parallel to developments within the Christian churches. The vaguely optimistic progressionism supported by an earlier generation of philosophers and literary figures had close links with a generalized theism that was surprisingly close to the liberalized Christianity of the Modernists. Both flourished on the hope that humans were the intended goal of the evolutionary process and were now capable of directing that process themselves. The ideas of creative and emergent evolution might make the goal of cosmic progress less predictable in detail, but about the overall result there could be no doubt, whether one believed that this was intended by a Creator or was the culmination of a universal process only now acquiring a sense of meaning for itself. In all too many cases the assumption that progress was inevitable implied a willingness to think in terms of authoritarian social policies such as eugenics. Once it was admitted that the goal was the future perfection of the human race, it made little difference in practice whether that goal was endorsed by the Deity or not. The optimistic progressionism offered by the non-Darwinian biologists still implied that life advanced through struggling against adversity. The danger of going down this road was illustrated by the enthusiasm for eugenics on one hand and the worship of individual self-expression on the other. Wyndham Lewis's attack on the ideology of progress through struggle paralleled C. S. Lewis's more traditional Christian fears of the same ideology. The all-too-evident parallels between Modernism and the explicitly anti-Christian visions of writers such as Shaw and Younghusband illustrated the dangers that many orthodox Christians had feared would be the consequence of unrestrained liberalization. There was a continuous spectrum linking Modernism via Lodge's very loosely Christian spiritualism to the more radical theology of creative evolutionism. Under these circumstances, there were many who wondered why they needed to come back to the church if its new vision had actually repudiated much of what Christianity had once stood for. Modernism failed to win converts for the churches because it was too much of a break with the past, and it was difficult to see the point of putting such new wine into the old bottles.

In the end, Christianity did best by reformulating its traditional message, and in the short term at least, this involved a rejection of natural theology along with the idea of progress. The study of nature became irrelevant to the need for salvation. Few of the theologians or literary figures involved in this reformulation seem to have been aware of the latest developments in science, even when those developments would have reinforced their suspicions. Belloc and Chesterton had insisted that Darwinism was dead, and Lewis seems to have had no better understanding of what was happening in biology when he too rejected progressionism. The fact that Lewis could articulate this return to the old Christian message in a way that could once again touch the lives of ordinary people suggests that this revival was a genuine reflection of wider public concerns in a time of growing social crisis, not a response to what was happening in science itself.

The growing polarization between the churches and science that marked the late 1930s represented the end of an episode in which a serious attempt had been made to overcome the divisions that had become apparent in the Victorian period. We have seen that the early decades of the twentieth century were marked by a widespread enthusiasm for reconciliation in some quarters. So great was this enthusiasm that it led to a rewriting of history, sweeping away memories of those Victorian scientists who had retained some form of religious belief and creating the illusion of an era dominated by materialism and Darwinism. Reacting against what they perceived as an age of unbelief, a generation of religious thinkers and popular writers argued that science had now abandoned both its temporary allegiance to materialism and its demand to be recognized as the only valid source of knowledge and values. A new, nonmaterialistic science had emerged that could serve as the basis for a natural theology appropriate for the modern age, in which evolution became a moral force under the control of minds that were agents for the Creator's purpose. There were many scientists at the turn of the century ready to endorse such a claim, and the rationalists within and without science seemed to be thrown onto the defensive.

Liberal church leaders from R. J. Campbell to the Anglican Modernists worked to purge Christianity of those aspects of traditional belief that were incompatible with the world of modern science. Original sin and the need for salvation were swept aside to make humans the driving force of progressive evolution, self-conscious agents fulfilling God's plan for the triumph of moral values in this world rather than the next. Science, too, seemed at first to have accepted the new role marked out for it. Even in the late nineteenth century there had been a reaction against mechanistic biology and a preference for non-Darwinian modes of evolution, which seemed to allow the purposeful behavior of living things to play a creative role in the history of life. Physics, too, had abandoned mechanism in favor of the vision of a harmonious cosmos unified by the all-pervading influence of the ether. In the hands of popularizers such as Oliver Lodge, creative evolution was linked with the new physics and with the apparent evidence that personal development continued in the next world. Here was an optimistic worldview in which science played a role in promoting the drive toward a better life—but a life that was spiritually better as well as freed from the hardships that had so far been humanity's lot. Those who saw progress as a merely utilitarian goal that would eliminate our need for spiritual values had been checkmated.

The liberal consensus was never one of unanimity. Not everyone accepted the evidence for the paranormal, and there were disagreements over the social policies that would best fulfill God's purpose. Some were prepared to take a heavy-handed view of what to do with those individuals who seemed to contaminate the human race with less than noble values, while others recognized the need for social reform to help those unfortunates improve themselves. Nevertheless, there seemed to be enough agreement on the basics to create a philosophy in which science would take its place in sustaining a modernized version of traditional values. Science would repair the damage it had done in the previous generation, and would thus help to stem the tide of materialism and secularism that had swept the country following its undermining of the original framework of belief.

The fragile nature of this synthesis was all too apparent to its many opponents. Evangelical and Catholic churchmen were appalled to see exponents of the New Theology or Anglican Modernists sharing platforms with outright critics of Christianity such as George Bernard Shaw. How could liberal Christians make common cause with radical thinkers who believed that the new spirituality should be based on principles quite different from those of the old religion? Did the theology offered by Lodge represent a genuine modernization of traditional religious insights, or had it effectively undermined all that the faith had stood for through previous centuries? There was little of the overt opposition to science that characterized the creationist movement in the United States, but there were many who felt than an evolutionism that reduced humans to nothing more than superior animals hoping for a better life in this world had gone too far.

It should also have been apparent that the new generation of scientists was far less positive about the synthesis than the old. Most biologists soon turned their backs on vitalism, while the emergence of genetics was rapidly showing that the Darwinian selection theory had a potential not recognized in the previous generation. Only in physics and cosmology did the revolutions of the early twentieth century seem to keep up the momentum of opposition to materialism, although few scientists felt comfortable with the interpretation that Jeans and Eddington put on the new theories. Yet the growing indifference of younger scientists to the "new" natural theology went largely unnoticed among religious thinkers and popular writers. Shaw, Belloc, and Chesterton trumpeted the death of Darwinism even while that theory was sweeping its opponents aside to become the dominant force in biology. The rationalists who rejected the new natural theology from the opposite perspective were only too eager to point out that its supporters were biologists from a previous generation or physicists with a highly idiosyncratic approach to the latest ideas. Their warnings went largely unheeded by the theologians, however. By the same token, the theologians' growing indifference to natural theology during the 1930s was fueled by a return to Christian orthodoxy rather than by a belated recognition that science was not, after all, following the course mapped out for it by the advocates of an antimechanistic philosophy of nature.

If the threat from changing ideas within science was recognized by only a few of the liberal religious thinkers, they were all aware of the changing attitudes in society at large that began to undermine the synthesis in the 1930s. As Europe lurched toward depression and the threat of war, few could still believe that humanity was able to save itself or would achieve a paradise on earth this side of Armageddon. The events of the 1930s achieved what the Great War had predicted: the final destruction of the nineteenth century's faith in progress. Without that underpinning, the liberal consensus crumbled, and with it, the need to prop up the creaking edifice of nonmaterialistic science. Those Christians who had held the fort against Modernism now leapt over the barricades to proclaim their faith to a public that was listening once again. They rejoiced that the facile optimism of their opponents had been undermined by events and relaunched the traditional message that humanity's only hope of salvation came from Christ. Far from stemming the tide of secularism, the Modernists' diluted spirituality had actually contributed to the destruction of the values that were the only bulwark against darkness. Even the old-style rationalists found themselves marginalized by the new dialectical materialism of the Marxists, as though the only hope of sustaining the philosophy of unbelief was to tie it to a rival gospel of salvation through revolution.

The tensions that had built up in the 1930s came to a head with the outbreak of war. Ready or not, Britain now faced the might of Nazi Germany, the embodiment of what many had feared as the forces threatening civilization. For a while things hung in the balance, with Britain alone in Western Europe still resisting the onslaught. Then the conflict widened to include other, potentially much greater powers, and the war was eventually won. It was won at a price, as far as Britain was concerned, because the country was impoverished by the effort and faced the imminent dissolution of its empire. For a while the hardships endured by the people were almost as great as in the war years themselves, yet the election of the Labour government with a clear majority and a farsighted reform program lent an air of optimism to their otherwise dark situation. By the 1950s a sense of normalcy and confidence had returned, although the threat posed by Soviet Communism was now evident to all and the Cold War had begun. Science had created new weapons more frightening than anything the world had seen before—yet science was a force that had to be used in the ongoing struggle against what was perceived as a new form of godless imperialism. Those who saw the Cold War as a further episode in the defense of Western civilization were no longer so open to the loss of faith in progress that had blighted the relations between science and religion in the immediately prewar years. One enemy had been defeated, and further technical progress was necessary to overcome the next challenge, even if it brought potentially greater threats. In this more positive climate of opinion, a new era of dialogue between religion and science could open.

For the churches and for religious thinkers generally, the postwar decade offered great opportunities. It was the only period in the twentieth century when it might have seemed to a realistic onlooker that the decline of organized religion could be halted and reversed. The enthusiasm for Christianity engendered by the writings of C. S. Lewis and others continued, and for the first time in a long while significant numbers of younger people were coming back to the churches.[1] Initially the attraction of Christianity depended, as it had before the war, on neo-orthodoxy. Bishop Barnes's Modernist ideas were now an embarrassment to the rest of his church, and in this climate of opinion a dialogue with science was unlikely to reopen. There were new threats from applied science, too. Ronald Knox produced the first detailed assessment of the challenge to morality posed by nuclear weapons, his *God and the*

---

1. Hastings, *History of English Christianity*, chap. 32.

Atom of 1945. But Knox was out of touch in at least one respect: few as yet felt a serious fear of global annihilation, and it was only in the 1950s that the campaign for nuclear disarmament began in earnest (under the leadership of Bertrand Russell).[2] The hostility to science that had characterized the neo-orthodoxy inspired by the theology of Karl Barth seemed increasingly out of place in a world where it was clear that the new enemy would have to be fought with weapons of mounting technical sophistication. The fact that the enemy claimed to base its political system on a scientific analysis of the human situation made it all the more necessary to open up a dialogue with our own scientists to see if their work could be integrated once again into a coherent foundation for Western culture that would preserve at least some of its ancient roots. The appeals by Alfred North Whitehead, Charles Raven, Joseph Needham, and others for a transformation of science that would reinstate it as a form of religious activity now took on a renewed sense of urgency, and were echoed by writers such as A. D. Lindsay and A. D. Ritchie.[3] Modernist ideas about a purposeful universe were once again to be heard—yet this was a muted Modernism, which no longer stressed the elimination of the miraculous and the consequent reassessment of the historical events on which Christianity is based.

The scientists, of course, were only too willing to argue that they should play a greater role in the nation's affairs. For public consumption, though, they were more likely to stress the industrial rather than the military effects of new technology. The British were already becoming rather touchy over claims that their inventiveness was being exploited for profit by others, as Robert Bud has shown in the case of penicillin.[4] Bud's characterization of the "new Elizabethans" shows that the effort to celebrate the achievements of science and technology in the era following the coronation of the Queen in 1953 built on a more general postwar ethos linking scientific and social progress. But as the Cold War intensified, it was vital to stress that the best way of making use of science was in a free-enterprise economy, not in a planned system of the kind now identified with the Soviets. J. D. Bernal and the few remaining Communists were increasingly isolated (some, of course, remained hidden) as the scientific community followed Michael Polanyi and those who insisted that research must be free of state control. Scientists had cooperated with government in a centralized war effort, but saw this as an abnormal

2. On the reception of *God and the Atom*, see Waugh, *Life of Ronald Knox*, 303–4.
3. Lindsay, *Religion, Science, and Society in the Modern World;* Ritchie, *Civilization, Science, and Religion.*
4. Bud, "Penicillin and the New Elizabethans."

situation. In normal circumstances, they wanted no part of Bernal's planned economy, and the Society for Freedom in Science was founded in the winter of 1940–1941 to counter the campaign from the Left.

Rationalism had not gone away, of course, and writers such as Russell and Julian Huxley still campaigned actively for a nonreligious foundation for human knowledge and morality. Huxley now favored the term "humanism" for this program. But those who wanted to sustain the movement back toward Christianity that had become visible just before the war had to reexamine their attitude toward science. The wave of hostility to natural theology fostered by neo-orthodoxy began to abate, allowing religious thinkers once again to talk of the study of nature as something that needed to be taken into account by religion. Some scientists were also interested in renewing the efforts at reconciliation—but the situation had changed, especially in biology, and it was now difficult for anyone to ignore the resurgence of Darwinism and the continued success of the mechanistic approach. The early-twentieth-century dialogue had been based on the assumption that science was turning its back on materialism. If the postwar theologians' renewed willingness to consider an interaction with science was to have any hope of gaining credibility, they could no longer afford to set limits on the kind of science that was acceptable, especially if their demands were out of touch with the latest trends in research.

One sign of a new attitude among religious thinkers was a growing willingness to join Raven in arguing against the rejection of natural theology expressed by neo-orthodox Christian thinkers such as Barth. The Barthian position had been endorsed by another Protestant theologian, Karl Heim, whose *Christian Faith and Natural Science* was translated in 1953. But there were now some neo-orthodox thinkers who thought that the distrust of science had been pushed too far. W. A. Whitehouse, a reader in divinity at the University of Durham, conceded that science might endorse theism, although he insisted that it could tell us nothing of the God who is associated with the carpenter of Nazareth.[5] A more active critique of the Barthian position came from Arthur Smethurst, canon of Salisbury, who had himself done seven years of work in science at Imperial College. Smethurst endorsed Raven's rejection of Barth and added a detailed account of Heim's work in which he tried to argue that neither theologian was as rigid in his opposition to natural theology as his followers pretended.[6] He also argued for the study of the

---

5. Whitehouse, *Christian Faith and the Scientific Attitude*, 83.

6. Smethurst, *Modern Science and Christian Belief*; for a detailed account of Heim's views, see appendix D.

paranormal as evidence that science could still offer a weapon against materialism. Raven himself was offered a platform to renew his appeals when he gave the Gifford Lectures in 1951–1952. He again stressed the need for Christianity to retain a sense of the Creator as immanent within nature, but he had little new to offer in his interpretation of science. He still felt that the trend toward materialism should be reversed, repeating his claim that biology would be stultified unless it adopted nonmechanistic and non-Darwinian theories.[7] He endorsed the theistic evolutionism of Pierre Teilhard de Chardin, with its vision of life advancing toward spiritual unity.[8] There was, in fact, a considerable wave of enthusiasm for Teilhard's thought in the later 1950s. Another Anglican, Greville Yarnold, argued that evolution was guided by the divine spirit, but opted for a saltation or jump at the point where the human spirit emerged.[9] The hope that evolution would turn out to have a spiritually significant trend built into it was thus still popular among some religious thinkers, imposing constraints on the possibility of reconciliation that many scientists would find unacceptable.

What, then, was on offer from science? There was certainly some effort being made by religious scientists, especially physicists, to hold out the hope of a reconciliation. The Eddington Memorial Lectures were founded at Cambridge and served as a vehicle for exploring Eddington's vision of how the latest developments in science could support a religious view of the world. The Oxford professor of mathematics E. A. Milne published his *Modern Cosmology and the Christian Idea of God* in 1950 to push Eddington's project ahead. A dissenting voice came from Fred Hoyle, who gave a series of radio talks in the same year stressing the alienation of humanity from a vast and indifferent cosmos and insisting that religion was a delusion.[10] Hoyle's steady-state theory of cosmology was introduced in order to eliminate the need for a point of creation, as presupposed by the rival "big bang" model.

Significantly, Milne had commented on biological evolution, suggesting that God intervenes with "deft touches" to steer mutations in the right direction.[11] And it was in the biological sciences that the most obvious resistance to the demands of religious thinkers became apparent. Julian Huxley's enthusiasm for the idea of progress led him to endorse Teilhard de Chardin's evolutionary vision, but the biologist Peter Medawar wrote a vitriolic review

7. Raven, *Natural Religion and Christian Theology*, vol. 1, chap. 10.
8. Raven, *Teilhard de Chardin: Scientist and Seer*.
9. Yarnold, *Spiritual Crisis of the Scientific Age*, 95–96.
10. Hoyle, *Nature of the Universe*, 115.
11. Milne, *Modern Cosmology and the Christian Idea of God*, 153.

of Teilhard's book, dismissing it as mysticism that had nothing to do with science.[12] Many evolutionists, now committed to the new Darwinian synthesis, would have shared Medawar's discomfort. It was certainly possible to link the theory of natural selection with a liberalized Christian theology, and R. A. Fisher came out openly to proclaim his vision of how this could be done.[13] But the evident difficulties of reconciling Darwinism with the Christian view of the human situation were displayed when Huxley's student, David Lack, was confirmed in the Anglican Church in 1947. Some time later Lack expressed his doubts in his *Evolutionary Theory and Christian Belief*, in which he not only explored the more complex situation with respect to design created by the resurgence of Darwinism, but also concluded that the human spirit would have to be explained by some force outside the realm of physical causation.[14] From the perspective of this eminent evolutionist, at least, the expectations of Raven and his supporters were not to be endorsed by the latest trends in science. A 1962 article by the Anglican theologian J. S. Habgood singled out this topic to proclaim that the truce between science and religion was at best uneasy. Medawar's attack on Teilhard might be aimed more at the style than the substance, but unless scientists became more willing to acknowledge the widespread distrust of reductionism, religion would find it hard to keep open the door to dialogue.[15]

Efforts to reestablish the kind of relationship between religion and science that had been the goal of the earlier generation of Modernist theologians were thus only marginally successful, at least as far as the biologists were concerned. The trend toward mechanism and Darwinism was now in full swing, and any efforts by theologians based on the hope that such a trend represented a blind alley in science were hardly likely to command respect among the scientists themselves. There were, however, more sophisticated moves afoot to redefine the form of the relationship in a way that made the dialogue less dependent on a particular set of scientific theories. Those scientists who wished to present their activity as something consistent with religious belief offered a way forward that was more in tune with Eddington's policy, in which the effort to create a new natural theology was replaced by a recognition that the search for scientific knowledge did not

12. Medawar, "Review of Teilhard de Chardin, *The Phenomenon of Man.*" Huxley had written an introduction to the English translation.

13. Fisher, *Creative Aspects of Natural Law.* These ideas had clearly sustained Fisher's thought throughout his career; see chap. 4 above.

14. Lack, *Evolutionary Theory and Christian Belief,* 87–89. Note the book's subtitle, *The Unresolved Conflict.*

15. Habgood, "Uneasy Truce between Science and Religion."

close off other avenues toward an understanding of the human situation. Most scientists now were suspicious of the Marxist view of their social obligations promoted by Bernal, and this offered a foundation for a very different epistemology more in tune with religion. Michael Polanyi, a founding member of the Society for Freedom in Science, thought that the search for truth—possible only if research was free from external constraint—had a higher purpose. His *Science, Faith, and Society* of 1946 suggested that scientific and moral progress went hand in hand, and implied that the very existence of progress showed that the process brought us closer to something absolute.[16] Polanyi went on to give the Gifford Lectures in 1951–1952, published as *Personal Knowledge*, in which he stressed the involvement of the observer in the creation of knowledge and the existence of unproven traditional beliefs in the foundations of all knowledge systems.

It was Milne's successor at Oxford, C. A. Coulson, who produced the most successful attempt to bring science and religion together in the postwar decade. His *Science and Christian Belief* of 1955 was widely and enthusiastically reviewed. Coulson shared the view of Raven and William Temple that all existence is a medium of revelation and saw science as a source of vitality in our civilization. Heim and the neo-orthodox theologians had tried to establish religion in a realm beyond science, but their views cut no ice with most scientists. Yet Coulson was suspicious of most conventional efforts to reconcile the two viewpoints. Milne's and Eddington's attempts to define how creation must have occurred would not do—we cannot restrict God with metaphysics. Nor could we accept a God who sneaks in through the loopholes left by statistical explanations of nature. Appeals to Heisenberg's uncertainty principle in this context were just nonsense. Milne's "deft touches" implied a God who just tinkered—but for Coulson He was either everywhere or nowhere. Invoking the paranormal was equally fruitless—if psychic phenomena were real (which Coulson did not dispute), they would tell us more about time and space than about the soul.[17]

Coulson's strategy for reopening the dialogue was based not on a survey of each area of science in turn, but on the argument that both science and religion were efforts to understand the wider significance of our existence. Some interwar writers had explored this methodological link between the two ways of seeking knowledge, but for Coulson it became central. Both ar-

16. M. Polanyi, *Science, Faith, and Society*, 64–65. On the significance of Polanyi's thought for religion, see Torrance, "The Transcendental Role of Wisdom in Science" and "Ultimate and Penultimate Beliefs in Science." On the Society for Freedom in Science, see McGucken, *Scientists, Society, and State*, chap. 9.

17. Coulson, *Science and Christian Belief*, 17–25.

eas required imagination, and science's willingness to challenge authority had a valuable role to play in encouraging freedom. Science was a religious activity because it studied the creation, but it was not a religion because it aimed at only a partial understanding, while religion demanded a total response to the universe in which we live.[18] Although he refrained from detailed exposition of particular points at which scientific thinking had been seen to interact with religious values, Coulson did make the point that the Christian should expect to find suffering to be an integral part of nature because it was central to our own relationship with God. Evolution was the travail of God's energy seeking to create humanity in his image, but it was shot through with pain and sacrifice.[19] Here was at least an indirect way of making peace with the biologists.

Coulson's book marked the start of a renewed dialogue that has gone on—not without opposition from the modern rationalists—through the present day. Are there any conclusions we can reach from the episode studied in this book that may be of use to those engaged in the current debates? We must beware of trying to impose models of the past on the present, but the situation in the early twentieth century was close enough to that which obtains today for comparisons to offer some useful insights. Some of the points highlighted above seem fairly obvious in retrospect. Then as now, physics and cosmology were the areas of science most easily reconciled with the claim that the universe has a rational Creator. Biology and psychology offered the greatest stumbling blocks once the mechanistic and Darwinian modes of explanation became well established. From the side of theology, it was the more liberal approach that offered the hand of reconciliation to science; what worked best as far as most scientists were concerned was a generalized theism in which evolution became the unfolding of a divine purpose. The specifically Christian view of humanity as a fallen species was more difficult to reconcile with the proposed consensus, as was the Christian emphasis on divine intervention in the physical world. The proposed reconciliation seemed possible because a generation of scientists who hoped to reverse the trend toward materialism coincided with a generation of theologians who were willing to liberalize their beliefs in a way that downplayed exactly those aspects of the Christian message that the scientists found it hardest to accept. Between them, they persuaded a significant number of thinking people that a new world of harmonious interaction had dawned. The proposed reconciliation remained plausible as long as the general cultural environment encouraged

18. Ibid., 83–86.
19. Ibid., 108–9.

the majority to retain their faith in progress as a key to understanding the natural world and the human situation.

If the basic outlines of the reconciliation seem predictable enough, the specific circumstances that made possible its rise and its eventual demise need careful analysis. Modernism made what many Christians thought were impossible demands on religion, but at the same time it imposed limits on what kind of scientific theorizing was permissible. It required that physics abandon (if it had ever seriously considered) a billiard-ball atomism in favor of a worldview in which everything was integrated in a coherent manner and in which it was possible for mental powers to have a real existence. Biology had to allow for the reality of both the life force and mental functions, and evolution had to be a teleological process shaped by these nonphysical forces and designed to enhance them. Psychology could not be reductionist, although it might accept the unconscious as a hidden source of mental energy. It was possible to argue that these positions were acceptable to science, although in fact they were mostly the products of turn-of-the-century science now increasingly out of date. In physics, Lodge's ether theory provided a suitable foundation, and although it was soon superseded, the new ideas of quantum mechanics at least sustained the impression that science was moving even further away from a simplistic mechanical model of reality. In biology, however, the antimechanistic and anti-Darwinian impulse of the decades around 1900 was soon overtaken by the emergence of genetics, the genetical theory of natural selection, and a host of mechanistic initiatives in biochemistry and physiology. To anyone in touch with the latest research in the life sciences, the Modernist program was already out of date by the mid-1920s, although few religious thinkers were aware of this.

It is also important for us to understand why the reconciliation proffered by the Modernists fell apart in the later 1930s. The thesis defended above is that the synthesis broke down not because developments in science undermined the teleology of the new natural theology, but because changing attitudes within the Christian churches destroyed the Modernist position and left its few remaining advocates isolated and powerless. Barnes was one of the few Modernists aware of Fisher's new basis for the Darwinian selection theory, and Raven was still maintaining that biology should abandon Darwinism in the 1950s. The Modernists saw themselves marginalized not by the new science, of which many remained unaware, but by changing values within the churches, which brought back a sense of human sinfulness and alienation from God incompatible with the idea of progress. In effect, the old traditions of Christianity reasserted themselves in a new form adapted to the sense of crisis in European culture and society. At one level, this was a shift in the balance of power within a purely theological debate, but it would be pointless to ignore the role of the wider issues. The new orthodoxy not only affected what was acceptable within the churches, but also allowed the churches as a whole to roll back, at least temporarily, the tide of indifference that had begun to marginalize them since the turn of the century. British culture as a whole responded to the crisis of the depression and the rise of Fascism by questioning for the first time the ethos of progressionism that had held sway, with only a temporary eclipse during the Great War, since the mid-nineteenth century. The destruction of the Modernist compromise was almost a by-product of this broader reaction, just as the reestablishment of a dialogue in the 1950s was the product of a new sense that science would have to be incorporated into the anti-Communist ethos of the Cold War. Whatever the intellectual or moral coherence of the various positions taken up in these debates, their relative strengths waxed and waned under the influence of broader social trends.

If the changing cultural and social environment influenced the debates in so striking a manner, the lessons offered by this episode need to be read with some care. Most important is the need to recognize that neither science nor organized religion are coherent entities, so there can be no single form of relationship between them. Each is subject to internal and external constraints that change the balance of power between competing interest groups and hence affect the possibilities of interaction between the two areas. Scientists have long differed among themselves over the broader implications of their enterprise, with some supporting rationalism, some a form of natural theology, and an increasing number becoming reluctant to commit themselves in public at all. Theologians represent their different churches' positions, but all must establish their relationship to the liberalizing trend that has been a conspicuous feature of modern religious thought. For every liberal who expresses enthusiasm for dialogue with science, there is a conservative who remains suspicious. To this extent there is a parallel to the early twentieth century today, although the balance between conservatives and liberals differs on either side of the Atlantic and has, of course, changed through time in response to both intellectual and cultural developments. Outside the churches, various forms of unorthodox spirituality have gained in strength, and here, too, the effect of social changes is apparent. Shaw's creative evolution lost its appeal in parallel with the Modernist movement in the churches as the pessimism of the 1930s took hold.

More controversially for the scientists themselves, there seems to have been a period in the early twentieth century when the direction of their theorizing was directly influenced by cultural pressures. In a reaction against the excesses of the Victorian scientific naturalists, antimechanistic values encouraged theoretical initiatives in both the physical and the life sciences. Most modern scientists would regard this as an aberration and would insist that whatever their views about religion, scientists put them aside at the laboratory door. Sociologists and historians of science doubt that such a clear separation is possible. Perhaps Michael Ruse is right when he depicts the twentieth century as a period in which this kind of overt ideological influence was gradually purged from science to be replaced with its own internal respect for the facts.[20] As Ruse insists, this does not mean that scientists are indifferent to or unaffected by outside beliefs, but when it comes to testing their theories, they are constrained by the results of their experiments, and their room to maneuver has been steadily circumscribed. The early decades of the twentieth century may indeed be the time at which this internal constraint began to dominate, symbolized in biology by the rapid collapse of the antimechanistic counterattack and the ability of figures with ideologies as diverse as R. A. Fisher and J. B. S. Haldane to contribute to the same theoretical perspective, the genetical theory of natural selection. If this interpretation is accepted, the growing reluctance of biologists, at least, to endorse a form of natural theology becomes an inevitable product of the internal development of science. Like Fisher, one could try to reconcile Darwinism with design, but it was an uphill struggle.

The present study has, however, been concerned more with the reception of science than with the internal development of science itself. Given that many scientists do not make public pronouncements about their beliefs, we need to exercise vigilance when claims are made about the consensus of scientists on a particular issue. The interaction between science and religion is shaped to some extent by perceptions of what each side believes and wants, and such perceptions can be manipulated by those with an interest in driving the debate in a particular direction. Theologians and literary figures through the first three decades of the twentieth century were persuaded that science as a whole had turned its back on Victorian materialism. This interpretation was not totally false, but neither did it present an accurate picture of either the contemporary situation or the immediate past. There was a strong move in some areas of science at the turn of the century to explore avenues of theorizing that were self-consciously a reaction against the perceived materialism of Huxley and Tyndall. But in biology, at least, this move was on the wane by the 1920s—yet neither the theologians nor the literary figures seemed to be aware of the transition.

20. Ruse, *Mystery of Mysteries*.

A point of interest to historians is the extent to which later generations' image of the late nineteenth century has been influenced by the model created just after 1900. In their anxiety to stress how much science had now swung back to a less materialistic way of thinking, writers from this period seem to have deliberately exaggerated the extent to which late-nineteenth-century scientists had become committed to Darwinism and a more generally mechanistic view of nature. Most "Darwinists" in the late Victorian era were just evolutionists, yet to read Shaw and his contemporaries, one would think that virtually every biologist of the previous generation had been a Darwinist in the strictest sense of the term. The image of the godless Victorians is at least in part a product of the debate that emerged in the next generation as those who sought to defend or transform religion tried to enlist the aid of the diminishing number of scientists who were interested in maintaining the dialogue.

The inability of theologians and literary figures to recognize what was actually going on in science in the 1920s raises another point of concern. A relatively small number of influential writers were able to present an interpretation of science that was almost certainly out of touch with what the majority of working scientists thought. Rationalists like Joseph McCabe railed against what they perceived (with some justification) to be a small clique of conservative scientists who had access to the means of publication and who kept up a barrage of literature designed to promote the nonmaterialistic version of science. Whether or not they intended to give this impression, their readers thought they were talking about the latest research trends. In fact, writers like Oliver Lodge, C. Lloyd Morgan, and J. A. Thomson were members of a previous generation, out of touch with what was happening at the research front, while Jeans and Eddington offered an interpretation of the new physics that would not have been accepted by the majority of working scientists at the time. There *were* religious scientists still, of course, but there were relatively fewer from the younger generation who were willing to take a stand in public, and the flood of literature on nonmechanistic biology dried up as the older generation died out. In biology especially, the writings of those who argued for a renewed dialogue between science and religion created a misleading impression that left most ordinary readers with an unrealistic expectation of what was about to emerge from current research. The growing power of the popular press and mass-market publishing created an opportunity for particular interest groups to manipulate what was presented to the public. Whatever its significance for the debate over science and religion, this is a point that needs to be borne in mind by anyone concerned with the way in which science is popularized and discussed today.

FRS stands for Fellow of the Royal Society. Ordination means into the Anglican Church unless otherwise stated. "Wrangler" indicates position in the Mathematics Tripos at Cambridge.

BALFOUR, Arthur James (1848–1930). Educated in Moral Sciences at Trinity College, Cambridge, 1866–1869. Conservative member of Parliament from 1879; appointed First Lord of the Treasury 1891 and served as Prime Minister 1902–1905. Noted writer on philosophy and opponent of materialism.

BARNES, Ernest William (1874–1953). Obtained mathematics scholarship to Trinity College, Cambridge; Second Wrangler 1896; subsequently fellow of Trinity. Elected FRS 1909 for work in applied mathematics. Ordained 1902. Appointed Master of the Temple 1915, Canon of Westminster 1920, and Bishop of Birmingham 1924. Leading Anglican Modernist, remembered for his support for evolution and rejection of miracles.

BARRETT, William Fletcher (1844–1925). Educated privately; served as assistant to physicist John Tyndall at Royal Institution, London, 1863. Science teacher, then lecturer in physics at Royal School of Naval Architecture 1869. Professor of physics, Royal College of Science, Dublin, 1873–1910. Anglican; supporter of spiritualism.

BELLOC, Joseph Hilaire Pierre René (1870–1953). Born in France; brought to England in 1872. Studied history at Balliol College, Oxford, 1893–1895; also active in Oxford Union. Supported himself as editor and author, writing on morality, history, and religion. Popular writings attacked rationalism and defended Roman Catholicism.

BERNAL, John Desmond (1901–1971). Born in Ireland. Studied mathematics and science and Emmanuel College, Cambridge, 1919–1923. Worked on crystallography under William Bragg at Royal Institution 1923–1927; lecturer at Cambridge 1927–1938. Elected FRS 1937. Professor of physics at Birkbeck College, London, 1938–1963 and of crystallography 1963–1968. Active Marxist from undergraduate days.

BETHUNE-BAKER, James Franklin (1861–1951). Studied classics and theology at Pembroke College, Cambridge; appointed fellow of Pembroke 1891. Ordained 1899. Lady Margaret Professor of Divinity, Cambridge, 1911–1935. Modernist theologian and New Testament scholar.

BLATCHFORD, Robert Peel Glanville (1851–1943). Apprenticed to a brushmaker, then served in the British Army, where he taught himself to write. Publisher of Socialist newspaper *The Clarion*, in which he wrote as "Nunquam." Opponent of religion, but later came to believe in spiritualism.

BRAGG, William Henry (1862–1942). Obtained mathematics scholarship to Trinity College, Cambridge, 1881; Third Wrangler 1884. Professor of mathematics and physics, University of Adelaide 1885–1904. Elected FRS 1907 for work on radium and radiation. Appointed Quain Professor of Physics, University College, London, 1915 and head of Royal Institution 1923. Knighted 1920.

BROOM, Robert (1866–1951). Obtained medical degree from Glasgow 1889. Practiced in Australia 1892–1897 and then in South Africa. Studied the anatomy of primitive mammals and fossil mammal-like reptiles; later worked on hominid fossils. Professor of zoology and geology, Victoria College, Stellenbosch, 1903–1910; Curator of Transvaal Museum, Pretoria, from 1934. Presbyterian; argued for evolution designed by God.

BURROUGHS, Edward Arthur (1882–1934). Educated in classics and philosophy at Balliol College, Oxford; fellow of Hertford College 1912–1920. Ordained 1908. Canon of Peterborough 1917–1922; chaplain of Trinity College, Oxford, 1920–1921; Dean of Bristol 1922–1926, then appointed Bishop of Ripon. Highlighted the social dangers of unrestrained science and technology.

CAMPBELL, Reginald John (1867–1956). Raised in Ulster; studied history and politics at Christ Church, Oxford, 1891–1895. Joined Congregational Church 1895; minister of Union Street Church, Brighton, 1895–1903, and of City Temple, London, 1903–1915. Published the very liberal *New Theology* in 1907. Received into the Anglican Church 1915, ordained the following year, and served as Anglican priest for the rest of his career.

CARR, Herbert Wildon (1857–1931). Privately educated; honorary fellow of King's College, London, 1914, then professor of philosophy 1918–1926. Moved to University of Southern California 1925. Leading advocate of Bergson's philosophy.

CHESTERTON, Gilbert Keith (1879–1936). Educated at St. Paul's School, London. Journalist, editor, and author; perhaps most popular for his *Father Brown* stories, which he began to publish in 1911. Converted to Roman Catholicism 1922; active defender of Christian values.

D'ARCY, Charles Frederick (1859–1938). Educated at Trinity College, Dublin. Ordained into Church of Ireland 1884. Vicar of Belfast 1900–1903, Bishop of Down and Connor 1903–1919, Archbishop of Dublin 1919–1929, Archbishop of Armagh 1929–1938.

DAVIDSON, Randall Thomas (1848–1930). Educated at Harrow and Trinity College, Oxford. Ordained 1875. Bishop of Rochester 1891–1895, Bishop of Winchester 1895–1903, Archbishop of Canterbury 1903–1928. Created baron 1928.

DEWAR, Douglas (1875–1957). Educated at Jesus College, Cambridge. Became barrister and served in Indian Civil Service 1898–1924. Expert on Indian wildlife and opponent of evolution theory; president of the Evolution Protest Movement.

EDDINGTON, Arthur Stanley (1882–1944). Studied at Owen's College, Manchester, and at Trinity College, Cambridge; First Wrangler 1904. Appointed Chief Assistant at Royal Observatory, Greenwich, 1906; Plumian Professor of Astronomy, Cambridge, 1912. Worked in astrophysics and on relativity (his study of the 1919 eclipse confirmed Einstein's predictions). Elected FRS 1914. Knighted 1930. Quaker; wrote popular books linking modern physics with an idealist philosophy that reinstated the plausibility of belief in God.

FISHER, Ronald Aylmer (1890–1962). Studied mathematics at Cambridge 1909–1912. Appointed statistician at Rothamstead Experimental Station 1919; Galton Professor of Eugenics, University College, London, 1933; Balfour Professor of Genetics, Cambridge, 1943. Elected FRS 1929 for work in statistics and population genetics. Knighted 1952 and spent last years in Adelaide, Australia. Anglican; in later life wrote to urge compatibility of Darwinism and religion.

FLEMING, John Ambrose (1849–1945). Graduated from University College, London, 1870; studied physics at Cambridge from 1877 and appointed demonstrator 1880. Professor of electrical technology, University College, London, 1885–1926. Worked on telegraphy and radio, patented thermionic valve for radio 1904. Evangelical Anglican; opponent of evolution.

GORE, Charles (1853–1932). Studied classics and philosophy at Balliol College, Oxford; elected fellow of Trinity College, Oxford, 1875 and appointed principal of Pusey House 1883. Ordained 1878. Appointed Canon of Westminster 1894; Bishop of Worcester 1901–1905; Bishop of Birmingham 1905–1911; Bishop of Oxford 1911–1919. Leading Anglo-Catholic theologian.

GRIFFITH-JONES, Ebenezer (1860–1942). Educated at Presbyterian College, Carmarthen, and University College, London. Congregationalist minister from 1885. Principal of Yorkshire United Independent College, Bradford, 1907–1932. Chair of Congregational Union of England and Wales 1918–1919; President of National Free Church Council 1931–1932.

HALDANE, John Burdon Sanderson (1892–1964). Son of John Scott Haldane. Educated at Eton and Oxford (although did not complete degree). Studied biochemistry at Cambridge and became reader in biochemistry there in 1921. Part-time appointment at John Innes Horticultural Institute 1927–1936. Appointed professor of genetics, University College, London, 1933. Marxist from 1933; wrote at popular level on science and society. Moved to India in 1957.

HALDANE, John Scott (1860–1936). Graduated from Edinburgh University in medicine 1884. Demonstrator in physiology at Oxford 1887; fellow of New College 1907. Elected FRS 1897 for work on physiology of respiration. Companion of Honour 1928. Wrote in support of an idealist philosophy of knowledge.

HENSON, Herbert Hensley (1863–1947). Educated at Oxford, elected fellow of All Souls 1884. Ordained 1888; Canon of Westminster 1900–1912; Dean of Durham 1912–

1918. Appointed Bishop of Hereford 1918 despite strong opposition by Anglo-Catholics because of his Modernist views. Bishop of Durham 1920–1939.

HOGBEN, Lancelot (1895–1975). Son of a Plymouth Brethren preacher; became a Quaker and later a Marxist. Studied medicine at Trinity College, Cambridge, 1912–1915. Imprisoned as conscientious objector in World War I. Lecturer in zoology at Imperial College, London, 1919; lecturer in experimental zoology at Edinburgh 1922; assistant professor of zoology at McGill University, Montreal, 1925; professor of zoology at Cape Town 1926; professor of social biology at London School of Economics 1929; professor of natural history at Aberdeen 1937; professor of zoology and later of medical statistics at Birmingham 1941–1961.

HUXLEY, Julian Sorell (1887–1975). Educated in natural sciences at Balliol College, Oxford. Scholar at Naples Zoological Station 1909–1910. Lecturer in zoology at Oxford 1910–1912; assistant professor at Rice Institution, Houston, Texas, 1916–1919; fellow of New College, Oxford, 1919–1925; professor of zoology at King's College, London, 1925–1927; professor of physiology at Royal Institution, London, 1926–1929; secretary of Zoological Society of London 1935–1942. Director General of UNESCO 1946–1948. Humanist scholar and writer.

INGE, William Ralph (1860–1954). Educated at Eton and King's College, Cambridge. Taught at Winchester and Eton. Fellow of Hertford College, Oxford, 1888. Ordained 1888. Lady Margaret Professor of Divinity, Cambridge, 1907–1911. Dean of St. Paul's, London, 1911–1934. Scholar dealing especially with Christian mysticism, also Modernist theologian. Wrote extensively for popular press expressing conservative social views.

JEANS, James Hopwood (1877–1946). Studied mathematics at Trinity College, Cambridge; Second Wrangler 1898. Fellow of Trinity 1901–1905; professor of applied mathematics at Princeton 1905–1907; Stokes lecturer in applied mathematics at Cambridge 1910–1912, retired to do independent research. Elected FRS 1907 for work on dynamical theory of gases and radiation. Popularized idealist view of knowledge and claim that intelligent life must be rare in the universe.

JOAD, Cyril Edwin Mitchinson (1891–1953). Studied classics and philosophy at Balliol College, Oxford. Served with Board of Trade 1914–1930. Head of philosophy department, Birkbeck College, London, 1930–1953; reader in philosophy from 1945. Wrote popular works on philosophy and appeared frequently on radio. Converted to Anglicanism in last years of his life.

KEITH, Arthur (1866–1955). Graduated from Aberdeen in medicine 1888, worked as doctor in Thailand, then returned to obtain Aberdeen M. D. in 1894. Senior demonstrator in anatomy at the London Hospital 1895, head of department from 1899. Conservator of museum of Royal College of Surgeons 1908. Elected FRS 1913 for work in anatomy and paleoanthropology. Knighted 1933 and in same year appointed master of Royal College of Surgeons research institute, Downe. Active supporter of rationalism.

KNOX, Ronald Arbuthnot (1888–1957). Studied classics and philosophy at Balliol College, Oxford; elected fellow of Trinity College, Oxford, 1910 and chaplain 1912.

Ordained 1912. Converted to Roman Catholicism in 1917 and ordained in 1919. Catholic chaplain at Oxford 1926–1939. Worked on translation of the Bible, published 1955.

LANKESTER, Edwin Ray (1847–1929). Educated in science at Downing College, Cambridge, and Christ Church, Oxford. Fellow of Exeter College, Oxford, 1872; professor of zoology at University College, London, 1874; Linacre Professor of Comparative Anatomy, Oxford, 1894; director of British Museum (Natural History) 1898–1907. Elected FRS 1875 for work on comparative anatomy and evolution. Knighted 1907. Active supporter of rationalism; wrote on science for popular press after retirement.

LEWIS, Clive Staples (1898–1963). Studied classics, philosophy, and English at University College, Oxford; appointed fellow of Magdalen College, Oxford, 1925. Professor of English medieval and Renaissance literature at Cambridge 1954–1963. Early skepticism eroded in late 1920s, became Anglican in 1931; wrote popular works on theology and on dangers of scientific rationalism.

LEWIS, Percy Wyndham (1882–1957). Educated at the Slade School of Art, London; avant-garde artist and writer working in Paris 1901–1909 and then in London.

LODGE, Oliver Joseph (1851–1940). Educated at the Royal College of Science and University College, London. Appointed demonstrator in physics at University College and obtained D.Sc. 1877. Professor of physics at University College of Liverpool 1881–1900; principal of Birmingham University 1900–1919. Elected FRS 1887; studied electromagnetism and did pioneering work on radio. Knighted 1902. Wrote extensively on spiritualism and on reconciliation of science and religion.

LUNN, Arnold (1888–1974). Educated at Balliol College, Oxford. President of the Ski Club of Great Britain and inventor of the slalom race. Received into the Roman Catholic Church in 1933 and wrote in its defense. Opponent of evolutionism. Knighted 1952.

MACBRIDE, Ernest William (1866–1940). Studied science at Queen's College, Belfast, and Cambridge, also at Zoological Station of Naples. Professor of zoology at McGill University, Montreal, 1897–1909; appointed lecturer in zoology at Imperial College, London, 1909, professor 1914–1934. Elected FRS 1905 for work on development and affinities of echinoderms. Leading neo-Lamarckian evolutionist.

MATTHEWS, Walter Robert (1881–1973). Educated at King's College, London. Lecturer in philosophy and theology at King's College 1908–1918, professor and dean 1918–1932. Chaplain to the King 1923–1931. Dean of Exeter 1931–1934.

MCDOUGALL, William (1871–1938). Educated at Owen's College, Manchester, and St. John's College, Cambridge; elected fellow of St. John's 1897. Wilde Reader in Mental Philosophy, Oxford, 1912–1920; professor of psychology at Harvard University 1920–1923 and at Duke University 1923–1938. Elected FRS 1912. Exponent of vitalism and neo-Lamarckian evolutionism.

MITCHELL, Peter Chalmers (1864–1945). Educated in natural science at Aberdeen and Christ Church, Oxford. Demonstrator in comparative anatomy, Oxford, 1888–1891, then lectured in anatomy at London hospitals. Secretary of the Zoological Society of London 1903–1935. Rationalist and opponent of vitalist biology.

MORGAN, Conwy Lloyd (1852–1936). Educated at School of Mines, London; taught in South Africa 1878–1883. Appointed professor of zoology at University College of Bristol 1883, principal 1887, and Vice-Chancellor when college made a university in 1909. Elected FRS 1899 for work on animal psychology and the evolutionary understanding of behavior. Originator of theory of emergent evolution.

NEEDHAM, Terrence Noel Joseph (1900–1995). Studied natural sciences at Gonville and Caius College, Cambridge, 1918–1922; obtained doctorate in 1924 and became fellow of Caius. Elected FRS 1941 for work in biochemistry and embryology, but following visit to China in 1942–1943 began work on history of Chinese science. Director of science UNESCO 1945–1948. Anglican and Marxist.

RAVEN, Charles Earle (1885–1964). Studied classics and theology at Gonville and Caius College, Cambridge. Assistant secretary for secondary education, Liverpool 1904–1909. Ordained 1909. Dean of Emmanuel College, Cambridge, 1909. Vicar of Blechingley, Surrey 1920–1922. Canon of Liverpool 1924–1932. Regius Professor of Divinity, Cambridge, 1932. Master of Christ's College, Cambridge, 1939–1950; Vice-Chancellor of the university 1947–1949. Modernist theologian, historian of science, also wrote popular books on natural history.

RAYLEIGH, John William Strutt, third baron (1842–1919). Studied mathematics at Cambridge. Fellow of Trinity College, Cambridge, 1866, then did independent research until appointed Cavendish Professor of Experimental Physics, Cambridge, 1879–1884. Nobel Prize for isolation of argon 1904. Chancellor of Cambridge 1908–1909.

ROMANES, George John (1848–1894). Educated in mathematics and natural sciences, Gonville and Caius College, Cambridge, 1867–1873. Worked in physiology at University College, London, 1874; then did independent research in psychology and evolution theory. Skeptic, but returned to Christianity shortly before death.

RUSSELL, Bertrand Arthur William (1872–1970). Educated at Trinity College, Cambridge; elected to fellowship 1895. Elected FRS 1908; lecturer in logic and philosophy of mathematics, Cambridge, 1910. Wrote extensively on philosophy from a humanist perspective.

RUSSELL, Edward Stuart (1887–1954). Educated in biology at Glasgow; studied briefly at Aberdeen and then worked as fisheries expert for Board of Agriculture and Fisheries. Argued for influence of mind in animal behavior.

SAYERS, Dorothy Leigh (1893–1957). Studied modern languages at Somerville College, Oxford, 1912–1915. Advertising copywriter 1916–1931, also novelist and writer on religious topics, best known for detective stories featuring Lord Peter Wimsey.

SHAW, George Bernard (1856–1950). Born and educated in Dublin; came to London in 1876 and established himself first as literary and music critic, then as playwright. Joined Fabian Society 1884. Exponent of vitalism, neo-Lamarckism, and creative evolutionism.

SHERRINGTON, Charles Scott (1857–1952). After medical training at St. Thomas's Hospital, London, studied physiology at Cambridge. Lecturer in physiology at St. Thomas's 1887. Holt Professor of Physiology, Liverpool, 1895–1912; Wayneflete Professor of Physiology, Oxford, 1913–1935. Knighted 1922. Nobel Prize in medicine for work on neurophysiology 1932.

SIMPSON, James Young (1873–1934). Educated in science at Edinburgh and Christ's College, Cambridge. Served with Political Intelligence Department in World War I, subsequently president of Latvian-Lithuanian Frontier Court of Arbitration. Professor of natural sciences at New College, Edinburgh.

SINGER, Charles Joseph (1876–1960). Studied medicine at Magdalen College, Oxford, then physician at Cancer Hospital and Dreadnaught Hospital, London, 1906–1914. Lecturer in history of medicine, University College, London, 1920–1931, professor 1931–1934.

SMUTS, Jan Christiaan (1870–1950). Trained in law in South Africa and at Cambridge; senior attorney of Transvaal 1898. Served as commander on Boer side in war of 1901–1902 and on peace commission. Colonial Secretary of South Africa 1907–1910. Commanded British forces in East Africa in World War I, member of British War Cabinet. Prime Minister of South Africa 1919–1924 and 1939–1948. Wrote on philosophy of holism while in opposition.

STREETER, Burnett Hillman (1874–1937). Studied humanities and theology at Queen's College, Oxford; elected fellow and dean 1905, chaplain 1928, provost 1933. Ordained 1899; Canon of Hereford 1915–1934. New Testament scholar and Modernist theologian.

TAYLOR, Frank Sherwood (1892–1956). Studied chemistry at Lincoln College, Oxford. Schoolteacher 1921–1933. Assistant lecturer in chemistry, Queen Mary College, London, 1933–1938. Curator of Science Museum, Oxford, 1940–1950; director of Science Museum, London, 1950–1956. Received into Roman Catholic Church 1941.

TEMPLE, William (1881–1944). Studied humanities at Balliol College, Oxford; fellow of Queen's College, Oxford, 1904–1910. Ordained 1909. Headmaster of Repton School 1910–1914. Rector of St. James, Piccadilly, London, 1914–1918; Canon of Westminster 1919–1921; Bishop of Manchester 1921–1929; Archbishop of York 1929–1943; Archbishop of Canterbury 1942–1944.

TENNANT, Frederick Robert (1866–1957). Educated in natural sciences at Gonville and Caius College, Cambridge. Science teacher 1891–1894. Curate of St. Matthews, Walsall, 1894–1897; chaplain of Caius College 1897; curate of St. Mary's, Cambridge, 1899. University lecturer in theology, Cambridge, 1913–1931.

THOMSON, John Arthur (1861–1933). Educated in biology at Edinburgh, Jena, and Berlin. Lecturer in medicine and biology, Edinburgh; Regius Professor of Natural History at Aberdeen 1899–1930. Knighted 1930. Evolutionist and exponent of nonmaterialistic approach to biology.

THOMSON, Joseph John (1856–1940). Educated in mathematics at Owen's College, Manchester, and Trinity College, Cambridge; Second Wrangler 1880. Fellow of Trinity College 1881. Cavendish Professor of Experimental Physics, Cambridge, 1884. Nobel Prize in physics 1906. Knighted 1908. Order of Merit 1912. Master of Trinity College 1918.

WALLACE, Alfred Russel (1825–1913). Early career as surveyor and schoolteacher; then collected zoological specimens in South America and the Malay archipelago. Elected FRS 1893 for work in evolutionary theory and biogeography. In later life supported himself by writing and lecturing. Strong advocate of spiritualism.

WARD, James (1843–1925). Studied moral sciences at Trinity College, Cambridge; elected fellow of Trinity 1875. Professor of mental philosophy and logic at Cambridge 1892–1925. Elected Fellow of the British Academy 1902.

WELLS, Herbert George (1866–1946). Studied at Normal School of Science, London, 1884–1887, then supported himself by teaching until he gained success as writer of novels and science fiction stories. Rationalist and campaigner for state-imposed social reform.

WHITEHEAD, Alfred North (1861–1947). Studied mathematics at Trinity College, Cambridge; Fourth Wrangler 1883. Elected fellow of Trinity College 1884; lecturer in mathematics at Cambridge 1884–1910. Professor of applied mathematics at Imperial College, London, 1914–1924. Professor of philosophy at Harvard 1924–1937. Sought reformulation of knowledge on nonmechanistic principles.

YOUNGHUSBAND, Francis (1863–1942). Educated at Sandhurst; joined Indian Army and later the Foreign Department; explored Tibet and Central Asia, later encouraged expeditions to Mount Everest. Knighted 1904. Founded World Congress of Faiths 1936.

This bibliography contains all the primary and secondary sources referred to in the text with the exception of anonymous newspaper reports. More substantial anonymous articles from magazines are included, alphabetized under the article title. Where known, the original date of publication of reprinted items is indicated in square brackets.

Abir-Am, Pnina G. "The Biotheoretical Gathering, Transdisciplinary Authority, and the Incipient Legitimation of Molecular Biology: New Perspectives on the Historical Sociology of Science." *History of Science* 25 (1987): 1–70.

———. "The Philosophical Background of Joseph Needham's Work in Chemical Embryology." In *A Conceptual History of Modern Embryology*, edited by Scott F. Gilbert, 159–80. Baltimore: Johns Hopkins University Press, 1991.

Ackerman, Robert. *J. G. Frazer: His Life and Work.* Cambridge: Cambridge University Press, 1987.

Acworth, Bernard. *This Bondage: A Study of the "Migration" of Birds, Insects, and Aircraft, with Some Reflections on "Evolution" and Relativity.* London: John Murray, 1929.

———. *This Progress: The Tragedy of Evolution.* London: Rich and Cowan, 1934.

Adami, J. George. *Medical Contributions to the Study of Evolution.* London: Duckworth, 1918.

———. "The Eternal Spirit of Nature as Seen by a Student of Science." *Modern Churchman* 19 (1926): 509–19.

———. *The Unity of Faith and Science.* London: Hodder and Stoughton, n.d.

Adami, Marie. *J. George Adami: Vice-Chancellor of the University of Liverpool, 1919–1926, Sometime Strathcona Professor of Pathology, McGill University, Montreal.* London: Constable, 1930.

Adams, Mary, ed. *Science in the Changing World.* London: Allen and Unwin, 1933.

Adeney, W. F. *A Century's Progress in Religious Life and Thought.* London: James Clarke, 1901.

Aldis, Arnold. "The Present Position of Evangelicals in Relation to Theology and Science." *Evangelical Quarterly*, October 1939, 336–44.

Alexander, Samuel. *Space, Time, and Deity: The Gifford Lectures at Glasgow, 1916–1918.* 2 vols. London: Macmillan, 1920.

Aliotta, Antonio. *The Idealist Reaction against Science.* Translated by Agnes McCaskill. London: Macmillan, 1914.

————. "Science and Religion in the Nineteenth Century." In *Science, Religion, and Reality*, edited by Joseph Needham, 149–86. London: Sheldon Press, 1925.

Allen, Garland E. *Life Science in the Twentieth Century.* New York: Wiley, 1975.

Allen, Grant. *The Woman Who Did.* Reprint, Oxford: Oxford University Press, 1995 [1895].

————. *The Evolution of the Idea of God: An Inquiry into the Origins of Religions.* New ed. London: Grant Richards, 1901 [1897].

Allen, H. S. "The Search for Truth." *Nature* 146 (1940): 637–40.

Alter, Peter. *The Reluctant Patron: Science and the State in Britain, 1850–1920.* Translated by Angela Davies. Oxford: Berg, 1987.

Altholz, Josef L. *Anatomy of a Controversy: The Debate over* Essays and Reviews, *1860–1864.* Aldershot: Scolar Press, 1994.

Annan, Noel. *Our Age: Portrait of a Generation.* London: Weidenfeld and Nicolson, 1990.

Archer-Shepherd, E. H. *Orthodox Religion in the Light of Today: Studies in Evolution, the Higher Criticism, Apologetics, Christology, and Other Subjects.* London: Rivingtons, 1929.

Armstrong, Henry E. "The Chemical Romance of the Green Leaf." In *The Great Design: Order and Progress in Nature*, edited by Frances Mason, 189–206. London: Duckworth, 1934.

Arrhenius, Svante. *Worlds in the Making: The Evolution of the Universe.* New York: Harper, 1908.

Ashton, J. "Evolution as a Theory of Ascent." *Dublin Review* 180 (1927): 53–72.

Askwith, Edward Harrison. "Sin, and the Need of Atonement." In *Essays on Some Theological Questions of the Day*, edited by H. B. Swete, 175–218. London: Macmillan, 1905.

Ayer, A. J. *Language, Truth, and Logic.* 2d ed. London: Victor Gollancz, 1964 [1936].

Balfour, Arthur J. *The Foundations of Belief: Being Notes Introductory to the Study of Theology.* London: Longmans, Green, 1895.

————. "Reflections Suggested by the New Theory of Matter." *Report of the British Association for the Advancement of Science* 1904: 3–14.

————. *Theism and Humanism: Being the Gifford Lectures Delivered at the University of Glasgow, 1914.* London: Hodder and Stoughton, 1915.

————. *Essays Speculative and Political.* London: Hodder and Stoughton, 1920.

————. "Introduction." In *Science, Religion, and Reality*, edited by Joseph Needham, 1–18. London: Sheldon Press, 1925.

Ball, Robert Stawell. *The Story of the Heavens.* New ed. London: Cassell, 1893 [1886].

Balmforth, Henry. *Is Christian Experience an Illusion? An Essay in the Philosophy of Religion.* London: Student Christian Movement, 1923.

Barnes, Barry, David Bloor, and John Henry. *Scientific Knowledge: A Sociological Analysis.* London: Athlone, 1996.

Barnes, Barry, and Steven Shapin, eds. *Natural Order: Historical Studies of Scientific Culture.* Beverly Hills: Sage Publications, 1979.

Barnes, Ernest William. *Religion and Science.* London: Hodder and Stoughton, 1923.

————. "Science and Modern Humanism." *School Science Review* 7 (1926): 145–53.

————. "Some Reflections on Eugenics and Religion." *Eugenics Review* 18 (1926): 7–14.

————. *Should Such a Faith Offend? Sermons and Addresses.* London: Hodder and Stoughton, 1927.

————. "Contributions to a British Association Discussion on the Evolution of the Universe." *Nature* 128 (1931): 719–22.

————. *Scientific Theory and Religion: The World Described by Science and Its Spiritual Interpretation.* Cambridge: Cambridge University Press, 1933.

————. "Christianity and Science." *Modern Churchman*, February 1936, 614–17.

————. Foreword to *The Church in the Twentieth Century*, edited by G. L. H. Harvey, ix–xviii. London: Macmillan, 1936.

————. *The Rise of Christianity.* London: Longmans, Green, 1947.

————. *Religion amid Turmoil.* Cambridge: Cambridge University Press, 1949.

Barnes, John. *Ahead of His Age: Bishop Barnes of Birmingham.* London: Collins, 1979.

Barrett, George S. "The Virgin Birth." In *The Old Faith and the New Theology*, edited by Charles H. Vine, 63–85. London: Sampson Low, Marston, 1907.

Barrett, William Fletcher. *Psychical Research.* London: Williams and Norgate, 1911.

————. *On the Threshold of the Unseen: An Examination of the Phenomena of Spiritualism and of the Evidence for Survival after Death.* London: Kegan Paul, 1917.

————. "The Deeper Issues of Psychical Research." *Contemporary Review* 113 (1918): 169–79.

————. "The Psychic Factor in Evolution." *Quest*, January 1918, 177–202.

————. *Death-Bed Visions.* London: Methuen, 1926.

————. *Personality Survives Death: Messages from Sir William Barrett.* Edited by Florence Barrett. London: Longmans, 1937.

————. *Creative Thought and the Problem of Evil.* 2d ed., rev. and enl. London: John M. Watkins, n.d.

Barry, F. R. *Christianity and Psychology: Lectures towards an Introduction.* London: Student Christian Movement, 1923.

Batchelor, John. *H. G. Wells.* Cambridge: Cambridge University Press, 1984.

Bateson, William. "Evolutionary Faith and Modern Doubts." *Science* 55 (1922): 55–61.

————. "The Revolt against the Teaching of Evolution in the United States." *Nature* 112 (1923): 313–14.

Baxendale, John, and Chris Pawling. *Narrating the Thirties: A Decade in the Making, 1930 to the Present.* London: Macmillan, 1996.

Beaman, Frank C. O. "The Church and Science." *Nineteenth Century and After* 89 (1921): 467–74.

Bebbington, D. W. *Evangelicalism in Modern Britain: A History from the 1730s to the 1980s.* London: Unwin Hyman, 1989.

———. "Science and Evangelical Theology in Britain from Wesley to Orr." In *Evangelicals and Science in Historical Perspective*, edited by David N. Livingstone, D. G. Hart, and Mark A. Noll, 120–39. New York: Oxford University Press, 1999.

Beeby, C. E. "Doctrinal Significance of a Miraculous Birth." *Hibbert Journal* 2 (1903): 125–40.

Bell, G. K. A. *Randall Davidson, Archbishop of Canterbury.* 2 vols. London: Oxford University Press, 1935.

Belloc, Hilaire. "A Few Words with Mr. Wells." *Dublin Review* 166 (1920): 182–202.

———. "Mr. Wells' *Outline of History.*" *London Mercury* 3 (1920): 43–62.

———. *A Companion to Mr. Wells'* Outline of History. London: Sheed and Ward, 1926.

———. "Is Darwinism Dead?" *Nature* 119 (1927): 277.

———. *Essays of a Catholic Layman in England.* London: Sheed and Ward, 1933 [1931].

———. "Man and the Machine." In *Science in the Changing World*, edited by Mary Adams, 240–52. London: Allen and Unwin, 1933.

———. *The Crisis of Our Civilization.* London: Cassell, 1937.

Bennett, J. H., ed. *Natural Selection, Heredity, and Eugenics: Including Selected Correspondence of R. A. Fisher with Leonard Darwin and Others.* Oxford: Clarendon Press, 1983.

Bergson, Henri. *Creative Evolution.* Translated by Arthur Mitchell. New York: Henry Holt, 1911.

Berman, David. *A History of Atheism in Britain from Hobbes to Russell.* London: Croom Helm, 1988.

Bernal, J. D. *The World, the Flesh, and the Devil: An Enquiry into the Future of the Three Enemies of the Rational Soul.* London: Kegan Paul, 1929.

———. "Irreligion." *Spectator*, 18 October 1930, 518–19.

———. *The Social Function of Science.* London: Routledge, 1939.

———. *Science in History.* 3d ed. 4 vols. Cambridge, MA: MIT Press, 1969 [1954].

Bethune-Baker, J. F. "Evolution and the New Theology." *Edinburgh Review* 243 (1926): 158–73.

———. *The Way of Modernism and Other Essays.* Cambridge: Cambridge University Press, 1927.

Bevan, Edwyn. "Bishop Barnes on Science and Superstition." *Hibbert Journal* 31 (1933): 176–88.

Birks, J. B. *Rutherford at Manchester.* London: Heywood, 1962.

Blatchford, Robert. *God and My Neighbour.* London: The Clarion Press, 1903.

———. *More Things in Heaven and Earth.* London: Methuen, 1925.

———. *My Eighty Years.* London: Cassell, 1931.

Blitz, David. *Emergent Evolution: Qualitative Novelty and the Levels of Reality.* Dordrecht: Kluwer, 1992.

Blyton, W. J. "Three Current Kinds of 'Dope': Ruthless Monopolists, Pseudo-Science, and Feather-Headed Neo-Psychology." *Catholic Gazette*, August 1930, 253–55.

Boakes, Robert. *From Darwin to Behaviourism: Psychology and the Mind of Animals.* Cambridge: Cambridge University Press, 1985.

Bonner, G. H. "The Case against 'Evolution.'" *Nineteenth Century* 102 (1927): 581–90.

———. "Evolution: A Reply to Sir Arthur Keith." *Nineteenth Century* 103 (1928): 375–90.

Booth, Bramwell. *These Fifty Years.* London: Cassell, 1929.

Boucher, David, and Andrew Vincent. *A Radical Hegelian: The Political and Social Philosophy of Henry Jones.* Cardiff: University of Wales Press, 1993.

Bouquet, A. C. *The Doctrine of God: Studies in the Divine Nature and Attributes with Chapters on the Philosophy of Worship.* Cambridge: Heffer, 1934.

Bower, Frederick O., et al. *Evolution in the Light of Modern Knowledge: A Collective Work.* London: Blackie, 1925.

Bowler, Peter J. *The Eclipse of Darwinism: Anti-Darwinian Evolution Theories in the Decades around 1900.* Baltimore: Johns Hopkins University Press, 1983.

———. "E. W. MacBride's Lamarckian Eugenics." *Annals of Science* 41 (1984): 245–60.

———. *Theories of Human Evolution: A Century of Debate, 1844–1944.* Baltimore: Johns Hopkins University Press; Oxford: Basil Blackwell, 1986.

———. *The Non-Darwinian Revolution: Reinterpreting a Historical Myth.* Baltimore: Johns Hopkins University Press, 1988.

———. *Life's Splendid Drama: Evolutionary Biology and the Reconstruction of Life's Ancestry, 1860–1940.* Chicago: University of Chicago Press, 1996.

———. "Evolution and the Eucharist: Bishop E. W. Barnes on Science and Religion in the 1920s and 1930s." *British Journal for the History of Science* 31 (1998): 453–67.

Box, Hubert S. *God and the Modern Mind.* With a foreword by A. E. Taylor. London: Society for the Promotion of Christian Knowledge, 1937.

Box, Joan Fisher. *R. A. Fisher: The Life of a Scientist.* New York: Wiley, 1978.

Bradshaw, David. *The Hidden Huxley: Contempt and Compassion for the Masses, 1920–36.* London: Faber and Faber, 1994.

Bragg, William Henry. *The World of Sound: Six Lectures Delivered before a Juvenile Audience at the Royal Institution, Christmas, 1919.* London: G. Bell, 1920.

———. "Science and the Worshipper." *Hibbert Journal* 38 (1940): 289–95.

———. *Science and Faith: Riddell Memorial Lectures, Thirteenth Series, Delivered before the University of Durham at King's College, Newcastle upon Tyne on 7 March 1941.* Oxford: Oxford University Press; London: Humphrey Milford, 1941.

Brandon, Ruth. *The Spiritualists: The Passion for the Occult in the Nineteenth and Twentieth Centuries.* London: Weidenfeld and Nicolson, 1983.

Bricknell, Edward John, "Sin and the Fall." In *Essays Catholic and Critical,* edited by E. G. Selwyn, 205–24. London: Society for the Promotion of Christian Knowledge, 1926.

Briggs, Asa. *The Birth of Broadcasting.* Vol. 1 of *The History of Broadcasting in the United Kingdom.* London: Oxford University Press, 1961.

————. *The Golden Age of Wireless.* Vol. 2 of *The History of Broadcasting in the United Kingdom.* London: Oxford University Press, 1965.

————. *The BBC: The First Fifty Years.* London: Oxford University Press, 1985.

Brinkworth, Guy. "Can Darwinism Be Revived?" *The Month,* November 1932, 395–401.

Broad, C. D. *The Mind and Its Place in Nature.* London: Kegan Paul, 1925.

————. "The Present Relations of Science and Religion." *Philosophy* 14 (1939): 131–54.

Brock, W. H., and R. M. MacLeod. "The Scientists' Declaration: Reflexions on Science and Belief in the Wake of *Essays and Reviews.*" *British Journal for the History of Science* 9 (1976): 39–66.

Broderick, J. "Evolution Still an Hypothesis." *The Month,* August 1926, 97–107.

Broks, Peter. *Media Science before the Great War.* London: Macmillan, 1996.

Brooke, John Hedley. "Natural Theology and the Plurality of Worlds: Observations on the Brewster-Whewell Debate." *Annals of Science* 34 (1977): 221–86.

————. *Science and Religion: Some Historical Perspectives.* Cambridge: Cambridge University Press, 1991.

Broom, Robert. *The Mammal-Like Reptiles of South Africa and the Origin of Mammals.* London: H. F. and G. Witherby, 1932.

————. *The Coming of Man: Was It Accident or Design?* London: H. F. and G. Witherby, 1933.

Brush, Stephen G. *The Temperature of History: Phases of Science and Culture in the Nineteenth Century.* New York: Burt Franklin, 1978.

Bud, Robert. "Penicillin and the New Elizabethans." *British Journal for the History of Science* 31 (1998): 305–33.

Budd, Susan. *Varieties of Unbelief: Atheists and Agnostics in English Society, 1850–1960.* London: Heinemann, 1970.

Burke, John Butler. "Mr. Wells and Modern Science." *Dublin Review* 169 (1921): 222–36.

Burt, Cyril, ed. *How the Mind Works.* London: Allen and Unwin, 1933.

Butterworth, G. W. "What the Scientists Forget: A Reply to Professor MacBride." *Hibbert Journal* 34 (1936): 368–77.

Campbell, Reginald John. "The Aim of the New Theology." *Hibbert Journal* 5 (1907): 481–98.

————. *The New Theology.* London: Chapman and Hall, 1907.

————. *A Spiritual Pilgrimage.* 2d imp. London: Williams and Norgate, 1916.

————. *The War and the Soul.* London: Chapman and Hall, 1916.

Caroe, G. M. *William Henry Bragg, 1862–1942: Man and Scientist.* Cambridge: Cambridge University Press, 1978.

Carpenter, James. *Gore: A Study in Liberal Catholic Thought.* London: Faith Press, 1960.

Carr, H. Wildon. *Henri Bergson: The Philosophy of Change.* London: T. C. and E. C. Jack, 1911.

————. *The Philosophy of Change: A Study of the Fundamental Principle of the Philosophy of Bergson.* London: Macmillan, 1914.

————. *Changing Backgrounds in Religion and Ethics: A Metaphysical Meditation.* London: Macmillan, 1927.

————. *The Unique Status of Man.* London: Macmillan, 1927.

Cecil, William Gascoyne. *Science and Religion.* London: Hodder and Stoughton, 1906.

————. "Darwinism and What It Implies." *Hibbert Journal* 27 (1929): 666–75.

Chadwick, Owen. *The Victorian Church.* London: A. & C. Black, 1966.

————. *Hensley Henson: A Study in the Friction between Church and State.* Oxford: Clarendon Press, 1983.

Cherry, Matt. "Science vs. Religion: War or Peace?" *Free Inquiry* 18 (1998): 14.

Chesterton, Gilbert K. *All Things Considered.* 5th ed. London: Methuen, 1909 [1908].

————. *George Bernard Shaw.* London: John Lane, The Bodley Head, 1914 [1909].

————. *Eugenics and Other Evils.* London: Cassell, 1922.

————. *The Everlasting Man.* London: Hodder and Stoughton, 1926 [1925].

————. *As I Was Saying: A Book of Essays.* London: Methuen, 1936.

Child, R. L. "Science, A Friend of Religion." *Baptist Quarterly,* January 1935, 202–8.

Clark, R. E. D. *The Universe and God: A Study of the Order of Nature in the Light of Modern Knowledge.* London: Hodder and Stoughton, 1939.

Clarke, A. H. T. "The Church of the Future: An Eirenicon." *Nineteenth Century* 86 (1919): 90–102.

————. "Evolution v. Creation: A Reply to Sir E. Ray Lankester." *Nineteenth Century* 90 (1921): 165–76.

————. "The Passing of the Nineteenth Century." *Nineteenth Century* 98 (1925): 933–41.

Clements, Keith W. *Lovers of Discord: Twentieth-Century Theological Controversies in England.* London: SPCK, 1988.

Clodd, Edward. *The Story of Creation: A Plain Account of Evolution.* Rev. ed. London: Longmans, Green, 1909 [1888].

————. *Grant Allen: A Memoir.* London: Grant Richards, 1900.

————. *Animism: The Seed of Religion.* London: Archibald Constable, 1905.

Clutton-Brock, A. "Spirit and Matter." In *The Spirit,* edited by B. H. Streeter, 313–46. London: Macmillan, 1919.

Cohen, Chapman. *Religion and Sex: Studies in the Pathology of Religious Development.* London: T. N. Foulis, 1919.

————. *God and Evolution.* London: Pioneer Press, 1925.

————. *God and the Universe: Eddington, Jeans, Huxley, and Einstein.* With a Reply by Professor A. S. Eddington. London: Pioneer Press, for the Secular Society, 1931.

Cohen, Chapman, and C. E. M. Joad. *Materialism: Has It Been Exploded? Verbatim Report of Debate between Chapman Cohen and C. E. M. Joad held at the Caxton Hall, Westminster, SW1, on Wednesday, September 26, 1928*. London: Watts, 1928.

Collingwood, R. G. *The Idea of Nature*. Oxford: Clarendon Press, 1945.

Collini, Stefan. *Liberalism and Sociology: L. T. Hobhouse and Political Argument in England, 1880–1914*. Cambridge: Cambridge University Press, 1979.

Collins, Peter. "The British Association as Public Apologist for Science, 1919–1946." In *The Parliament of Science: The British Association for the Advancement of Science, 1831–1981*, edited by Roy MacLeod and Peter Collins, 211–36. Northwood, Middlesex: Science Reviews, 1981.

Conn, J. C. *The Menace of the New Psychology*. London: Inter-Varsity Fellowship, 1939.

Coren, Michael. *The Invisible Man: The Life and Liberties of H. G. Wells*. New York: Atheneum, 1993.

Cornforth, Maurice. *Science versus Idealism: An Examination of "Pure Empiricism" and Modern Logic*. London: Lawrence and Wishart, 1946.

Coulson, C. A. *Science and Christian Belief*. London: Oxford University Press, 1955.

Courtney, W. L., ed. *Do We Believe? A Record of a Great Correspondence in the* Daily Telegraph, *October, November, December, 1904*. London: Hodder and Stoughton, 1905.

Crichton-Miller, H. *The New Psychology and the Preacher*. London: Jarrolds, 1924.

Crook, D. P. *Darwinism, War, and History: The Debate over the Biology of War from the* Origin of Species *to the First World War*. Cambridge: Cambridge University Press, 1994.

Crookes, William. "President's Address." *Report of the British Association for the Advancement of Science* 1898: 3–38.

Cross, F. Leslie. *Religion and the Reign of Science*. London: Longmans, Green, 1930.

———. "Science and Religion in Contemporary Culture." *Church Quarterly Review*, October 1933, 132–44.

Crowther, J. G. *The Social Relations of Science*. London: Macmillan, 1941.

———, ed. *Science Today: The Scientific Outlook on World Problems Explained by Leading Exponents of Modern Scientific Thought. Planned and Arranged by the late Sir J. Arthur Thomson*. London: Eyre and Spottiswoode, 1934.

Crowther, James Arnold. "Radiation." In *The Great Design: Order and Progress in Nature*, edited by Frances Mason, 41–61. London: Duckworth, 1934.

Curran, James, and Jean Seaton. *Power without Responsibility: The Press and Broadcasting in Britain*. 3d ed. London: Routledge, 1988.

Dale, Alzina Stone. *The Outline of Sanity: A Biography of G. K. Chesterton*. Grand Rapids, MI: Eerdmans, 1982.

Dampier-Whetham, William Cecil Dampier. *A History of Science and Its Relations with Philosophy and Religion*. Cambridge: Cambridge University Press, 1929.

D'Arcy, Charles F. "Love and Omnipotence." In *God and the Struggle for Existence*, edited by B. H. Streeter, 15–65. London: Student Christian Movement, 1919.

———. *Science and Creation: The Christian Interpretation.* London: Longmans, Green, 1925.

———. *Providence and the World-Order.* London: Hodder and Stoughton, 1932.

———. *The Adventures of a Bishop: A Phase of Irish Life: A Personal and Historical Narrative.* London: Hodder and Stoughton, 1934.

D'Arcy, M. C. "Science and Theology." In *Science Today*, edited by J. G. Crowther, 173–98. London: Macmillan, 1934.

Darwin, Charles. *The Origin of Species.* 6th ed. London: John Murray, 1872.

Darwin, Francis, ed. *The Life and Letters of Charles Darwin.* 3 vols. London: John Murray, 1887.

Davies, A. Morley. *Evolution and Its Modern Critics.* London: Thomas Murby, 1937.

Davies, Lewis Merson. *The Bible and Modern Science.* 3d ed. London: Pickering and Inglis, 1934 [1925].

Dawson, W. Bell. *The Bible Confirmed by Modern Science.* London: Marshall, Morgan and Scott, 1936.

Dendy, A. *Animal Life and Human Progress.* London: Constable, 1919.

Den Otter, Sandra M. *British Idealism and Social Explanation: A Study in Late Victorian Thought.* Oxford: Clarendon Press, 1996.

Desmond, Adrian. *Huxley: The Devil's Disciple.* London: Michael Joseph, 1994.

———. *Huxley: Evolution's High Priest.* London: Michael Joseph, 1996.

Dewar, Douglas. *Difficulties of the Evolution Theory.* London: Edward Arnold, 1931.

———. "The Limitations of Organic Evolution." *Journal of the Transactions of the Victoria Institute* 64 (1932): 120–43.

———. "A Critical Examination of the Supposed Fossil Links between Man and the Lower Animals." *Journal of the Transactions of the Victoria Institute* 67 (1934): 157–83.

———. *More Difficulties of the Evolution Theory.* London: Thynne, 1938.

———. *Man: A Special Creation.* London: Thynne, n.d.

Dewar, Douglas, L. Merson Davies, and J. B. S. Haldane. *Is Evolution a Myth?* London: Evolution Protest Movement, 1957.

Dick, Steven J. *The Biological Universe: The Twentieth-Century Extraterrestrial Life Debate and the Limits of Science.* Cambridge: Cambridge University Press, 1996.

Dickinson, G. Lowes. *Religion: A Criticism and a Forecast.* London: Brimley Johnson and Ince, 1905.

Di Gregorio, Mario. "Entre Mephistopheles et Luther: Ernst Haeckel et la réforme de l'univers." In *Darwinism et société*, edited by Patrick Tort, 237–84. Paris: PUF, 1992.

Dilliston, F. W. *Charles Raven: Naturalist, Historian, Theologian.* London: Hodder and Stoughton, 1975.

Dingle, Herbert. "Physics and God." *Hibbert Journal* 27 (1928): 35–46.

————. "Physics and Reality." *Nature* 126 (1930): 799–800.

————. *Through Science to Philosophy.* Oxford: Clarendon Press, 1937.

————. *The Sources of Eddington's Philosophy.* Cambridge: Cambridge University Press, 1954.

Dingle, Reginald J. "What of the New Physics?" *The Month,* January 1931, 37–44.

Dival, Colin. "From a Victorian to a Modern: Julian Huxley and the English Intellectual Climate." In *Julian Huxley: Biologist and Statesman of Science,* edited by C. Kenneth Waters and Albert Van Helden, 31–44. Houston: Rice University Press, 1992.

Donkin, Bryan. "Science and Psychical Research." *Nature* 118 (1926): 658–59.

Dorlodot, Henri de. *Darwinism and Catholic Thought.* Translated by E. Messenger. New York: Benziger Brothers, 1925.

Dougal, Lily. "The Survival of the Fittest." In *God and the Struggle for Existence,* edited by B. H. Streeter, 66–109. London: Student Christian Movement, 1919.

Douglas, A. Vibart. *The Life of Arthur Stanley Eddington.* London: Nelson, 1956.

Downey, Richard. *Some Errors of H. G. Wells: A Catholic's Criticism of* The Outline of History. London: Burnes Oates and Washbourne, 1921.

Doyle, Arthur Conan. *The New Revelation.* London: Hodder and Stoughton, 1918.

————. *The Coming of the Fairies.* London: Hodder and Stoughton, 1922.

————. *Memories and Adventures.* 2d ed. London: John Murray, 1930.

Draper, John William. *History of the Conflict between Religion and Science.* London: Kegan Paul, 1882 [1875].

Drawbridge, C. L. *The Religion of Scientists: Being Recent Opinions Expressed by Two Hundred Fellows of the Royal Society on the Subject of Religion and Theology.* London: Ernest Benn, 1932.

Driesch, Hans. *The Science and Philosophy of the Organism: The Gifford Lectures Delivered before the University of Aberdeen in the Years 1907/08.* 2 vols. London: A. and C. Black, 1908.

————. *Mind and Body: A Criticism of Psychological Parallelism.* Translated by Theodore Besterman. London: Methuen, 1927.

————. "The Breakdown of Materialism." In *The Great Design: Order and Progress in Nature,* edited by Frances Mason, 283–303. London: Duckworth, 1934.

Dronamraju, Krishna R. *If I Am to Be Remembered: The Life and Work of Julian Huxley, with Selected Correspondence.* Singapore: World Scientific, 1993.

Drummond, Henry. *Natural Law in the Spiritual World.* London: Hodder and Stoughton, 1885.

————. *The Ascent of Man.* New York: James Pott, 1904 [1894].

Duckworth, Wynfrid Laurence Henry. "Man's Origin, and His Place in Nature." In *Essays on Some Theological Questions of the Day,* edited by H. B. Swete, 147–74. London: Macmillan, 1905.

Dunne, J. W. *An Experiment with Time.* London: Faber and Faber, 1947 [1927].

————. *The Serial Universe.* London: Faber and Faber, 1934.

Durant, John. "Darwinism and Divinity: A Century of Debate." In *Darwinism and Divinity: Essays on Evolution and Religious Belief*, edited by John Durant, 9–39. Oxford: Basil Blackwell, 1985.

Eagle, Albert. *The Philosophy of Religion versus the Philosophy of Science: An Exposure of the Worthlessness and Absurdity of Some Conventional Conclusions of Modern Science*. Printed for private circulation, 1935.

Eddington, Arthur Stanley. "The Domain of Physical Science." In *Science, Religion, and Reality*, edited by Joseph Needham, 189–218. London: Sheldon Press, 1925.

———. *The Nature of the Physical World*. Cambridge: Cambridge University Press, 1928.

———. *Science and the Unseen World*. London: Allen and Unwin, 1929.

———. *The Expanding Universe*. Cambridge: Cambridge University Press, 1933.

———. *New Pathways in Science*. Cambridge: Cambridge University Press, 1947 [1935].

———. *The Philosophy of Physical Science*. Cambridge: Cambridge University Press, 1941 [1939].

———. *Fundamental Theory*. Cambridge: Cambridge University Press, 1946.

Einstein, Albert. "What I Believe." *Forum* 84 (1930): 193–94.

———. "Science and Religion." *Nature* 146 (1940): 605–7.

Eisendrath, Craig R. *The Unifying Moment: The Psychological Philosophy of William James and Alfred North Whitehead*. Cambridge, MA: Harvard University Press, 1971.

Eksteins, Modris. *Rites of Spring: The Great War and the Birth of the Modern Age*. Boston: Houghton Mifflin, 1989.

Elder, Gregory P. *Chronic Vigour: Darwin, Anglicans, Catholics, and the Development of a Doctrine of Providential Evolution*. Lanham, MD: University Press of America, 1996.

Elliot, Hugh S. R. *Modern Science and the Illusions of Professor Bergson*. Preface by E. Ray Lankester. London: Longmans, Green, 1912.

———. *Modern Science and Materialism*. London: Longmans, Green, 1919.

Emmet, Dorothy. "The Philosopher." In *William Temple, Archbishop of Canterbury: His Life and Letters*, by F. A. Iremonger, 521–39. London: Oxford University Press, 1948.

———. *Whitehead's Philosophy of Organism*. 2d ed. London: Macmillan, 1966.

"Energy and Atoms." *Nature* 122 (1928): 555–56.

Ensor, R. C. K. *England, 1870–1914*. Oxford: Clarendon Press, 1931.

Eve, A. S., ed. *Rutherford: Being the Life and Letters of the Rt. Hon. Lord Rutherford, O. M.* Cambridge: Cambridge University Press, 1939.

Everdel, William R. *The First Moderns: Profiles in the Origins of Twentieth-Century Thought*. Chicago: University of Chicago Press, 1997.

Farley, John. *The Spontaneous Generation Controversy from Descartes to Oparin*. Baltimore: Johns Hopkins University Press, 1977.

Farmer, H. H. "Science and God." *Spectator*, 7 February 1931, 173–75.

Fawcett, Edward Douglas. *The World as Imagination* (series I). London: Macmillan, 1916.

Ferngren, Gary B., and Ronald L. Numbers. "C. S. Lewis on Creation and Evolution: The Acworth Letters, 1944–1960." *Perspectives on Science and Christian Faith* 48 (1996): 28–33.

Ferrar, W. J. "The Gloom of Dean Inge." *Hibbert Journal* 20 (1921–22): 317–23.

Figgis, John Neville. *Civilisation at the Cross Roads: Four Lectures Delivered before Harvard University in the Year 1911 on the William Beldon Noble Foundation.* London: Longmans, Green, 1912.

Findlay, G. H. *Dr. Robert Broom, F. R. S.: Palaeontologist and Physician, 1866–1951: A Biography, Appreciation, and Bibliography.* Cape Town: A. A. Balkema, 1972.

Fisch, Menachim, and Simon Schaffer. *William Whewell: A Composite Portrait.* Oxford: Clarendon Press, 1991.

Fisher, Ronald Aylmer. *The Genetical Theory of Natural Selection.* Oxford: Clarendon Press, 1930.

———. "Indeterminism and Natural Selection." *Philosophy of Science* 1 (1934): 99–117.

———. "The Renaissance of Darwinism." *The Listener* 37 (1947): 1001.

———. *Creative Aspects of Natural Law: The Eddington Memorial Lecture.* Cambridge: Cambridge University Press, 1950.

———. "Science and Christianity." *The Friend,* 21 October 1955, 995–96.

Fleming, John Ambrose. *Evolution and Revelation.* London: Religious Tract Society, 1926.

———. "Evolution and Revelation." *Journal of the Transactions of the Victoria Institute* 59 (1927): 11–40.

———. "Evolution: A Reply to Sir Arthur Keith." *Nineteenth Century* 103 (1928): 510–15.

———. "Truth and Error in the Doctrine of Evolution." *Nineteenth Century* 103 (1928): 76–87.

———. *Memories of a Scientific Life.* London: Marshall, Morgan and Scott, 1934.

———. "Modern Anthropology versus Biblical Statements." *Journal of the Transactions of the Victoria Institute* 67 (1934): 15–42.

———. *Modern Anthropology versus Biblical Statements on Human Origins.* London: Victoria Institute, 1935.

———. *The Origin of Mankind: Viewed from the Standpoint of Revelation and Research.* London: Marshall, Morgan and Scott, 1935.

———. *Evolution or Creation?* London: Marshall, Morgan and Scott, n.d.

Flint, Robert. *Theism: Being the Baird Lecture for 1876.* Edinburgh: William Blackwood, 1877.

Formby, C. W. *Re-Creation: A New Aspect of Evolution.* London: Williams and Norgate, 1907.

———. *The Unveiling of the Fall.* London: Williams and Norgate, 1923.

Forsyth, P. T. "Immanence and Incarnation." In *The Old Faith and the New Theology,* edited by Charles H. Vine, 47–61. London: Sampson Low, Marston, 1907.

Fournier d'Albe, E. E. *The Life of Sir William Crookes, O. M., F. R. S.* London: T. Fisher Unwin, 1923.

Fowler, J. Cyril. *Psychological Studies of Religious Questions.* London: Williams and Norgate, 1924.

Fox, Adam. *Dean Inge.* London: John Murray, 1960.

Frank, P. *Einstein: His Life and Times.* New York: Knopf, 1947.

Frazer, James George. *The Golden Bough: A Study in Magic and Religion.* Abridged ed. London: Macmillan, 1924.

French, Patrick. *Younghusband: The Last Great Imperial Adventurer.* London: Harper Collins, 1994.

Freud, Sigmund. *The Standard Edition of the Complete Psychological Works of Sigmund Freud.* Edited by James Strachey. 24 vols. London: Hogarth Press, 1959–1974.

Friend, Julius W., and James Feibleman. *Science and the Spirit of Man: A New Ordering of Existence.* London: Allen and Unwin, 1933.

Fulton, William. *Nature and God: An Introduction to Theistic Studies with Special Reference to the Relation of Science and Religion.* Edinburgh: T. and T. Clark, 1927.

Galton, Francis. *English Men of Science: Their Nature and Nurture.* London: Macmillan, 1874.

Gasman, Daniel. *Haeckel's Monism and the Birth of Fascist Ideology.* New York: Peter Lang, 1998.

Geddes, Patrick, and J. Arthur Thomson. *Biology.* London: Williams and Norgate, 1925.

Gibson, W. R. Boyce. "The Problem of Freedom in Its Relation to Psychology." In *Personal Idealism,* edited by H. Sturt, 134–92. London: Macmillan, 1902.

Gilbert, Alan D. *The Making of Post-Christian Britain: A History of the Secularization of Modern Society.* London: Longmans, 1980.

Gill, H. V. "Science Gets Religion!" *The Month,* April 1931, 310–19.

Glazebrook, M. G. *The Faith of a Modern Churchman.* London: John Murray, 1918.

Goldsmith, Maurice. *Joseph Needham: 20th-Century Renaissance Man.* Paris: UNESCO, 1995.

Goodrich, Albert. "The Immanence of God and the Divinity of Christ." In *The Old Faith and the New Theology,* edited by Charles H. Vine, 15–27. London: Sampson Low, Marston, 1907.

Gore, Charles. *The New Theology and the Old Religion: Being Eight Lectures, together with Five Sermons.* London: John Murray, 1907.

———. *Can We Then Believe? Summary of Volumes on "Reconstruction of Belief" and Reply to Criticisms.* London: John Murray, 1926.

———. *The Reconstruction of Belief: Belief in God. Belief in Christ. The Holy Spirit and the Church.* London: John Murray, 1926.

———, ed. *Lux Mundi: A Series of Studies in the Religion of the Incarnation.* 4th ed. London: John Murray, 1890 [1889].

Grant, John W. *Free Churchmanship in England, 1870–1940: With Special Reference to Congregationalism.* London: Independent Press, n.d.

Grant, Malcolm. *A New Argument for God and Survival: And a Solution to the Problem of Supernatural Events.* London: Faber and Faber, 1934.

Gray, J. "The Mechanistic View of Life." *Report of the British Association for the Advancement of Science* 1933: 81–92.

Green, Peter. *The Problem of Evil: Being an Attempt to Shew that the Existence of Sin and Pain in the World Is Not Inconsistent with the Goodness and Power of God.* London: Longmans, Green, 1920.

Greene, John C. "The Interaction of Science and World View: Sir Julian Huxley's Evolutionary Biology." *Journal of the History of Biology* 23 (1990): 39–55.

Greene, S. J. D. *Religion in an Age of Decline: Organization and Experience in Industrial Yorkshire, 1870–1920.* Cambridge: Cambridge University Press, 1996.

Greenwood, William Osborne. *Biology and Christian Belief.* London: Student Christian Movement, 1938.

———. *Christianity and the Mechanists.* London: Eyre and Spottiswoode, 1941.

Gregory, Frederick. "The Impact of Darwinian Evolution on Protestant Theology in the Nineteenth Century." In *God and Nature,* edited by David C. Lindberg and Ronald L. Numbers, 369–90. Berkeley: University of California Press, 1986.

Gregory, Richard. *Discovery: Or the Spirit and Service of Science.* London: Macmillan, 1917 [1916].

———. "Religion in Science." *Nature* 143 (1939): 68–70.

———. *Religion in Science and Civilization.* London: Macmillan, 1940.

Grensted, L. W. *Psychology and God: A Study of the Implications of Recent Psychology for Religious Belief and Practice.* London: Longmans, Green, 1930.

Griffith-Jones, Ebenezer. *The Ascent through Christ: A Study of the Doctrine of Redemption in the Light of the Theory of Evolution.* London: James Bowen, 1900 [1899].

———. *Providence—Divine and Human: A Study of the World-Order in the Light of Modern Thought.* Vol. 1, *Some Problems of Divine Providence.* London: Hodder and Stoughton, 1925.

———. *Providence—Divine and Human: A Study of the World-Order in the Light of Modern Thought.* Vol. 2, *The Dominion of Man (Some Problems in Human Providence).* London: Hodder and Stoughton, 1926.

Grist, Charles J. *Science and the Bible.* London: Skeffington, n.d. [1941].

Grogin, R. C. *The Bergsonian Controversy in France, 1900–1914.* Calgary: University of Calgary Press, 1988.

Gruber, Jacob. *A Conscience in Conflict: A Life of St. George Jackson Mivart.* New York: Columbia University Press, 1960.

Habgood, John Stapylton. "The Uneasy Truce between Science and Religion." In *Soundings: Essays Concerning Christian Understanding,* edited by A. R. Vidler, 21–41. Cambridge: Cambridge University Press, 1962.

Haeckel, Ernst. *The Riddle of the Universe at the Close of the Nineteenth Century.* Translated by Joseph McCabe. London: Watts, 1900.

Haeckel, Ernst, et al. *Evolution in Modern Thought.* New York: Boni and Liveright, 1917.

Haldane, J. B. S. *Daedalus: Or Science and the Future.* London: Kegan Paul, 1924.

———. *Possible Worlds and Other Essays.* London: Chatto and Windus, 1930 [1927].

———. *Science and Ethics.* London: Watts, 1928.

————. *The Causes of Evolution.* London: Longmans, Green, 1932.

————. *The Inequality of Man and Other Essays.* London: Chatto and Windus, 1932.

————. *Science and Everyday Life.* Harmondsworth: Pelican, 1941 [1939].

————. *Science and Life: Essays of a Rationalist.* Edited by J. Maynard Smith. London: Pemberton, 1968.

————. *Haldane's Daedalus Revisited.* Edited by Krishna R. Dronamraju. Oxford: Oxford University Press, 1995.

Haldane, John Scott. *Mechanism, Life, and Personality: An Examination of the Mechanistic Theory of Life and Mind.* London: John Murray, 1913.

————. *Organism and Environment as Illustrated by the Physiology of Breathing.* New Haven: Yale University Press, 1916.

————. "Natural Science and Religion." *Hibbert Journal* 21 (1921–22): 417–35.

————. "Biology and Religion." *Nature* 114 (1924): 468–71.

————. *The Sciences and Philosophy: Gifford Lectures, University of Glasgow, 1927–28.* London: Hodder and Stoughton, 1928.

————. *Materialism.* London: Hodder and Stoughton, 1932.

————. "The Bishop of Birmingham's Gifford Lectures." *The Listener,* supplement, 3 May 1933: 3–4.

————. *The Philosophy of a Biologist.* Oxford: Clarendon Press, 1935.

Haldane, R. B. *The Reign of Relativity.* London: John Murray, 1921.

Haldane, R. B., and J. S. Haldane. "The Relation of Philosophy to Science." In *Essays in Philosophical Criticism,* edited by Andrew Seth Pringle Pattison and R. B. Haldane, 41–66. London: Longmans, 1883.

Hall, A. Daniel. "The Faith of a Man of Science." *Nineteenth Century* 110 (1931): 717–22.

Halliday, W. Fearon. *Psychology and Religious Experience.* London: Hodder and Stoughton, 1929.

Hamilton, Floyd E. "The Present Status of Evolutionary Faith." *Evangelical Quarterly,* July 1929, 268–73.

————. *The Basis of Evolutionary Faith: A Critique of the Theory of Evolution.* London: James Clarke, 1931.

Hand, J. E., ed. *Ideas of Science and Faith: Essays by Various Authors.* London: George Allen, 1904.

Haraway, Donna. *Crystals, Fabrics, and Fields: Metaphors of Organicism in Twentieth-Century Developmental Biology.* New Haven: Yale University Press, 1976.

Hardman, O., ed. *Psychology and the Church.* London: Macmillan, 1925.

Hardwick, John Charlton. *Religion and Science from Galileo to Bergson.* London: Society for the Promotion of Christian Knowledge, 1920.

————. *Religion and Science.* Papers in Modern Churchmanship, 7. London: Longmans, Green, 1925.

————. *The Bible and Science.* London: John Murray, 1938.

Harper, George Mills. *Yeats's Golden Dawn.* London: Macmillan, 1974.

———, ed. *Yeats and the Occult.* London: Macmillan, 1975.

Harrington, Anne. *Reenchanted Science: Holism in German Culture from Wilhelm II to Hitler.* Princeton: Princeton University Press, 1996.

Harris, G. H. *Vernon Faithful Storr: A Memoir.* London: Society for the Promotion of Christian Knowledge, 1943.

Harvey, G. L. H., ed. *The Church in the Twentieth Century.* London: Macmillan, 1936.

Hastings, Adrian. *A History of English Christianity, 1920–1985.* London: Collins, 1986.

Hayward, F. H., and Arnold Freeman. *The Spiritual Foundations of Reconstruction: A Plea for New Educational Methods.* London: P. S. King, 1919.

Heard, Gerald. *The Source of Civilization.* London: Jonathan Cape, 1935.

Heim, Karl. *Christian Faith and Natural Science.* Translated by Neville Horton Smith. London: Student Christian Movement, 1953.

Heimann, P. M. "The Unseen Universe: Physics and the Philosophy of Nature in Victorian Britain." *British Journal for the History of Science* 6 (1972): 73–79.

Helmstadter, Richard J., and Bernard Lightman, eds. *Victorian Faith in Crisis: Continuity and Change in Nineteenth-Century Religious Belief.* London: Macmillan; Stanford: Stanford University Press, 1990.

Henslow, George. *The Proofs of the Truths of Spiritualism.* London: Kegan Paul, 1919.

———. *The Religion of the Spirit World: Written by the Spirits Themselves.* London: Kegan Paul, 1920.

Henson, Herbert Hensley. *Retrospect of an Unimportant Life.* 3 vols. London: Oxford University Press, 1942–1950.

———. *Letters of Herbert Hensley Henson.* Edited by Evelyn Fox Braley. London: Society for the Promotion of Christian Knowledge, 1950.

Heseltine, G. C. "Professor Haldane and Evolution." *English Review* 55 (1932): 11–18.

Hicks, G. Dawes. "Professor Eddington's Philosophy of Nature." *Proceedings of the Aristotelian Society* 29 (1928–29): 275–300.

Hiebert, Erwin N. "Modern Physics and Christian Faith." In *God and Nature,* edited by David C. Lindberg and Ronald L. Numbers, 427–47. Berkeley: University of California Press, 1986.

Hinchcliff, Peter. *God and History: Aspects of British Theology, 1875–1914.* Oxford: Clarendon Press, 1992.

Hobhouse, L. T. *Morals in Evolution: A Study in Comparative Ethics.* 2 vols. London: Chapman and Hall, 1906.

———. *Development and Purpose: An Essay towards a Philosophy of Evolution.* London: Macmillan, 1913.

Hobson, E. W. *The Domain of Natural Science: The Gifford Lectures Delivered in the University of Aberdeen in 1921 and 1922.* Cambridge: Cambridge University Press, 1923.

Hodge, M. J. S. "Biology and Philosophy (Including Ideology): A Study of Fisher and Wright." In *The Founders of Evolutionary Genetics*, edited by Sahotra Sarker, 231–93. Dordrecht: Kluwer, 1992.

Hogben, Lancelot. *The Nature of Living Matter.* London: Kegan Paul, 1930.

———. *Lancelot Hogben, Scientific Humanist: An Unauthorized Autobiography.* Edited by Adrian and Anne Hogben. Woodbridge, Suffolk: Merlin Press, 1998.

Holmes, Edmond G. A. "The Idea of Evolution and the Idea of God." *Hibbert Journal* 21 (1921–22): 227–47.

Holroyd, Michael. *Bernard Shaw.* 4 vols. London: Chatto and Windus, 1988–92.

Hopkinson, Austin. "Relativity and Revelation." *Hibbert Journal* 21 (1922): 53–64.

Howe, Ellic. *The Magicians of the Golden Dawn: A Documentary History of a Magical Order.* London: Routledge, 1972.

Hoyle, Fred. *The Nature of the Universe: A Series of Broadcast Lectures.* Oxford: Blackwell, 1950.

Huxley, Aldous. *Crome Yellow.* Harmondsworth: Penguin, 1936 [1921].

———. *Brave New World.* Harmondsworth: Penguin, 1955 [1932].

Huxley, Julian Sorrell. *The Individual in the Animal Kingdom.* Cambridge: Cambridge University Press, 1912.

———. "Progress: Biological and Other." *Hibbert Journal* 21 (1921–22): 436–60.

———. *Essays of a Biologist.* London: Chatto and Windus, 1923.

———. "Science and Religion." In *Science and Civilization*, edited by F. S. Marvin, 279–329. Oxford: Oxford University Press; London: Humphrey Milford, 1923.

———. *Religion without Revelation.* London: Ernest Benn, 1927.

———. "Progress Shown in Evolution." In *Creation by Evolution*, edited by Frances Mason, 227–29. New York: Macmillan, 1928.

———. *Science, Religion, and Human Nature.* London: Watts, 1930.

———. *What Dare I Think? The Challenge of Modern Science to Human Action and Belief.* London: Chatto and Windus, 1931.

———. "Man and Reality." In *Science in the Changing World*, edited by Mary Adams, 186–98. London: Allen and Unwin, 1933.

———. *Evolution: The Modern Synthesis.* London: Allen and Unwin, 1942.

———. *Essays of a Humanist.* London: Chatto and Windus, 1964.

———. *Memories.* New York: Harper and Row, 1970.

———, ed. *The Humanist Frame.* London: Allen and Unwin, 1961.

Huxley, Thomas Henry. *Methods and Results.* Vol. 1 of *Collected Essays.* London: Macmillan, 1893.

———. *Darwiniana.* Vol. 2 of *Collected Essays.* London: Macmillan, 1893.

———. *Discourses Biological and Geological.* Vol. 7 of *Collected Essays.* London: Macmillan, 1894.

————. "Evolution and Ethics." In *Evolution and Ethics.* Vol. 9 of *Collected Essays.* London: Macmillan, 1894.

Hynes, Samuel. *The Edwardian Turn of Mind.* Princeton: Princeton University Press, 1968.

————. *The Auden Generation: Literature and Politics in England in the 1930s.* London: Faber, 1976.

Inge, William Ralph. "The Person of Christ." In *Contentio Veritatis,* by H. Rashdall et al., 59–104. London: John Murray, 1902.

————. "The Sacraments." In *Contentio Veritatis,* by H. Rashdall et al., 270–310. London: John Murray, 1902.

————. "Survival and Immortality." *Hibbert Journal* 15 (1917): 585–97.

————. *Outspoken Essays.* London: Longmans, Green, 1921 [1919].

————. *Outspoken Essays (Second Series).* London: Longmans, Green, 1922.

————. "Conclusion." In *Science, Religion, and Reality,* edited by Joseph Needham, 347–89. London: Sheldon Press, 1925.

————. *Lay Thoughts of a Dean.* London: Putnam, 1926.

————. *Science and Ultimate Truth: Fison Memorial Lecture, 1926.* London: Longmans, Green, 1926.

————. *The Church in the World: Collected Essays.* London: Longmans, Green, 1927.

————. *Wit and Wisdom of Dean Inge.* Selected by Sir James Marchant. London: Longmans, Green, 1927.

————. *Christian Ethics and Modern Problems.* London: Hodder and Stoughton, 1930.

————. *More Lay Thoughts of a Dean.* London: Putnam, 1931.

————. *The Eternal Values.* Oxford: Oxford University Press; London: Humphrey Milford, 1933.

————. *God and the Astronomers: Containing the Warburton Lectures 1931–33.* London: Longmans, Green, 1933.

————. *Vale.* London: Longmans, Green, 1934.

————. *Diary of a Dean: St. Paul's, 1911–1934.* London: Hutchinson, 1949.

Iremonger, F. A. *William Temple, Archbishop of Canterbury: His Life and Letters.* London: Oxford University Press, 1948.

Jacks, L. P. *Sir Arthur Eddington: Man of Science and Mystic.* Cambridge: Cambridge University Press, 1949.

Jacyna, L. S. "Science and Social Order in the Thought of A. J. Balfour." *Isis* 71 (1980): 11–34.

James, E. O. "The Origin and Fall of Man." *Theology* 3 (1921): 16–22 and 78–89.

————. "Science and Theology." *Theology* 4 (1922): 339–48.

————. "Evolution and the Faith." *Theology* 16 (1928): 2–9.

Jammer, Max. *Einstein and Religion: Physics and Theology.* Princeton: Princeton University Press, 1999.

Jeans, James. "Cosmogony." In *Evolution in the Light of Modern Knowledge*, by F. O. Bower et al., 1–30. London: Blackie, 1925.

———. "The Physics of the Universe." *Nature* 128 (1928): 689–700.

———. *The Universe around Us.* Cambridge: Cambridge University Press, 1929.

———. *The Mysterious Universe.* Cambridge: Cambridge University Press, 1930. Revised ed. 1932.

———. *The New Background of Science.* Cambridge: Cambridge University Press, 1933.

———. *Physics and Philosophy.* Cambridge: Cambridge University Press, 1943 [1942].

Joad, C. E. M. *Matter, Life, and Mind.* London: Oxford University Press, 1929.

———. "Modern Science and Religion." *Proceedings of the Aristotelian Society* 31 (1930–31): 55–86.

———. *Philosophical Aspects of Modern Science.* London: Allen and Unwin, 1932.

———. *Under the Fifth Rib: A Belligerent Autobiography.* London: Faber and Faber, 1932.

———. *Decadence: A Philosophical Inquiry.* London: Faber and Faber, 1947.

———. *Shaw.* London: Victor Gollancz, 1949.

———. *The Recovery of Belief: A Restatement of Christian Philosophy.* London: Faber and Faber, 1952.

Johnstone, J. "Emergent Evolution." *Nature* 127 (1931): 55.

Jolly, W. P. *Sir Oliver Lodge.* London: Constable, 1974.

Jones, Frederic Wood. "The Origin of Man." In *Animal Life and Human Progress*, ed. A. Dendy, 99–131. London: Constable, 1919.

———. *Man's Place among the Mammals.* London: Edward Arnold, 1929.

———. *Design and Purpose.* London: Kegan Paul, 1942.

———. *Habit and Heritage.* London: Kegan Paul, 1943.

———. *Hallmarks of Mankind.* London: Balliere, Tyndall and Cox, 1948.

———. *Trends of Life.* London: Edward Arnold, 1953.

Jones, Greta. *Social Hygiene in Twentieth-Century Britain.* London: Croom Helm, 1986.

———. "Eugenics in Ireland: The Belfast Eugenics Society, 1911–1915." *Irish Historical Studies* 28 (1992): 81–95.

Jones, Henry. *A Faith That Enquires: The Gifford Lectures Delivered in the University of Glasgow in the Years 1920 and 1921.* London: Macmillan, 1922.

Jones, R. Tudur. *Congregationalism in England, 1662–1962.* London: Independent Press, 1962.

Jones, W. Tudor. *The Reality of the Idea of God.* London: Williams and Norgate, 1929.

Joseph, H. W. B. "Professor Eddington on 'The Nature of the Physical World.'" *Hibbert Journal* 27 (1929): 406–23.

Keating, Joseph. "Where Does Adam Come In?" *The Month*, October 1927, 340–52.

Keith, Arthur. "Is Darwinism at the Dusk or Dawn?" *Nineteenth Century* 92 (1922): 165–76.

———. "Why I Am a Darwinist." *RPA Annual* 1922: 11–14.

———. *Concerning Man's Origin: Being the Presidential Address Given at the Meeting of the British Association Held in Leeds on August 31, 1927, and Recent Essays on Darwinian Subjects.* London: Watts, 1927.

———. "Is Darwinism Dead?" *Nature* 119 (1927): 75–76.

———. *Darwinism and What It Implies.* London: Watts, 1928.

———. "Evolution and Its Modern Critics." *Nineteenth Century* 103 (1928): 225–37.

———. "What I Believe." *Forum* 84 (1930): 220–25.

———. *Darwinism and Its Critics.* London: Watts, 1935.

———. *Essays on Human Evolution.* London: Scientific Book Club, 1944.

———. *An Autobiography.* London: Watts, 1950.

"Keith versus Moses." *New Statesman,* 10 September 1927, 672–73.

Kent, John. *From Darwin to Blatchford: The Role of Darwinism in Christian Apologetics, 1875–1910.* London: Dr. Williams Trust, 1966.

Kevles, Daniel. *In the Name of Eugenics: Genetics and the Uses of Human Heredity.* New York: Knopf, 1985.

Kilmister, C. W. *Eddington's Search for a Fundamental Theory: A Key to the Universe.* Cambridge: Cambridge University Press, 1994.

Kitchen, Paddy. *A Most Unsettling Person: An Introduction to the Ideas and Life of Patrick Geddes.* London: Gollancz, 1975.

Knight, James. "The Evolution Theory Today." *Evangelical Quarterly,* January 1933, 3–13.

Knox, Ronald A. *Some Loose Stones: Being a Consideration of Certain Tendencies in Modern Theology Illustrated by Reference to the Book called "Foundations."* London: Longmans, Green, 1913.

———. *Caliban in Grub Street.* London: Sheed and Ward, 1930.

———. *Broadcast Minds.* London: Sheed and Ward, 1932.

———. *God and the Atom.* London: Sheed and Ward, 1945.

Koestler, Arthur. *The Case of the Midwife Toad.* London: Hutchinson, 1971.

Kottler, Malcolm. "Alfred Russel Wallace, the Origin of Man, and Spiritualism." *Isis* 65 (1974): 145–92.

Kropotkin, Peter. *Mutual Aid: A Factor of Evolution.* London: Heinemann, 1908 [1902].

Lacey, T. A. *Mr. Britling's Finite God: Two Lectures Delivered at All Saints', Margaret Street.* London: A. R. Mowbray, 1917.

Lack, David. *Evolutionary Theory and Christian Belief: The Unresolved Conflict.* London: Methuen, 1957.

Laing, Samuel. *Modern Science and Modern Thought.* London: Chapman and Hall, 1893 [1885].

———. *Problems of the Future: And Essays.* London: Chapman and Hall, 1892.

———. *Human Origins.* London: Chapman and Hall, 1894.

The Lambeth Conference, 1930: Encyclical Letter from the Bishops with Resolutions and Reports. London: Society for the Promotion of Christian Knowledge, 1930.

Lang, W. D. "Human Origins and Christian Doctrine." *Nature* 136 (1935): 168–70.

Langdon-Brown, Walter. *Thus We Are Men.* London: Kegan Paul, 1938.

Langdon-Davies, John. *Science and Common Sense.* London: Hamish Hamilton, 1931.

———. "Science and God." *Spectator*, 31 January 1931, 137–38.

Lankester, E. Ray. *Degeneration: A Chapter in Darwinism.* London: Macmillan, 1880.

———. *Zoological Articles Contributed to the Encyclopaedia Britannica.* London: A. and C. Black, 1891.

———. *The Kingdom of Man.* London: Constable, 1907.

———. *Science from an Easy Chair.* London: Methuen, 1910.

———. "Will Orthodox Christianity Survive the World War?" *RPA Annual* 1917: 13–18.

———. "The Church of the Future: A Protest." *Nineteenth Century* 89 (1921): 844–48.

———. "Evolution v. Creation." *Nineteenth Century* 90 (1921): 165–76.

———. "A Matter of Fact." *Nineteenth Century* 90 (1921): 544–50.

———. "Is There a Revival of Superstition?" *RPA Annual* 1922: 3–10.

———. "Evolution and Intellectual Freedom." *Nature* 116 (1925): 71–72.

Larson, Edward J. *Summer for the Gods: The Scopes Trial and America's Continuing Debate over Science and Religion.* New York: Basic Books; Cambridge, MA: Harvard University Press, 1998.

Larson, Edward J., and Larry Witham. "Scientists Are Still Keeping the Faith." *Nature* 386 (1997): 435–36.

———. "Leading Scientists Still Reject God." *Nature* 394 (1998): 313.

Latourelle, Kenneth Scott. *Christianity in a Revolutionary Age: A History of Christianity in the Nineteenth and Twentieth Centuries.* Vol. 4, *The Twentieth Century in Europe: The Roman Catholic, Protestant, and Eastern Churches.* London: Eyre and Spottiswoode, 1961.

Laurence, Dan H. *Bernard Shaw: A Bibliography.* 2 vols. Oxford: Clarendon Press, 1983.

Lawrence, D. H. *Lady Chatterley's Lover.* Introduction by Richard Hoggart. Harmondsworth: Penguin, 1961 [1928].

LeConte, Joseph. *Evolution: Its Nature, Its Evidences, and Its Relation to Religious Thought.* 2d ed. Reprint. New York: Kraus, 1970 [1899].

Leighton, Gerald. *The Greatest Life.* London: Duckworth, 1908.

Lester, Joseph. *E. Ray Lankester and the Making of Modern British Biology.* Edited and with additional material by Peter J. Bowler. BSHS monographs, 9. British Society for the History of Science, 1995.

Leuba, James H. *The Psychological Origin and the Nature of Religion.* London: Archibald Constable, 1909.

Levy, Hyman. "What Is Science?" In *Science in the Changing World,* ed. Mary Adams, 30–185. London: Allen and Unwin, 1933.

Lewin, P. D. "Embryology and the Evolutionary Synthesis: Waddington, Development, and Genetics." Ph.D. thesis, University of Leeds, 1998.

Lewis, Clive Staples. *The Problem of Pain.* London: Geoffrey Bles, 1940.

———. *The Screwtape Letters.* London: Geoffrey Bles, 1942.

———. *The Abolition of Man.* Oxford: Oxford University Press; London: Humphrey Milford, 1943.

———. *Miracles: A Preliminary Study.* London, Geoffrey Bles, 1947.

———. *Surprised by Joy: The Shape of My Early Life.* London: Geoffrey Bles, 1955.

———. *Letters of C. S. Lewis.* Edited by W. H. Lewis. London: Geoffrey Bles, 1966.

———. *Of Other Worlds: Essays and Stories.* Edited by Walter Hooper. London: Geoffrey Bles, 1966.

———. *The Cosmic Trilogy: Out of the Silent Planet; Perelandra; That Hideous Strength.* London: Bodley Head/Pan, 1989.

Lewis, John, Karl Polanyi, and Donald K. Kitchin, eds. *Christianity and the Social Revolution.* Reissue with additional preface by John Lewis. London: Left Book Club, 1937 [1935].

Lewis, Wyndham. *Time and Western Man.* London: Chatto and Windus, 1927.

Lidgett, J. Scott. "Contrasted Cosmologies." *Contemporary Review,* November 1933, 550–58.

Lightman, Bernard. *The Origins of Agnosticism: Victorian Unbelief and the Limits of Knowledge.* Baltimore: Johns Hopkins University Press, 1987.

———. "Ideology, Evolution, and Late-Victorian Agnostic Popularizers." In *History, Humanity, and Evolution,* edited by J. R. Moore, 289–309. Cambridge: Cambridge University Press, 1989.

———. "'Fighting Even with Death': Balfour, Scientific Naturalism, and Thomas Henry Huxley's Final Battle." In *Thomas Henry Huxley's Place in Science and Letters,* edited by Alan P. Barr, 323–50. Athens: University of Georgia Press, 1997.

Lindberg, David C., and Ronald L. Numbers, eds. *God and Nature: Historical Essays on the Encounter between Christianity and the Sciences.* Berkeley: University of California Press, 1986.

Lindsay, A. D. *Religion, Science, and Society in the Modern World.* London: Oxford University Press, 1943.

Livingstone, David. *Darwin's Forgotten Defenders: The Encounter between Evangelical Theology and Evolutionary Thought.* Edinburgh: Scottish Universities Press; Grand Rapids, MI: Eerdmans, 1987.

———. "Science and Religion: Towards a New Cartography." *Christian Scholars Review* 26 (1997): 270–92.

Livingstone, David, D. G. Hart, and Mark A. Noll, eds. *Evangelicals and Science in Historical Perspective.* New York: Oxford University Press, 1999.

Lloyd, Roger. *The Church of England: 1900–1965.* London: Student Christian Movement, 1966.

Lockhart-Mummery, J. Percy. *After Us: Or the World as It Might Be.* London: Stanley Paul, 1936.

———. *Nothing New under the Sun.* London: Andrew Melrose, n.d. [1947].

Lodge, Oliver. "The Outstanding Controversy between Science and Faith." *Hibbert Journal* 1 (1902): 46–61.

———. "Suggestions towards the Reinterpretation of Christian Dogma." *Hibbert Journal* 2 (1904): 461–75.

———. *Life and Matter: A Criticism of Professor Haeckel's* Riddle of the Universe. London: Williams and Norgate, 1905.

———. *The Substance of Faith Allied with Science: A Catechism for Parents and Teachers.* London: Methuen, 1907.

———. *Man and the Universe: A Study of the Influence of the Advance in Scientific Knowledge upon our Understanding of the Christian Faith.* London: Methuen, 1908.

———. *Raymond: Or Life and Death.* London: Methuen, 1916.

———. "The Effort of Evolution." *Hibbert Journal* 21 (1921–22): 461–73.

———. "Psychic Science." In *The Outline of Science: A Plain Story Simply Told,* vol. 2, edited by J. Arthur Thomson, 401–20. London: G. Newnes, 1921–1922.

———. *Making of Man: A Study in Evolution.* London: People's Library, 1929 [1924].

———. *Evolution and Creation.* London: Hodder and Stoughton, 1926.

———. *Science and Human Progress.* London: Allen and Unwin, 1926.

———. *Phantom Walls.* London: Hodder and Stoughton, 1929.

———. "The Philosophy of Evolution." *Review of the Churches,* January 1929, 86–91.

———. *Past Years: An Autobiography.* London: Hodder and Stoughton, 1931.

———. "Religion and the New Knowledge." *Hibbert Journal* 30 (1932): 204–19.

———. *My Philosophy: Representing My Views on the Many Functions of the Ether of Space.* London: Ernest Benn, 1933.

———. "On the Conflict between Religion and Science." *Philosophy* 8 (1933): 44–51.

Longford, Frank. "A. J. Ayer's Tour de Force, *Language, Truth and Logic.*" *Sunday Times,* 18 February 1996, 7/7.

Lotsy, J. P. "Science and Psychical Research." *Nature* 118 (1926): 370.

Lowe, Victor. *Alfred North Whitehead: The Man and His Work.* 2 vols. Baltimore: Johns Hopkins University Press, 1985–1990.

Lunn, Arnold. *The Flight from Reason: A Criticism of the Dogmas of Popular Science.* London: Eyre and Spottiswoode, 1932.

———. *Within That City.* London: Sheed and Ward, 1936.

Lunn, Arnold, and J. B. S. Haldane. *Science and the Supernatural.* London: Eyre and Spottiswoode, 1935.

Lunn, Arnold, and C. E. M. Joad. *Is Christianity True? A Correspondence between Arnold Lunn and C. E. M. Joad.* London: Eyre and Spottiswoode, 1932.

MacBride, Ernest William. "The Present Position of Organic Evolution." *Journal of the Transactions of the Victoria Institute* 47 (1915): 93–124.

————. "Evolution and Intellectual Freedom." *Nature* 116 (1925): 72–73.

————. "Evolution as Shown by the Development of the Individual." In *Creation by Evolution*, edited by Frances Mason, 49–61. New York: Macmillan, 1928.

————. "The Oneness and Uniqueness of Life." In *The Great Design: Order and Progress in Nature*, edited by Frances Mason, 134–58. London: Duckworth, 1934.

————. "The Scientific Atmosphere and the Creeds of the Christian Church." *Hibbert Journal* 34 (1935): 206–18.

MacFie, Ronald Campbell. *Metanthropos: Or the Body of the Future*. London: Kegan Paul, 1928.

————. *Science Rediscovers God: Or the Theodicy of Science*. London: T. and T. Clark, 1930.

————. *The Theology of Evolution*. London: Unicorn Press, 1933.

————. *The Complete Poems of Ronald Campbell MacFie*. London: Humphrey Toulmin, 1937.

MacGregor-Morris, J. T. *The Inventor of the Valve: A Biography of Sir Ambrose Fleming*. London: Television Society, 1954.

Machin, G. I. T. *Politics and the Churches in Great Britain, 1869 to 1921*. Oxford: Clarendon Press, 1987.

Mackenzie, Norman, and Jean Mackenzie. *The Time Traveller: The Life of H. G. Wells*. London: Weidenfeld and Nicolson, 1973.

MacLeod, Roy, and Peter Collins, eds. *The Parliament of Science: The British Association for the Advancement of Science, 1831–1981*. Northwood, Middlesex: Science Reviews, 1981.

MacPherson, Hector. *The Church and Science: A Study of the Inter-relation of Theological and Scientific Thought*. London: James Clarke, 1927.

Magnello, Eileen. "Pearson, Karl." In *Encyclopedia of Biostatistics*, edited by Peter Armitage and Theodor Colton, vol. 4, 3308–15. New York: Wiley, 1998.

Major, H. D. A. *English Modernism: Its Origins, Methods, Aims. Being the William Belden Noble Lectures Delivered in Harvard University, 1925–1926*. Cambridge, MA: Harvard University Press, 1927.

Manwaring, Randle. *From Controversy to Co-existence: Evangelicals in the Church of England, 1914–1980*. Cambridge: Cambridge University Press, 1985.

Marrin, Albert. *The Last Crusade: The Church of England and the First World War*. Durham, NC: Duke University Press, 1974.

Marshall, L. H. "Religion and Science." *Baptist Quarterly*, July 1938, 148–58.

Marvin, F. S. "Science and Human Affairs." In *Science and Civilization*, edited by F. S. Marvin, 330–50. Oxford: Oxford University Press; London: Humphrey Milford, 1923.

————, ed. *Science and Civilization*. Oxford: Oxford University Press; London: Humphrey Milford, 1923.

Mason, Frances, ed. *Creation by Evolution: A Consensus of Present-Day Knowledge as Set Forth by Leading Authorities in Non-Technical Language that All May Understand*. New York: Macmillan, 1928.

————, ed. *The Great Design: Order and Progress in Nature*. London: Duckworth, 1934.

Massey, Charles C. *Thoughts of a Modern Mystic: A Selection from the Writings of the Late C. C. Massey.* Edited by W. F. Barrett. London: Kegan Paul, 1909.

Matczak, Sebastian A. *Karl Barth on God: The Knowledge of the Divine Existence.* New York: St. Paul Publications, 1962.

Matheson, P. E. *The Life of Hastings Rashdall, D. D.* Oxford: Oxford University Press; London: Humphrey Milford, 1928.

Matthews, W. R. "The Psychological Standpoint and Its Limitations." In *Psychology and the Church,* edited by O. Hardman, 3–15. London: Macmillan, 1925.

———. *God and Evolution.* London: Longmans, Green, 1926.

Mayer, Anna-Katherina. "Moralizing Science: The Uses of Science's Past in National Education in the 1920s." *British Journal for the History of Science* 30 (1997): 51–70.

———. "'A Combative Sense of Duty': Englishness and the Scientists." Unpublished manuscript, 1999.

Mays, W. *The Philosophy of Whitehead.* London: Allen and Unwin, 1959.

McCabe, Joseph. *The Evolution of Mind.* Revised ed. London: Watts, 1921.

———. *The Existence of God.* Revised ed. London: Watts, 1933.

———. *The Riddle of the Universe Today.* London: Watts, 1934.

McDougall, William. *Body and Mind: A History and Defence of Animism.* 7th ed. London: Methuen, 1928 [1911].

———. *The Group Mind.* Cambridge: Cambridge University Press, 1920.

———. "Mental Evolution." In F. O. Bower et al., *Evolution in the Light of Modern Knowledge,* 321–54. London: Blackie, 1925.

———. "An Experiment for Testing the Hypothesis of Lamarck." *British Journal of Psychology* 17 (1927): 267–304.

———. *Modern Materialism and Emergent Evolution.* London: Methuen, 1929.

———. "William McDougall." In *A History of Psychology in Autobiography,* edited by Carl Murchison, vol. 1, 191–223. Worcester, MA: Clark University Press, 1934.

———. *The Riddle of Life: A Survey of Theories.* London: Methuen, 1938.

McDowall, Stewart A. *Evolution and Spiritual Life.* Cambridge: Cambridge University Press, 1915.

———. *Evolution and the Doctrine of the Trinity.* Cambridge: Cambridge University Press, 1918.

———. *Evolution, Knowledge, and Revelation: Being the Hulsean Lectures Delivered before the University of Cambridge, 1923–24.* Cambridge: Cambridge University Press, 1924.

McGucken, William. *Scientists, Society, and State: The Social Relations of Science Movement in Great Britain, 1931–1947.* Columbus: Ohio State University Press, 1984.

McKerrow, J. C. *Evolution without Natural Selection.* London: Longmans, Green, 1937.

McKibbin, Ross. *Classes and Culture: England, 1918–1951.* Oxford: Oxford University Press, 1999.

McLaughlin, Brian P. "The Rise and Fall of British Emergentism." In *Emergence or Reduction? Essays on the Prospects of Nonreductive Physicalism,* edited by Ansgar Beckerman, Hans Flohr, and Jaegwon Kim, 49–93. Berlin: Walter de Gruyter, 1992.

McLeod, Hugh. *Religion and Society in England, 1850–1914.* London: Macmillan, 1996.

McOuat, Gordon, and Mary P. Winsor. "J. B. S. Haldane's Darwinism in Its Religious Context." *British Journal for the History of Science* 28 (1995): 227–31.

McTaggart, John McTaggart Ellis. *Studies in Hegelian Cosmology.* Cambridge: Cambridge University Press, 1901.

———. *Some Dogmas of Religion.* London: Edward Arnold, 1906.

Medawar, Peter. "Review of Teilhard de Chardin, *The Phenomenon of Man.*" *Mind* 70 (1961): 99–106.

Medcalf, Stephen. "Cries in the Wilderness." *Times Literary Supplement,* 23 July 1999, 3–4.

Meer, Jitse M. van der, ed. *Facets of Faith and Science.* Vol. 1, *Historiography and Modes of Interaction.* Lanham, MD: University Press of America, 1996.

Mercier, Charles. "Sir Oliver Lodge and the Scientific World." *Hibbert Journal* 15 (1917): 598–613.

Messenger, Ernest. *Evolution and Theology: The Problem of Man's Origin.* London: Burns Oates and Washbourne, 1931.

Metchnikoff, Elie. *The Nature of Man: Studies in Optimistic Philosophy.* English translation edited by P. Chalmers Mitchell. London: Heinemann, 1903.

Micklem, Nathaniel. *What Is the Faith?* London: Hodder and Stoughton, 1936.

Midgley, Mary. *Science as Salvation: A Modern Myth and Its Making.* London: Routledge, 1992.

Milburn, R. Gordon. *The Logic of Religious Thought: An Answer to Professor Eddington.* London: Williams and Norgate, 1929.

Millikan, Robert Andrews. *Evolution in Science and Religion.* New Haven: Yale University Press, 1927.

Milne, E. A. *Modern Cosmology and the Christian Idea of God: Being the Edward Cadbury Lectures in the University of Birmingham for 1950.* Oxford: Clarendon Press, 1952.

———. *Sir James Jeans: A Biography.* Cambridge: Cambridge University Press, 1952.

Milum, J. Parton. *Evolution and the Spirit of Man: Being an Indication of Some Paths Leading to the Reconquest of the 'Eternal Values' through the Present Knowledge of Nature.* London: The Epworth Press, 1928.

———. "The Science of Man and of God." *London Quarterly Review,* April 1937, 172–82.

Mitchell, Peter Chalmers. *Evolution and the War.* London: John Murray, 1915.

———. *Materialism and Vitalism in Biology: The Herbert Spencer Lecture Delivered at Oxford, 3 June 1930.* Oxford: Clarendon Press, 1930.

Mivart, St. George Jackson. *On the Genesis of Species.* London: Macmillan, 1871 [1870].

Monk, Ray. *Bertrand Russell: The Spirit of Substance.* London: Jonathan Cape, 1996.

Moore, George. *The Brook Kerith: A Syrian Story.* Edinburgh: T. Werner Laurie, 1916.

Moore, James R. *The Post-Darwinian Controversies: A Study of the Protestant Struggle to Come to Terms with Darwinism in Britain and America, 1870–1900.* Cambridge: Cambridge University Press, 1979.

———. "Evangelicals and Evolution: Henry Drummond, Herbert Spencer, and the Naturalization of the Spiritual World." *Scottish Journal of Theology* 38 (1985): 383–417.

———, ed. *History, Humanity, and Evolution: Essays for John C. Greene.* Cambridge: Cambridge University Press, 1989.

Moorehead, Caroline. *Bertrand Russell: A Life.* London: Sinclair and Stevenson, 1992.

Morgan, Conwy Lloyd. *Instinct and Experience.* London: Methuen, 1912.

———. *Emergent Evolution: The Gifford Lectures Delivered in the University of St. Andrews in the Year 1922.* London: Williams and Norgate, 1923.

———. "Biology." In *Evolution in the Light of Modern Knowledge*, edited by F. O. Bower et al., 107–62. London: Blackie, 1925.

———. *Life, Mind, and Spirit: Being the Second Course of Gifford Lectures Delivered in the University of St. Andrews in the Year 1923 under the General Title of* Emergent Evolution. London: Williams and Norgate, 1926.

———. "Mind in Evolution." In *Creation by Evolution*, edited by Frances Mason, 340–54. New York: Macmillan, 1928.

———. *The Animal Mind.* London: Edward Arnold, 1930.

———. *The Emergence of Novelty.* London: Williams and Norgate, 1933.

———. "The Ascent of Mind." In *The Great Design: Order and Progress in Nature*, edited by Frances Mason, 115–32. London: Duckworth, 1934.

———. "C. Lloyd Morgan." In *A History of Psychology in Autobiography*, edited by Carl Murchison, vol. 2, 237–64. Worcester, MA: Clark University Press, 1934.

Morris, Kevin L. *Mgr Ronald Knox: A Great Teacher.* London: CTS, 1995.

Morrison, J. H. *Christian Faith and the Science of Today.* London: Hodder and Stoughton, 1936.

Morton, Harold C. *The Bankruptcy of Evolution.* London: Marshall Brothers, 1925.

Moulton, James Hope. *Is Christianity True? How God Prepared for Christianity.* London: Charles H. Kelly, 1904.

Mozley, John Kenneth. "The Incarnation." In *Essays Catholic and Critical*, edited by E. G. Selwyn, 178–202. London: Society for the Promotion of Christian Knowledge, 1926.

———. *Some Tendencies in British Theology: From the Publication of* Lux Mundi *to the Present.* London: Society for the Promotion of Christian Knowledge, 1951.

Mulliner, H. G. *Arthur Burroughs: A Memoir.* London: Nisbett, 1936.

Murchison, Carl, ed. *A History of Psychology in Autobiography.* 2 vols. Worcester, MA: Clark University Press, 1934.

Murray, John Owen Farquhar. "The Spiritual and Historical Evidence for Miracles." In *Essays on Some Theological Questions of the Day*, edited by H. B. Swete, 307–40. London: Macmillan, 1905.

Needham, Joseph. "Mechanistic Biology and the Religious Consciousness." In *Science, Religion, and Reality*, edited by Joseph Needham, 221–57. London: Sheldon Press, 1925.

———. "The Limitations of Optick Glasses: Some Observations on Science and Religion." *Hibbert Journal* 24 (1926): 463–80.

———. "Biochemistry and Mental Phenomena." Appendix to *The Creator Spirit*, by Charles Raven, 285–303. London: Martin Hopkinson, 1927.

———. "Organicism in Biology." *Journal of Philosophy* 3 (1928): 29–40.

———. *The Sceptical Biologist: Ten Essays*. London: Chatto and Windus, 1929.

———. *A History of Embryology*. 2d ed. New York: Abelard-Schuman, 1959 [1931].

———. "Religion and the Scientific Mind." *Criterion* 39 (1931): 233–63.

———. "Religion in a Scientific Age." *Nineteenth Century* 110 (1931): 580–93.

———. "Laud, the Levellers, and the Virtuosi." In *Christianity and the Social Revolution*, edited by J. Lewis, K. Polanyi and D. K. Lewis, eds., 180–205. Reprint. London: Left Book Club, 1937 [1935].

———. "Science, Religion, and Socialism." In *Christianity and the Social Revolution*, edited by J. Lewis, K. Polanyi and D. K. Lewis, eds., 416–41. Reprint. London: Left Book Club, 1937 [1935].

———. *Time, the Refreshing River: Essays and Addresses, 1932–1942*. London: Allen and Unwin, 1943.

———. *History Is on Our Side: A Contribution to Political Religion and Scientific Faith*. London: Allen and Unwin, 1946.

———. *Integrative Levels: A Revaluation of the Idea of Progress: The Herbert Spencer Lecture Delivered at Oxford, 27 May 1937*. Reprint for the author, n.p., n.d.

———, ed. *Science, Religion, and Reality*. London: Sheldon Press, 1925.

Needham, Joseph, and Walter Pagel, eds. *Background to Modern Science: Ten Lectures at Cambridge arranged by the History of Science Committee*. Cambridge: Cambridge University Press, 1938.

Neeson, Hugh. "The Educational Work of Sir Bertram Windle, FRS (1858–1929)." M. A. thesis, Queen's University of Belfast, 1962.

Newsholme, H. P. *Evolution and Redemption*. London: Williams and Norgate, 1933.

Noyes, Alfred. "Radio and the Master-Secret." *Radio Times*, 18 September 1925, 549–55.

———. *The Unknown God*. London: Sheed and Ward, 1934.

Numbers, Ronald L. *The Creationists*. New York: Alfred Knopf, 1992.

———. *Darwinism Comes to America*. Cambridge, MA: Harvard University Press, 1998.

O'Hara, C. W. Untitled. In *Science and Religion: A Symposium*, 107–16. London: Gerald Howe, 1931.

Oldham, J. A. *Christianity and the Race Problem*. London: SCM, 1924.

Oman, John. "The Sphere of Religion." In *Science, Religion, and Reality*, edited by Joseph Needham, 261–99. London: Sheldon Press, 1925.

————. *The Natural and the Supernatural.* Cambridge: Cambridge University Press, 1931.

Oppenheim, Janet. *The Other World: Spiritualism and Psychic Research in England, 1850–1914.* Cambridge: Cambridge University Press, 1985.

Orchard, W. E. *The Present Crisis in Religion.* London: Cassell, 1929.

————. *From Faith to Faith: An Autobiography of Religious Development.* London and New York: Putnam, 1933.

Orr, James. *God's Image in Man: And Its Defacement in the Light of Modern Denials.* London: Hodder and Stoughton, 1905.

Otto, Rudolph. *Naturalism and Religion.* Translated by J. Arthur Thomson and Margaret R. Thomson. London: Williams and Norgate, 1913.

Palladino, Paolo. "On the Contradictions of Humanism: The Medical and Other Writings of Percy Lockhart-Mummery." Paper delivered to a Symposium on Science, Humanism, and Citizenship in the Interwar Period, Wellcome Institute for the History of Medicine, London, 1996.

Passmore, John. *A Hundred Years of Philosophy.* Harmondsworth: Pelican, 1968.

Pattison-Muir, M. M. "The Vain Appeal of Dogma to Science." *Hibbert Journal* 10 (1911): 824–34.

Peacocke, Arthur. *Theology for a Scientific Age.* Enl. ed. London: Student Christian Movement Press, 1993.

Peake, Harold. *The Flood: New Light on an Old Story.* London: Kegan Paul, 1930.

Pearce, Joseph. *Literary Converts: Spiritual Inspiration in an Age of Unbelief.* London: Harper Collins, 1999.

Pearson, E. S. *Karl Pearson: An Appreciation of Some Aspects of His Life and Work.* Cambridge: Cambridge University Press, 1938.

Pearson, Karl. *The Grammar of Science.* 2d ed. London: A. and C. Black, 1900 [1892].

Pegg, Mark. *Broadcasting and Society, 1918–1939.* London: Croom Helm, 1983.

Platt, William. "Religion and Sir Arthur Keith." *Nineteenth Century* 104 (1928): 712–15.

Polanyi, Michael. *Science, Faith, and Society.* Reprint. Chicago: University of Chicago Press, 1964 [1946].

————. *Personal Knowledge: Towards a Post-Critical Philosophy.* Reprint. New York: Harper, 1964 [1958].

Prestige, G. L. *The Life of Charles Gore: A Great Englishman.* London: Heinemann, 1935.

Price, George McCready, and Joseph McCabe. *Is Evolution True? Verbatim Report of a Debate between George McCready Price and Joseph McCabe held at the Queen's Hall, Langham Place, London W., on September 6 1925.* London: Watts, 1925.

Price, H. H. *Some Aspects of the Conflict between Science and Religion: The A. S. Eddington Memorial Lecture, 1953.* Cambridge: Cambridge University Press, 1953.

Profeit, W. *The Creation of Matter: Or Material Elements, Evolution, and Creation.* Edinburgh: T. and T. Clark, 1903.

"Propaganda and Philosophy." *Nature* 118 (1926): 543–45.

Pym, T. W. *Psychology and the Christian Life.* London: Student Christian Movement, 1921.

Ramsey, Arthur Michael. *From Gore to Temple: The Development of Anglican Theology between* Lux Mundi *and the Second World War, 1889–1939.* London: Longmans, 1960.

Randall, H. J. "The Intellectual Revolution in Nineteenth-Century England." *Edinburgh Review* 249 (1929): 41–61.

Rashdall, Hastings. "Personality: Human and Divine." In *Personal Idealism,* edited by H. Sturt, 369–93. London: Macmillan, 1902.

———. "The Ultimate Basis of Theism." In *Contentio Veritatis,* by H. Rashdall et al., 1–58. London: John Murray, 1902.

Rashdall, Hastings, et al. *Contentio Veritatis: Essays on Constructive Theology by Six Oxford Tutors.* London: John Murray, 1902.

Raven, Charles E. *In Praise of Birds: Pictures of Bird Life.* London: Martin Hopkinson, 1925.

———. *The Eternal Spirit: An Account of the Church Congress held at Southport, October 1926.* Liverpool: Diocesan Publishing Co.; London: Hodder and Stoughton, 1926.

———. *The Creator Spirit: A Survey of Christian Doctrine in the Light of Biology, Psychology, and Mysticism.* London: Martin Hopkinson, 1927.

———. *The Ramblings of a Bird Lover.* London: Martin Hopkinson, 1927.

———. *A Wanderer's Way.* London: Martin Hopkinson, 1928.

———. *Bird Haunts and Bird Behaviour.* London: Martin Hopkinson, 1929.

———. *Looking Forward (Towards 1940).* London: Nisbett, n.d. [1931].

———. *Musings and Memories.* London: Martin Hopkinson, 1931.

———. "Introduction." In *Christianity and the Social Revolution,* edited by J. Lewis, K. Polanyi and D. K. Lewis, eds., 15–27. Reprint. London: Left Book Club, 1937 [1935].

———. *Evolution and the Christian Concept of God.* Oxford: Oxford University Press; London: Humphrey Milford, 1936.

———. *John Ray, Naturalist: His Life and Work.* Cambridge: Cambridge University Press, 1942.

———. *Science, Religion, and the Future: A Course of Eight Lectures.* Cambridge: Cambridge University Press, 1943.

———. *English Naturalists from Neckam to Ray.* Cambridge: Cambridge University Press, 1947.

———. *Natural Religion and Christian Theology: The Gifford Lectures, 1951/1952.* 2 vols. Cambridge: Cambridge University Press, 1953.

———. *Teilhard de Chardin: Scientist and Seer.* London: Collins, 1962.

Reardon, Bernard M. G. *From Coleridge to Gore: A Century of Religious Thought in Britain.* London: Longmans, 1973.

Reeman, Edmund H. *Do We Need a New Idea of God?* London: Hurst and Blackett, 1918.

Reinheimer, Hermann. "Cooperation among Species." *Hibbert Journal* 21 (1921–22): 158–77.

———. "Science and Religion." *London Quarterly Review,* October 1925, 239–43.

————. "The Passing of Darwinism." *Quest*, January 1928, 144–53.

"Religion and Evolution." *Review of Reviews*, June 1932, 39–42.

Richards, I. A. *Science and Poetry.* London: Kegan Paul, 1926.

Richards, Robert J. *Darwin and the Emergence of Evolutionary Theories of Mind and Behavior.* Chicago: University of Chicago Press, 1987.

Richardson, L. De C. "Bishop Barnes on Science and Superstition." *Hibbert Journal* 31 (1933): 358–71.

Ritchie, A. D. *Civilization, Science, and Religion.* Harmondsworth: Penguin Books, 1945.

————. *Reflections on the Philosophy of Sir Arthur Eddington.* Introduction by C. E. Raven. Cambridge: Cambridge University Press, 1948.

Robbins, Keith. "The Spiritual Pilgrimage of the Rev. R. J. Campbell." *Journal of Ecclesiastical History* 30 (1979): 261–76.

Robbins, Peter. *The British Hegelians, 1875–1925.* New York: Garland, 1982.

Roberts, Harry, and Lord Thomas Horder. *The Philosophy of Jesus.* London: Dent, 1945.

Robinson, Arthur Williamson. "Prayer, in Relation to the Idea of Law." In *Essays on Some Theological Questions of the Day,* edited by H. B. Swete, 263–306. London: Macmillan, 1905.

Robinson, W. "Christianity and Evolution." *Modern Churchman,* February 1929, 669–79.

Rodgers, John H. *The Theology of P. T. Forsyth: The Cross of Christ and the Revelation of God.* London: Independent Press, 1965.

Rogers, Ben. *A. J. Ayer: A Life.* London: Chatto and Windus, 1999.

Roll-Hansen, Nils. "E. S. Russell and J. H. Woodger: The Failure of Two 20th-Century Opponents of Mechanistic Biology." *Journal of the History of Biology* 17 (1984): 399–428.

Romanes, George John. *Thoughts on Religion.* Edited by Charles Gore. London: Longmans, Green, 1905 [1895].

Rothblatt, Sheldon. *The Revolution of the Dons: Cambridge and Society in Victorian England.* Reprint. Cambridge: Cambridge University Press, 1981 [1968].

Royle, Edward. *Radicals, Secularists, and Republicans: Popular Free Thought in Britain, 1865–1915.* Manchester: Manchester University Press, 1980.

Ruse, Michael. *Monad to Man: The Concept of Progress in Evolutionary Biology.* Cambridge, MA: Harvard University Press, 1996.

————. *Mystery of Mysteries: Is Evolution a Social Construct?* Cambridge, MA: Harvard University Press, 1999.

Russell, Alexander Smith. "The Dynamic of Science." In *Adventure: The Faith of Science and the Science of Faith,* edited by B. H. Streeter, 1–20. London: Macmillan, 1927.

Russell, Bertrand. *Philosophical Essays.* London: Longmans, Green, 1910.

————. *The Philosophy of Bergson: With a Reply by Mr. H. Wildon Carr.* Cambridge: Bowes and Bowes, 1914.

————. *Mysticism and Logic: And Other Essays.* London: Longmans, Green, 1918.

————. *Icarus: Or The Future of Science.* London: Kegan Paul, 1924.

————. *Sceptical Essays.* London: Allen and Unwin, 1928.

————. *The Scientific Outlook.* London: Allen and Unwin, 1931.

————. *Religion and Science.* London: Oxford University Press, 1961 [1935].

————. *Portraits from Memory and Other Essays.* London: Allen and Unwin, 1956.

————. *The Basic Writings of Bertrand Russell.* Edited by Lester E. Dennon and Robert E. Egner. London: Allen and Unwin, 1961.

Russell, Colin A. *The Earth, Humanity, and God: The Templeton Lectures, Cambridge, 1993.* London: University College, London, Press, 1994.

Russell, Edward Stuart. *Form and Function: A Contribution to the History of Animal Morphology.* London: John Murray, 1916.

————. *The Study of Living Things: Prolegomena to a Functional Biology.* London: Methuen, 1924.

————. "The Study of Behaviour." *Report of the British Association for the Advancement of Science* 1934: 83–98.

————. *The Behaviour of Animals: An Introduction to Its Study.* 2d ed. London: Edward Arnold, 1938.

Russell, Leonard. "A. S. Eddington, 'The Nature of the Physical World.'" *Journal of Philosophical Studies* 4 (1929): 252–55.

Saleeby, C. W. *Evolution: The Master-Key.* London and New York: Harper, 1906.

Sandhurst, B. G. *How Heathen Is Britain?* Rev. and enl. ed. London: Collins, 1948 [1946].

Sapp, Jan. *Evolution by Association: A History of Symbiosis.* New York: Oxford University Press, 1994.

Sarkar, Sahotra, ed. *The Founders of Evolutionary Genetics: A Centenary Reappraisal.* Dordrecht: Kluwer, 1992.

Sayers, Dorothy L. *Lord Peter Views the Body.* Reprint. London: New English Library, 1992 [1928].

————. *The Unpleasantness at the Bellona Club.* Reprint. London: New English Library, 1993 [1928].

————. *The Mind of the Maker.* London: Methuen, 1959 [1941].

————. *The Man Born to Be King: A Play-Cycle on the Life of Our Lord and Saviour Jesus Christ.* London: Victor Gollancz, 1943.

————. *Unpopular Opinions.* London: Victor Gollancz, 1946.

————. *The Letters of Dorothy L. Sayers.* Vol. 2. Edited by Barbara Reynolds. Swaversy, Cambridge: The Dorothy L. Sayers Society, 1997.

Scannell, Paddy, and David Cardiff. *A Social History of British Broadcasting.* Vol. 1, *1922–1939: Serving the Nation.* Oxford: Blackwell, 1991.

"Science and Faith." *Nature* 148 (1941): 181–83.

"Science and Religion." *Nature* 118 (1926): 577–78.

Science and Religion: A Symposium. London: Gerald Howe, 1931.

"Science and Values." *Nature* 134 (1934): 233–34.

"Scientific Theory and Religion." *Nature* 132 (1933): 79–81.

Searle, G. R. *Eugenics and Politics in Britain, 1900–1914.* Leiden: Noordhoff International, 1976.

———. "Eugenics and Politics in Britain in the 1930s." *Annals of Science* 36 (1979): 159–69.

Selbie, W. B. *The Psychology of Religion.* Oxford: Clarendon Press, 1924.

———. *Christianity and the New Psychology.* London: Centenary Press, 1939.

Sell, Alan P. F. *Philosophical Idealism and Christian Belief.* Cardiff: University of Wales Press, 1995.

Selwyn, Edward Gordon, ed. *Essays Catholic and Critical: By Members of The Anglican Communion.* London: Society for the Promotion of Christian Knowledge, 1926.

Seth Pringle-Pattison, Andrew. *The Idea of God in the Light of Recent Philosophy: The Gifford Lectures Delivered in the University of Aberdeen in the Years 1912 and 1913.* Oxford: Clarendon Press, 1917.

———. "Immanence and Transcendence." In *The Spirit: God and His Relations to Man Considered from the Standpoint of Philosophy, Psychology, and Art,* edited by B. H. Streeter, 1–22. London: Macmillan, 1919.

Seth Pringle-Pattison, Andrew, and R. B. Haldane, eds. *Essays in Philosophical Criticism.* London: Longmans, 1883.

Shaw, George Bernard. *The Heretics: Mr. Bernard Shaw: May 29, 1911.* Cambridge: The Heretics, 1911.

———. *Back to Methuselah: A Metabiological Pentateuch.* London: Constable, 1921.

———. *Too True to Be Good, Village Wooing, and On the Rocks.* London: Constable, 1934.

———. *The Religious Speeches of Bernard Shaw.* Ed. Warren Sylvester Smith. University Park: Pennsylvania State University Press, 1963.

———. *Shaw on Religion.* Ed. Warren Sylvester Smith. London: Constable, 1967.

———. *The Bodley Head Bernard Shaw.* 7 vols. London: Bodley Head, 1970–1974.

Shearman, J. N. *The Natural Theology of Evolution.* London: Allen and Unwin, 1915.

Sherrington, Charles. Introduction to *Creation by Evolution,* edited by Frances Mason. New York: Macmillan, 1928.

———. Introduction to *The Great Design,* edited by Frances Mason. London: Duckworth, 1934.

———. *Man on His Nature.* Cambridge: Cambridge University Press, 1940.

———. *The Endeavour of Jean Fernel: With a List of the Editions of His Writings.* Cambridge: Cambridge University Press, 1946.

Simpson, James Young. *The Spiritual Interpretation of Nature.* New and rev. ed. London: Hodder and Stoughton, 1923 [1912].

———. *Landmarks in the Struggle between Science and Religion.* London: Hodder and Stoughton, 1925.

————. *Nature: Cosmic, Human, and Divine.* New Haven: Yale University Press, 1929.

————. *The Garment of the Living God: Studies in the Relations of Science and Religion. The Sprunt Lectures Delivered in Richmond Theological Seminary in 1934. With a Memoir by G. F. Barbour.* London: Hodder and Stoughton, 1934.

Singer, Charles Joseph. "Ancient Medicine." In *Science and Civilization,* edited by F. S. Marvin, 43–71. Oxford: Oxford University Press; London: Humphrey Milford, 1923.

————. "The Dark Ages." In *Science and Civilization,* edited by F. S. Marvin, 112–50. Oxford: Oxford University Press; London: Humphrey Milford, 1923.

————. "Historical Relations of Science and Religion." In *Science, Religion, and Reality,* edited by Joseph Needham, 85–148. London: Sheldon Press, 1925.

————. *Religion and Science Considered in Their Historical Relations.* London: Ernest Benn, 1928.

————. *A History of Biology to about the Year 1900.* 3d ed. New York: Abelard Schuman, 1959 [1931].

————. *A Short History of Biology: A General Introduction to the Study of Living Things.* Oxford: Clarendon Press, 1931.

————. *A Short History of Science to the Nineteenth Century.* Oxford: Clarendon Press, 1941.

————. *The Christian Failure.* London: Victor Gollancz, 1943.

Smethurst, Arthur F. *Modern Science and Christian Belief.* London: James Nisbett, 1955.

Smith, C. W., and M. N. Wise. *Energy and Empire: A Biographical Study of Lord Kelvin.* Cambridge: Cambridge University Press, 1989.

Smith, David C. *H. G. Wells: Desperately Moral.* New Haven: Yale University Press, 1986.

Smith, Roger. "The Embodiment of Value: C. S. Sherrington and the Cultivation of Science." *British Journal for the History of Science* 33 (2000): 283–312.

Smocovitis, V. Betty. *Unifying Biology: The Evolutionary Synthesis and Evolutionary Biology.* Princeton: Princeton University Press, 1996.

Smuts, Jan Christiaan. *Holism and Evolution.* Reprint. New York: Viking, 1961 [1926].

————. "The Scientific World-Picture of Today." *Report of the British Association for the Advancement of Science,* 1931: 1–18.

Sollas, W. J. *Ancient Hunters and Their Modern Representatives.* 2d ed. London: Macmillan, 1915.

Sommer, Dudley. *Haldane of Cloan: His Life and Times, 1856–1928.* London: Allen and Unwin, 1960.

Speaight, Robert. *The Life of Hilaire Belloc.* London: Hollis and Carter, 1957.

Spencer, Frederick A. M. *The Meaning of Christianity.* London: T. Fisher Unwin, 1912.

————. *The Revival of Christianity and the Return of Christ.* Oxford: Blackwell, 1919.

————. "Darwinism and Christianity." *Contemporary Review,* May 1932, 640–44.

————. *The Future Life: A New Interpretation of Christian Doctrine.* London: Hamish Hamilton, 1935.

Spinks, G. Stephen, E. L. Allen, and James Parkes. *Religion in Britain since 1900*. London: Andrew Dakers, 1952.

Spurway, Neil, ed. *Humanity, Environment, and God: Glasgow Centenary Gifford Lectures*. Oxford: Oxford University Press, 1993.

Standing, Herbert F. *Spirit in Evolution: From Amoeba to Spirit*. London: Allen and Unwin, 1930.

Starbuck, Edwin Diller. *The Psychology of Religion: An Empirical Study of the Growth of Religious Consciousness*. 4th ed. London: Walter Scott, 1914.

Stebbing, L. Susan. *Philosophy and the Physicists*. London: Methuen, 1937.

Stenger, Victor J. *Physics and Psychics: The Search for a World Beyond the Senses*. Buffalo, NY: Prometheus Books, 1990.

Stephens, Lester G. *Joseph LeConte: Gentle Prophet of Evolution*. Baton Rouge: Louisiana State University Press, 1982.

Stephenson, Alan M. G. *The Rise and Decline of English Modernism*. London: Society for the Promotion of Christian Knowledge, 1984.

Stephenson, Thomas. "Emergent Evolution." *London Quarterly Review*, January 1928, 59–68.

Steward, Fred. "Political Formation." In *J. D. Bernal: A Life in Science and Politics*, edited by Brenda Swann and Francis Aprahamian, 37–77. London: Verso, 1999.

[Stewart, Balfour, and Peter Guthrie Tait.] *The Unseen Universe: Or Speculations on a Future State*. 6th ed. London: Macmillan, 1876 [1875].

———. *Paradoxical Philosophy: A Sequel to The Unseen Universe*. London: Macmillan, 1878.

Stokes, George Gabriel. *Natural Theology: The Gifford Lectures Delivered before the University of Edinburgh in 1891*. London: A. and C. Black, 1891.

———. *Natural Theology: The Gifford Lectures Delivered before the University of Edinburgh in 1893*. London: A. and C. Black, 1893.

Storr, Vernon F. *Development and Divine Purpose*. London: Methuen, 1906.

———. *God in the Modern Mind*. London: James Nisbett, n.d. [1931].

Streeter, Burnett Hillman. "The Defeat of Pain." In *God and the Struggle for Existence*, edited by B. H. Streeter, 157–207. London: Student Christian Movement, 1919.

———. *Reality: A New Correlation of Science and Religion*. London: Macmillan, 1927.

———. *The Church and Modern Psychology: The Nineteenth Annual Hale Memorial Sermon, Delivered March 26, 1934*. Evanston, IL: Seabury-Western Theological Seminary, 1934.

———, ed. *Foundations: A Statement of Christian Belief in Terms of Modern Thought. By Seven Oxford Men*. London: Macmillan, 1912.

———, ed. *God and the Struggle for Existence*. London: Student Christian Movement, 1919.

———, ed. *The Spirit: God and His Relations to Man Considered from the Standpoint of Philosophy, Psychology, and Art*. London: Macmillan, 1919.

———, ed. *Adventure: The Faith of Science and the Science of Faith*. London: Macmillan, 1927.

Strick, James. "Darwinism and the Origin of Life: The Role of H. C. Bastian in the British Spontaneous Generation Debates, 1868–1873." *Journal of the History of Biology* 32 (1999): 51–92.

Strutt, Robert John [4th Baron Rayleigh]. *John William Strutt, Third Baron Rayleigh, O. M., FRS: Sometime President of the Royal Society and Chancellor of the University of Cambridge.* London: Edward Arnold, 1924.

———. *The Life of Sir J. J. Thomson, O. M., Sometime Master of Trinity College, Cambridge.* Cambridge: Cambridge University Press, 1942.

Sturdy, Steve. "Biology as Social Theory: John Scott Haldane and Physiological Regulation." *British Journal for the History of Science* 21 (1988): 315–40.

Sturt, Henry, ed. *Personal Idealism: Philosophical Essays by Eight Members of the University of Oxford.* London: Macmillan, 1902.

Sullivan, J. W. N. *The Bases of Modern Science.* London: Ernest Benn, 1928.

———. *Limitations of Science.* London: Chatto and Windus, 1934 [1933].

Swann, Brenda, and Francis Aprahamian, eds. *J. D. Bernal: A Life in Science and Politics.* London: Verso, 1999.

Swete, Henry Barclay, ed. *Essays on Some Theological Questions of the Day: By Members of the University of Cambridge.* London: Macmillan, 1905.

Swetlitz, Mark. "Julian Huxley and the End of Evolution." *Journal of the History of Biology* 28 (1995): 181–217.

Swinton, A. A. Campbell. "Science and Psychical Research." *Nature* 118 (1926): 299–30 and 533.

Synge, Ann. "Early Years and Influences." In *J. D. Bernal: A Life in Science and Politics,* edited by Brenda Swann and Francis Aprahamian, 1–16. London: Verso, 1999.

Tabrum, Arthur H., ed. *Religious Beliefs of Scientists: Including One Hundred Hitherto Unpublished Letters on Science and Religion from Eminent Men of Science.* Introduction by C. L. Drawbridge. London: Hunter and Longhurst, for North London Christian Evidence League, 1910.

Taylor, F. Sherwood. *The Century of Science.* London: Scientific Book Club, 1943 [1941].

———. *The Fourfold Vision: A Study of the Relations of Science and Religion.* London: Chapman and Hall, 1945.

———. *Two Ways of Life: Christian and Materialist.* London: Burnes Oates, 1947.

———. *The Alchemists: Founders of Modern Chemistry.* London: Heinemann, 1951.

———. *Man and Matter: Essays Scientific and Christian.* London: Chapman and Hall, 1951.

Taylor, M. "Catholics and Biology." *The Month,* August 1934, 148–56.

Taylor, Monica, ed. *Sir Bertram Windle: A Memoir.* London: Longmans, Green, 1932.

Taylor, R. O. P. *The Meeting of the Roads: A Scientific View of God and Man.* London: Nisbet, 1931.

———. *Does Science Leave Room for God?* London: Hodder and Stoughton, 1933.

Taylor, Sally J. *The Great Outsiders: Northcliffe, Rothermere, and the Daily Mail.* London: Weidenfeld and Nicolson, 1996.

Teilhard de Chardin, Pierre. *The Phenomenon of Man.* Introduction by Julian Huxley. London: Collins, 1959.

Temple, William. *The Faith and Modern Thought: Six Lectures.* London: Macmillan, 1913 [1910].

———. *Mens Creatrix: An Essay.* London: Macmillan, 1923 [1917].

———. *Christus Veritas: An Essay.* London: Macmillan, 1954 [1924].

———. *Nature, Man, and God: Being the Gifford Lectures Delivered at the University of Glasgow in the Academical Years 1932–33 and 1933–34.* London: Macmillan, 1935.

Tennant, Frederick Robert. *The Origin and Propagation of Sin: Being the Hulsean Lectures Delivered before the University of Cambridge, 1901–02.* Cambridge: Cambridge University Press, 1902.

———. "The Being of God in the Light of Physical Science." In *Essays on Some Theological Questions of the Day,* edited by H. B. Swete, 55–100. London: Macmillan, 1905.

———. *The Concept of Sin.* Cambridge: Cambridge University Press, 1912.

———. "The Present Relations of Science and Theology." *Constructive Quarterly,* December 1921 and June 1922, 578–95 and 274–89.

———. *Miracle and Its Philosophical Presuppositions: Three Lectures Delivered in the University of London, 1924.* Cambridge: Cambridge University Press, 1925.

———. *Philosophical Theology.* 2 vols. Cambridge: Cambridge University Press, 1928–30.

Thompson, D'Arcy Wentworth. *On Growth and Form.* Cambridge: Cambridge University Press, 1917.

Thompson, J. M. *Miracles in the New Testament.* London: Edward Arnold, 1911.

Thompson, Laurence. *Robert Blatchford: Portrait of an Englishman.* London: Victor Gollancz, 1951.

Thomson, J. Arthur. "Biological Philosophy." *Nature* 87 (1911): 475–77.

———. *The System of Animate Nature: The Gifford Lectures Delivered in the University of St. Andrews in the Years 1915 and 1916.* 2 vols. London: Williams and Norgate, 1920.

———. "The Influence of Darwinism on Thought and Life." In *Science and Civilization,* edited by F. S. Marvin, 203–20. Oxford: Oxford University Press; London: Humphrey Milford, 1923.

———. *Science and Religion.* London: Methuen, 1925.

———. "Why We Must Be Evolutionists." In *Creation by Evolution,* edited by Frances Mason, 13–23. New York: Macmillan, 1928.

———. *Purpose in Evolution: Riddell Memorial Lectures, 4th ser., Delivered before the University of Durham at Armstrong College, Newcastle-on-Tyne on November 4, 5, 6, 1931.* Oxford: Oxford University Press; London: Humphrey Milford, 1932.

———. "Introduction." In *The Great Design: Order and Progress in Nature,* edited by Frances Mason, 11–16. London: Duckworth, 1934.

———. *The Gospel of Evolution.* London: George Newnes, n.d.

———, ed. *The Outline of Science: A Plain Story Simply Told.* 2 volumes. London: G. Newnes, 1921–1922.

Thomson, J. Arthur, and Patrick Geddes. "A Biological Approach." In *Ideals of Science and Faith*, edited by J. E. Hand, 49–80. London: George Allen, 1904.

———. *Life: Outlines of General Biology.* 2 vols. London: Williams and Norgate, 1931.

Thomson, Joseph John. "Inaugural Address" [British Association meeting, Winnipeg]. *Nature* 81 (1909): 248–57.

Thorne, Guy. *When It Was Dark: The Story of a Great Conspiracy.* London: Stanley Paul, 1925 [1905].

Thornton, Lionel Spencer. "The Christian Concept of God." In *Essays Catholic and Critical*, edited by E. G. Selwyn, 121–50. London: Society for the Promotion of Christian Knowledge, 1926.

———. *The Incarnate Lord: An Essay Concerning the Doctrine of the Incarnation in Its Relations to Organic Conceptions.* London: Longmans, Green, 1928.

———. "The Sciences and Religion in Recent Gifford Lectures." *Review of the Churches*, October 1929, 536–43.

Tilby, A. Wyatt. "General Smuts' Philosophy." *Nineteenth Century* 101 (1927): 242–54.

Tillyard, R. J. "Science and Psychical Research." *Nature* 118 (1926): 147–49.

Tisdall, W. St. Clair. "Dr. Barnes and the Fall." *The Record*, 2 September 1920, 692–93.

———. "Canon Barnes Again." *The Record*, 9 September 1920, 708.

Torrance, Thomas F. "The Transcendental Role of Wisdom in Science." In *Facets of Faith and Science*, vol. 1, *Historiography and Modes of Interaction*, edited by J. M. van der Meer, 131–49. Lanham, MD: University Press of America, 1996.

———. "Ultimate and Penultimate Beliefs in Science." In *Facets of Faith and Science*, vol. 1, *Historiography and Modes of Interaction*, edited by J. M. van der Meer, 151–76. Lanham, MD: University Press of America, 1996.

Turner, Frank Miller. *Between Science and Religion: The Reaction to Scientific Naturalism in Late Victorian England.* New Haven: Yale University Press, 1974.

———. "The Victorian Conflict between Science and Religion: A Professional Dimension." *Isis* 69 (1978): 356–76.

———. "Public Science in Britain, 1880–1919." *Isis* 71 (1980): 589–608.

Tyndall, John. *Address Delivered before the British Association Assembled at Belfast: With Additions.* London: Longmans Green, 1874.

Underhill, G. E. "The Limits of Evolution." In *Personal Idealism*, edited by H. Sturt, 193–220. London: Macmillan, 1902.

Vialleton, Louis. *L'origine des êtres vivants: L'illusion transformiste.* Paris: Librairie Plon, 1929.

Vidler, Alec R. *The Church in an Age of Revolution: 1798 to the Present Day.* London: Penguin, 1961.

———. *20th Century Defenders of the Faith: Some Theological Fashions Considered in the Robertson Lectures for 1964.* London: Student Christian Movement Press, 1965.

———. "Bishop Barnes: A Centenary Retrospect." *Modern Churchman* 18 (1975): 87–98.

———. *Scenes from Clerical Life: An Autobiography.* London: Collins, 1977.

————, ed. *Soundings: Essays Concerning Christian Understanding.* Cambridge: Cambridge University Press, 1962.

Vine, Charles H. *The Old Faith and the New Theology: A Series of Sermons and Essays on Some of the Truths Held by Evangelical Christians, and the Difficulties of Accepting Much of What Is Called the "New Theology."* London: Sampson Low, Marston, 1907.

Waddington, C. H. "The Relations between Science and Ethics." *Nature* 148 (1941): 270–74.

————. *The Scientific Attitude.* 2d ed. West Drayton, Sussex: Penguin Books, 1948 [1941].

————. *Science and Ethics: An Essay by C. H. Waddington Together with a Discussion between the Author and the Right Rev. E. W. Barnes [and Others].* London: Allen and Unwin, 1942.

————. "The Practical Consequences of Metaphysical Beliefs on a Biologist's Work: An Autobiographical Note." In *Towards a Theoretical Biology,* vol. 2, *Sketches,* edited by C. H. Waddington, 72–81. Edinburgh: Edinburgh University Press, 1969.

Waggett, Philip Napier. *The Scientific Temper in Religion and Other Addresses.* London: Longmans, Green, 1905.

————. "The Influence of Darwin upon Religious Thought." In E. Haeckel et al., *Evolution in Modern Thought,* 223–45. New York: Boni and Liveright, 1917.

Walker, Leslie J. *Science and Revelation.* London: Burns Oates and Washbourne, 1932.

Wallace, Alfred Russel. *On Miracles and Modern Spiritualism: Three Essays.* London: James Burns, 1875.

————. *Man's Place in the Universe: A Study of the Results of Scientific Research in Relation to the Unity or Plurality of Worlds.* 4th ed. London: Chapman and Hall, 1904 [1903].

————. *The World of Life: A Manifestation of Creative Power, Directive Mind, and Ultimate Purpose.* London: Chapman and Hall, 1910.

Ward, James. *Naturalism and Agnosticism: The Gifford Lectures Delivered before the University of Aberdeen in the Years 1896–1898.* 2 vols. London: A. and C. Black, 1899.

————. *The Realm of Ends or Pluralism and Theism: Being the Gifford Lectures Delivered in the University of St. Andrews in the Years 1907–1910.* 3d ed. Cambridge: Cambridge University Press, 1920.

Ward, Keith. *God, Chance, and Necessity.* Oxford: One World Publications, 1996.

Warschauer, J. *The New Evangel: Studies in the "New Theology."* London: James Clarke, 1907.

Washington, Peter. *Madame Blavatsky's Baboon: Theosophy and the Emergence of the Western Guru.* London: Secker and Warburg, 1993.

Waterhouse, E. S. *Psychology and Religion: A Series of Broadcast Talks.* London: Elkin Mathews and Marrot, 1930.

Waters, C. Kenneth, and Albert Van Helden, eds. *Julian Huxley: Biologist and Statesman of Science.* Houston: Rice University Press, 1992.

Watkin, E. "Professor Haldane on Science and Religion." *Dublin Review* 182 (1928): 116–31.

Waugh, Evelyn. *Decline and Fall.* Harmondsworth: Penguin, 1960 [1927].

————. *The Life of the Right Reverend Ronald Knox.* London: Chapman and Hall, 1959.

Weismann, August. *Essays upon Heredity and Kindred Biological Problems.* 2 vols. Oxford: Oxford University Press, 1891–1892.

Wells, G. P. "Lancelot Thomas Hogben." *Biographical Memoirs of the Fellows of the Royal Society* 24 (1978): 183–221.

Wells, Herbert George. *Mr. Britling Sees It Through.* London: Cassell, 1916.

————. *God the Invisible King.* London: Cassell, 1917.

————. *The Outline of History: Being a Plain History of Life and Mankind.* 4th revision. 2 vols. London: Cassell, 1925 [1920].

————. *The New Teaching of History: With a Reply to Some Recent Criticisms of* The Outline of History. London: Cassell, 1921.

————. *Mr. Belloc Objects to* The Outline of History. London: Watts, 1926.

————. *The Short Stories of H. G. Wells.* London: Ernest Benn, 1927.

————. *The Shape of Things to Come: The Ultimate Revolution.* London: Hutchinson, 1933.

————. *Experiment in Autobiography: Discoveries and Conclusions of a Very Ordinary Brain (since 1866).* 2 vols. London: Faber and Faber, 1984 [1934].

Wells, H. G., Julian Huxley, and G. P. Wells. *The Science of Life.* London: Cassell, 1931.

Welsby, Paul A. *A History of the Church of England, 1945–1980.* Oxford: Oxford University Press, 1984.

Werskey, Gary. *The Visible College: A Collective Biography of British Scientists and Socialists of the 1930s.* London: Free Association, 1988.

West, Anthony. *H. G. Wells: Aspects of a Life.* London: Hutchinson, 1984.

White, Andrew Dickson. *A History of the Warfare of Science with Theology in Christendom.* 2 vols. Reprint. New York: Dover, 1960 [1896].

Whitehead, Alfred North. *Religion in the Making.* Cambridge: Cambridge University Press, 1926.

————. *Science and the Modern World.* Cambridge: Cambridge University Press, 1933 [1926].

————. *Process and Reality: An Essay in Cosmology.* Cambridge: Cambridge University Press, 1929.

————. *Nature and Life.* Cambridge: Cambridge University Press, 1934.

Whitehouse, W. A. *Christian Faith and the Scientific Attitude.* Edinburgh: Oliver and Boyd, 1952.

Whittaker, Edmund. *A History of Theories of Aether and Electricity.* 2 vols. London: Nelson, 1951 [1910].

————. *The Beginning and End of the World: Riddell Memorial Lectures, Fourteenth Series, Delivered before the University of Durham at King's College, Newcastle-upon-Tyne in February 1942.* Oxford: Oxford University Press; London: Humphrey Milford, 1942.

————. *Space and Spirit: Theories of the Universe and the Arguments for the Existence of God.* London: Thomas Nelson, 1946.

Whyte, A. Gowans. *The Story of the RPA, 1899–1949.* London: Watts, 1949.

Wilkinson, Alan. *The Church of England and the First World War.* London: SPCK, 1978.

Williams, Norman Powell. *The Ideas of the Fall and of Original Sin: A Historical and Critical Study.* London: Longmans, Greene, 1929.

Williams, T. Rhondda. *The Evangel of the New Theology: Sermons.* London: W. Daniel; Bradford: Percy Lund, Humphries, 1905.

———. *How I Found My Faith.* London: Cassell, 1938.

Wilson, A. N. *C. S. Lewis: A Biography.* London: Collins, 1990.

Wilson, David B. "The Thought of Late Victorian Physicists: Oliver Lodge's Etherial Body." *Victorian Studies* 15 (1977): 29–48.

———. "A Physicist's Alternative to Materialism: The Religious Thoughts of George Gabriel Stokes." *Victorian Studies* 28 (1984): 69–96.

———. "On the Importance of Eliminating *Science* and *Religion* from the History of Science and Religion: The Cases of Oliver Lodge, J. H. Jeans, and A. S. Eddington." In *Facets of Faith and Science,* vol. 1, *Historiography and Modes of Interaction,* edited by J. M. van der Meer, 27–47. Lanham, MD.: University Press of America, 1996.

Wilson, James. "The Religious Effect of the Idea of Evolution." In *Evolution in the Light of Modern Knowledge,* by F. O. Bower et al., 477–515. London: Blackie, 1925.

Wimpenny, R. S. "Russell, Edward Stuart." In *Dictionary of National Biography 1951–1960,* ed. E. T. Williams and Helen M. Parker, 856–58. Oxford: Oxford University Press, 1971.

Windle, Bertram. *Science and Morals: And Other Essays.* London: Burns Oates and Washbourne, 1919.

———. *Vitalism and Scholasticism.* London: Sands, 1920.

———. *On Miracles and Some Other Matters.* London: Burns Oates and Washbourne, 1924.

———. *The Catholic Church and Its Relations with Science.* London: Burns Oates and Washbourne, 1927.

———. *The Evolutionary Problem as It Is Today.* New York: Joseph F. Wagner, 1927.

Wohl, Robert. *The Generation of 1914.* London: Weidenfeld and Nicolson, 1980.

Wolfe, John. *God and Greater Britain: Religion and National Life in Britain and Ireland, 1843–1945.* London: Routledge, 1994.

Wolfe, Kenneth M. *The Churches and the British Broadcasting Corporation, 1922–1956: The Politics of Broadcast Religion.* London: Student Christian Movement Press, 1984.

Wood, Neal. *Communism and British Intellectuals.* London: Victor Gollancz, 1959.

Woodger, J. H. *Biological Principles: A Critical Study.* London: Routledge and Kegan Paul, 1967 [1929].

Worrall, B. G. *The Making of the Modern Church: Christianity in England since 1800.* London: Society for the Promotion of Christian Knowledge, 1988.

Wynne, Bryan. "Physics and Psychics: Science, Symbolic Action and Social Control in Late Victorian England." In *Natural Order: Historical Studies of Scientific Culture*, edited by Barry Barnes and Steven Shapin, 167–87. Beverly Hills: Sage Publications, 1979.

Yarnold, Greville Dennis. *Christianity and Physical Science*. London: A. R. Mowbray, 1950.

———. *The Spiritual Crisis of the Scientific Age*. London: Allen and Unwin, 1959.

Yellowlees, David. *Psychology's Defence of the Faith*. London: Student Christian Movement, 1930.

Young, Robert M. "The Historiographical and Ideological Context of the Nineteenth-Century Debate on Man's Place in Nature." In *Changing Perspectives in the History of Science*, edited by M. Teich and R. M. Young, 344–438. London: Heinemann, 1973.

———. *Darwin's Metaphor: Nature's Place in Victorian Culture*. Cambridge: Cambridge University Press, 1985.

Younghusband, Francis E. *The Heart of a Continent: A Narrative of Travels in Manchuria, across the Gobi Desert, through the Himalayas, the Pamirs, and Chitral, 1884–1894*. London: John Murray, 1896.

———. *Within: Thoughts during Convalescence*. London: Williams and Norgate, 1912.

———. *Mutual Influence: A Re-Review of Religion*. London: Williams and Norgate, 1915.

———. *Life in the Stars: An Exposition of the View that on Some Planets of Some Stars Exist Beings Higher than Ourselves, and on One a World-Leader, the Supreme Embodiment of the Eternal Spirit which Animates the Whole*. London: John Murray, 1927.

———. "The Destiny of the Universe." *Hibbert Journal* 31 (1933): 161–75.

———. *The Living Universe*. London: John Murray, 1933.

———. "The Mystery of Nature." In *The Great Design: Order and Progress in Nature*, edited by Frances Mason, 237–55. London: Duckworth, 1934.

Yoxen, Edward. "Form and Strategy in Biology: Reflections on the Career of C. H. Waddington." In *A History of Embryology*, edited by T. J. Horder, J. A. Witkowski, and C. C. Wylie, 309–29. Cambridge: Cambridge University Press, 1986.